Julius Thomsen
A Life in Chemistry and Beyond

Abstract

The Danish chemist Julius Thomsen (1826-1909) was a pioneer of thermochemistry, a highly reputed member of the European chemical community, and an admired leader of Danish science. He spanned the period between two of the country's scientific luminaries, H. C. Ørsted and Niels Bohr. Apart from his important contributions to pure chemistry Thomsen also engaged successfully in Danish chemical industry by developing a method of producing soda from the Greenlandic mineral cryolite. This volume offers for the first time a comprehensive account of Thomsen's life and work intended for an international readership. Structured in eight chapters the book focuses on Thomsen's scientific work but also deals extensively with his life and career, his administrative and cultural activities, and his personality and relations to contemporary scientists. Based in part on hitherto unexplored archival sources it offers a novel picture of Thomsen, the man and the chemist, and places his work in the context of international history of science. Among other things, the book presents in detail Thomsen's late work on the composite atom and the meaning of the periodic system of the chemical elements.

Julius Thomsen

A Life in Chemistry and Beyond

By Helge Kragh

Scientia Danica. Series M, Mathematica et physica · vol. 2

DET KONGELIGE DANSKE VIDENSKABERNES SELSKAB

Submitted to the Academy May 2016
Published November 2016

Contents

Preface

Julius Thomsen was a prominent nineteenth-century Danish chemist who acquired fame for his systematic and meticulous measurements of the thermal effects associated with chemical processes. Apart from his position as one of the founders of classical thermochemistry he is also remembered for his early method of producing soda from the Greenlandic mineral cryolite, a discovery that resulted in a major industrial enterprise. Keenly interested in the electrical sciences he constructed a new type of battery which for a while attracted attention. Last but not least, he made significant contributions to the understanding of the periodic system in terms of the hypothesis of composite atoms – and he ended his career by questioning if argon was really a chemical element.

Although Thomsen once was an internationally celebrated scientist, today his name and work has largely fallen into oblivion even in his native country. On the occasion of his death in 1909 several obituaries and memorial articles were written about him, but apparently no-one thought of writing a full-scale biography of the man and his science. The present work offers a comprehensive account of Thomsen's life with an emphasis on his contributions to thermochemistry and other branches of science. A dominant actor on the scene of Danish science in the period beteen the two giants H. C. Ørsted and N. Bohr, Thomsen deserves attention from the point of view of national as well as international history of science. He was admittedly not an innovator of the same scale as Ørsted and Bohr, but in other respects his importance for Danish science was almost as great. Many years ago I wrote a couple of papers on some of Thomsen's major contributions to pure and applied science (see the bibliography, items Kragh 1982, 1984, and 1995). Since then nothing substantial has been published on him or his science.

In what follows I have attempted to present a full picture of Thomsen and the period in which he lived, meaning not only his science but also his significant contributions to Danish society in general. These contributions were in part of an industrial-technological nature (such as the cryolite business) and in part related to institutions of higher learning (such as the University of Copenhagen and the Polytechnic College). All the same, the focus of the book is on Thomsen's scientific work rather than his administrative and political activities. While the former is of interest to international history of science, and of course to history of chemistry in particular, the latter will mostly be of interest within a local or national context. The picture I present of Thomsen is in many respects novel, in particular because I call attention to his relatively broad interests and work in areas outside experimental thermochemistry. This work, covering physics, electrochemistry, structural organic chemistry, and atomic theory, has not previously attracted much attention. Unfortunately, the scarcity of primary sources which authentically illuminate his personality makes my description of Thomsen's personal life somewhat incomplete. On the other hand, it is less incomplete than earlier descriptions.

Most archival material concerning Thomsen's life and work is kept at the Royal Library, Copenhagen. There are also a few letters from Danish scientists (L. V. Lorenz, A. F. Paulsen, P. K. Prytz, J. Thomsen) located at the Uppsala University Library. The rich sources at the Royal Library consist of three parts of which one contains family letters and another is a collection of travel diaries, economic account books, and some letters and manuscripts. The third part is a large collection of letters relating to Thomsen's scientific work, most of them from foreigners. I shall refer to the latter source as "Royal Library, TSC" (Thomsen Scientific Correspondence). Although I have made use of the available archival sources I have done it selectively and not gone through the entire material systematically or in detail. The list of correspondents includes many of the period's most important scientists and illustrates Thomsen's central position in nineteenth-century chemistry. Among the correspondents are J. C. Poggendorff, H. Kolbe, A. Naumann, L. Pfaundler, J. Cooke, C. Guldberg, J. Volhard, J. Wislicenus, C. Blomstrand, P. Cleve, L. Meyer, V. Meyer, G. Wiedemann, W. Ramsay, D. I. Mendeleev, W. Ostwald, H. Kopp, A. W. Williamson, A. Horstmann, F. W. Clarke, and H. Jahn.

The book is structured in eight chapters followed by three appendices and an extensive bibliography covering primary as well as secondary sources. One of the appendices considers Thomsen's impressive scientific productivity from a quantitative point of view. In Chapter 1 I describe Thomsen's life from his birth in 1826 up to about 1866, when he was appointed full university professor of chemistry. Parts of this period is more fully described in the following two chapters which include his innovative work on cryolite soda (Chapter 2) and his no less innovative treatise with L. A. Colding on the cholera epidemic in Copenhagen (Chapter 3). Thomsen's seminal work on experimental and theoretical thermochemistry is the subject of Chapter 4 and to some extent also of Chapter 5. The latter chapter focuses on the many scientific controversies in which Thomsen was involved and of which a priority controversy with Marcellin Berthelot concerning thermochemistry was the most serious one. Although his experimental data survived the emergence of physical chemistry in the 1880s, his theory did not. This is described in Chapter 6, which also contains discussion of some of Thomsen's lesser known contributions to physics and chemistry. We meet another side of Thomsen's thinking in Chapter 7, which contains a detailed exposition of his – in some cases prophetic – ideas of atoms, atomic constitution, and the periodic system. While chapters 4-7 are predominantly of a scientific nature, in the final Chapter 8 I review aspects of Thomsen's life in so far that it concerned Danish culture, scientific institutions, and society generally. In addition, in this chapter I attempt to form a picture of what kind of person the somewhat reclusive scientist was.

It should be noted that some of the chapters are relatively technical and for this reason may not be easily followed by readers without an elementary knowledge of chemistry and physics. However, many of the technicalities are innocent and those which are less so are necessary to understand what Thomsen's science was all about. One potential problem is the chemical nomenclature which in the period was not only somewhat different from the present one but also included Danish words for chemical compounds that will not be recognisable today. In most cases I have used words and symbols which can be easily understood by a modern reader or supplied the obsolete word with a kind of non-anachronistic translation.

Helge Kragh

CHAPTER I

Early years

H. P. J. Julius Thomsen became one of Denmark's greatest scientists, internationally reputed for his work in thermochemistry and nationally celebrated as an icon for progress in both science and industry – and, not least, the close connection between the two. His unique status is illustrated by two large *fin de siècle* paintings by the prominent painter P. S. Krøyer, who portrayed the elite of Danish scientists and also the elite of Danish industrialists and technical leaders (see also Section 8.4). In both of these paintings Thomsen figures in the centre and is one of only two persons found worthy to appear in both groups. But Julius Thomsen had to fight hard for this unique position, for by birth, education and class he was in no way privileged. The wealthy professor, businessman, honorary doctor, and Privy Councillor of the early twentieth century was born in a lower middle-class family which had only recently moved to Copenhagen from the German-speaking part of southern Denmark.

In this chapter we follow young Thomsen struggling to obtain a position in Danish science, a struggle that started when he entered the Polytechnic College in 1843 and was crowned with success twenty-three years later with his appointment as university professor of chemistry. In between he had invented an industrial process to manufacture soda from the Greenlandic mineral cryolite and, in 1860, had become a member of the Royal Danish Academy of Sciences and Letters. To understand the circumstances under which Thomsen created his career one must be aware of Denmark's scientific-technical environment and infrastructure in the period from about 1830 to the 1860s. Much of this environment was marked by the powerful presence and no less powerful shadow of H. C. Ørsted, who directly and indirectly played an important role for Thomsen. This was the case not least with regard to the Polytechnic College, but also other institutions were instrumental in shaping his career. One of the sections sketches the institutional landscape of Danish science and another section deals more specifically with the early

Polytechnic College. Although these institutions may be said to have shaped Thomsen's career, the actual way in which his career was created was very much due to his own efforts and not the product of either institutions or influential patrons.

1.1. Childhood and adolescence

The death of Julius Thomsen in early 1909 was followed by a series of obituaries and memorial addresses, some national and others international. They will be considered in Section 7.3. The prominent British chemist Sir Edward Thorpe, a former president of the London Chemical Society, began a memorial lecture as follows: "Among the Danes whose names are inscribed as men of science on the eternal bead-roll of fame, that of Julius Thomsen stands preeminent – linked indeed with that of Oersted."[1] No higher praise could be given.

Always known as and referring to himself as just Julius Thomsen, his full name was Hans Peter Jürgen Julius Thomsen. The German "Jürgen" rather than the Danish "Jørgen" reveals his family background in Southern Schleswig in the German-speaking part of Denmark. Indeed, his grandfather, a farmer from this part of the country, was named Jürgen Thomsen. In the early 1820s Jürgen's son Thomas Thomsen (1787-1862) went to Copenhagen together with his wife Jensine Frederikke Thomsen, née Lund (1798-1862). As a young man Thomas had attended lectures in chemistry given by Christoph Pfaff at the University of Kiel and also, in 1814, followed lectures by H. C. Ørsted and the astronomer Christian Olufsen at the University of Copenhagen.[2] Thomas Thomsen eventually became an accountant at the National Bank and ended up as a Counsellor (Kammerråd), but for quite some time the family's economic situation was strained and scarcely adequate for the growing

---

1. Thorpe (1910), p. 161. Reprinted in E. Thorpe, *Essays in Historical Chemistry* (London: Macmillan, 1924), pp. 515-532. Also another prominent chemist, J. N. Brønsted, compared Thomsen to Ørsted: "It would be difficult to decide which of the two has cast a greater lustre over Danish science." Brønsted (1932), p. 147.

2. Archival material concerning the Thomsen family, Royal Library, Copenhagen, boxes NKS 4306-4°.

number of children. On 16 February 1826 Jensine gave birth to Julius, the first of four sons and the longest-living of them. He had a two year older sister, Franzisca Helleonora Marie, and two younger sisters, Meta Catharine (b. 1828) and Caroline Auguste (b. 1830). Two of Julius' brothers, Carl August Thomsen (1834-1894) and Thomas Gottfried Thomsen (1841-1901), became seriously involved in the chemical sciences, while Sigismund Gotthelf Thomsen (1831-1903) became an attorney.[3]

Julius Thomsen's two brothers August and Thomas both graduated from the Polytechnic College in "applied science," as Julius had done before them, and they followed careers in pure and applied chemistry. While Julius never became an "academic citizen," both of his brothers were students from a reputed Copenhagen high school ("Borgerdydskolen" on Christianshavn founded in 1795) and had passed the traditional entrance exam to the university known as *examen philosophicum*. In 1871 August was appointed a docent (lecturer) in technical chemistry at the Polytechnic College, a position which was changed into a full professorship shortly before his death in 1894. At several occasions he collaborated with his older brother, in particular as the founding editors of the *Tidsskrift for Physik og Chemie* (Journal of Physics and Chemistry) to be mentioned later in this chapter.

Also Thomas Thomsen junior worked with and under his brother Julius, in part as an assistant at the cryolite mines in Greenland and in part at the chemical laboratory belonging to the University of Copenhagen. He clearly had a talent for science, such as indicated by a series of publications on the optical rotation of organic substances that he wrote 1880-1881. While Thomas Thomsen was an internationally oriented research chemist, August Thomsen only published a few Danish works of a popular and technical nature.[4] A chemical treatise of 1885 dealing with the chemical equilibrium in

---

3. Copenhagen census of 1840, see http://www.onlinearkivalier.dk/cid4738638. According to Bjerrum (1909), Julius was the second of eight children. The census mentions a Carl Aage Thomsen born in 1834, the same year as Carl August. See also Appendix C.

4. For the research of the two Thomsen brothers, see Veibel (1943), and, for their careers, *Dansk Biografisk Leksikon*.

aqueous solutions was valuable but also the last one from Thomas Thomsen's pen. In stark contrast to his famous brother, Thomas was deeply religious and at the time he published his memoir on aqueous equilibria he decided to quit chemistry for the sake of theology. From 1888 to his death in 1901 he served as a priest in the Danish Lutheran Church. When the Royal Danish Academy of Sciences and Letters was informed about Thomas' decision, three members of its scientific class expressed their regret that a promising and original mind had been lost for science. On their recommendation the Academy chose to honour Thomas Thomsen with a gold medal "in recognition and gratitude for his beautiful contributions to science."[5] About thirty years earlier, Julius, Thomas's senior by fifteen years, had been awarded a silver medal from the Academy.

But back to Julius Thomsen, who grew up in a lower middle-class family with interest in science and culture but no tradition for higher education. He was sent to a German-Danish church school in central Copenhagen (Skt. Petri Kirkeskole) and subsequently to the von Westen Institute, a higher gymnasium school founded in 1799 from which it was possible to take a student's exam. The school offered a course in elementary physics for the higher classes, but no chemistry.[6] Like most other schools at the time its focus was classical education, meaning Greek and Latin. At any rate, although young Julius was recognised to be a bright kid he never passed the exam that would qualify him for studies at the University. He left the school in late September 1841, probably with no more than a rudimentary knowledge of Latin – a deficiency which mattered in those days. What interested Julius were neither Latin nor literature, but science generally and chemistry in particular.

In the years 1841-1843 Julius was allowed to work as a laboratory assistant to Edvard August Scharling, associate professor of chemistry and lecturer at the Polytechnic College (and H. C. Ørsted's son in law). Scharling valued his young assistant who, remarkably, made

---

5. Letter of 5 May 1885, signed by C. T. Barfoed, C. Christiansen, and S. M. Jørgensen. Royal Danish Academy, Main Archive 1882-1897.
6. For the von Westen Institute and its modest science teaching in the 1840s, see Riis Larsen (1991), pp. 175-178.

236   *Scharling, Versuche über die Quantität*

Stunden nach dem Essen und einem kurzen Spaziergang 89, 3 Stunden nach Tisch 80, und 4 Stunden nach dem Mittagsessen wieder 72 Schläge.

Vergleicht man nach meinen Versuchen das Verhältnifs zwischen der in der Nacht und der am Tage ausgeathmeten Kohlenstoff-Menge, so ergiebt sich folgende Tabelle:

Nr. 1. Scharling . . . . . 1 : 1,237
Nr. 2. Thomsen . . . . . 1 : 1,235
Nr. 3. Der Soldat . . . . . 1 : 1,42
Nr. 4. Das erwachsene Mädchen 1 : 1,24
Nr. 5. Der Knabe . . . . . 1 : 1,266
Nr. 6. Das kleine Mädchen . . 1 : 1,225.

Nimmt man aus diesen 6 Verhältnifszahlen das Mittel, so wird das Verhältnifs der in der Nacht ausgeathmeten Kohlenstoff-Menge zu der, welche am Tage ausgeathmet wird, wie 1 : 1,237; oder es wird am Tage, im wachenden Zustande ungefähr ¼ Kohlenstoff mehr ausgeathmet, als Nachts während des Schlafes.

Figure 1.1. Thomsen's debut in chemistry. Source: Scharling (1843).

his entry in the international chemical literature at the tender age of 16 (Figure 1.1). In 1842 Scharling performed systematic measurements of the amount of carbon dioxide exhaled by humans. Using himself and five other test persons he measured the emitted carbon dioxide at various conditions and for various ages. One of the guinea pigs was Thomsen, "a 16-year-old man who weighed 115.5 pounds and in 24 hours exhaled 224.37 g."[7] Not only did Thomsen figure by name in Scharling's paper, the Danish chemist also acknowledged some helpful suggestions made by his youthful assistant.

7. Scharling (1843). In the Danish version of the paper published in the Royal Academy's *Skrifter* Scharling referred to Thomsen as "a young man who most diligently studies chemistry under my supervision." 1 pound = 0.5 kg.

The experience in Scharling's laboratory undoubtedly wetted Thomsen's appetite for making a career as a chemist. He might have become the apprentice of a pharmacist, such as several notable Danish chemists had started their careers, including Ørsted, W. C. Zeise and Scharling. But at the time there was another alternative for a youngster without a student's exam.[8] The Polytechnic College founded in 1829 did not require this exam. One could enter the College either with the *examen philosophicum* (as August and Thomas did) or by passing an entrance exam. Admission was not easy, as it required knowledge of "first principles of algebra ... and the equations of first and second degree, including geometry, stereometry, and plane geometry; history and geography; experience in writing Danish; and as much knowledge of French and German which is necessary to understand scientific works in these languages."[9]

Not only did admission to the Polytechnic College require a passed entrance exam, it also required fees for tuition and the Thomsen family was short of money. Fortunately the College had access to a few free places paid by the Society for the Dissemination of Science. On 29 October 1843 Julius' father addressed the Society with a request that his son received one of the free places. The letter provides a rare insight in the life of Julius Thomsen as a teenager:

> My oldest son has throughout shown great desire and talent for chemical and mechanical subjects. Already when he was 12 years old he mastered logarithms. He was always the best in his class in mathematics, and after attending school he eagerly read books from the school library about the physics of the Earth, about electricity and about the steam engine, whose constitution he explained to me. Moreover, a little before confirmation he built an electrifying machine which worked quite well. I would rather have preferred that he followed training as an office clerk, for in this way it appeared to me

---

8. According to Veibel (1939), p. 202 and also Vinding (1941), p. 31, Thomsen worked briefly as an apprentice to a pharmacist. However, there is no documentation for this. Apparently his father thought of the possibility and tried perhaps, if so unsuccessfully, to persuade Julius. See Christensen (1910), p. 156.

9. See Lundbye (1929), p. 39. The textbooks used at the Polytechnic College were mostly in Danish, but some teaching material was in German or French.

that he could support himself and thus ease my heavy burden of educating seven children. After his confirmation I asked him if he might wish to study the dead languages [Latin, Greek]; but he replied with a firm no, stating that he was determined to become a polytechnic. So, on Michaelmas Day [29 September] I took him out of the second-highest class of von Westen's Institute. However, it was too late to be admitted to the polytechnic school, which would take another two years. For this reason I am most thankful to Mr. Privy Ørsted that he managed to provide my son with work at Professor Scharling's laboratory for two years.[10]

Thomsen senior's request of a free place for his son was granted on the assumption that he passed the entrance exam – which he did. Julius would later pay the debt he owed to Ørsted and the Society for the Dissemination of Science.

Julius prepared for the entrance exam while working for Scharling, and in November 1843 he entered the Polytechnic College as a student in the class of "applied science," essentially chemistry or chemical engineering. Little is known about his studies at the College, except that he must have worked hard and followed the required courses in chemistry, physics, mathematics and workshop training. After two and a half year of study, in the spring of 1846 he graduated with the highest grade (*Laudabilis*) at the age of twenty.[11] It should be noted that the number of College students at the time was very small; only four of the eights students that commenced their study on the applied science line in 1843 actually completed it with a polytechnic exam. Far from being anomalous, the small numbers were typical for the period. Thus, of the three students that passed the entrance exam in 1844, the year after Julius, not a single one graduated as a polytechnic candidate! During the 1840s

10. Translated from Harding (1924), p. 106. Since Ørsted was the head of both institutions, the Polytechnic College and the Society for Dissemination of Science, it was essentially his decision to accept Julius as a student. Notice that Thomsen senior does not mention anything about Julius as the apprentice of a pharmacist.

11. Thomsen's average grade was 7.08 on a scale invented by H. C. Ørsted where the top grade was 8 and the lowest – 23. For a complete list of graduates from the Polytechnic College 1829-1890, see Voigt (1890).

the number of graduates of both lines, mechanics and applied science, fluctuated between one and eight per year.[12]

Absorbed in his studies, the ambitious Thomsen – as I shall call him from hereon – had neither time nor desire (nor, perhaps, money) to socialise with his few fellow students and take part in their activities. He diligently and with great determination focused on his study plan, giving other aspects of a Copenhagen students' life very little attention. Concerning Thomsen's early years, one of his biographers offered the following characterisation: "Thomsen, always a reserved and taciturn man, talked little about himself even to his intimate friends – and least of all about the days of his youth. It was known to a few that these days had not been smooth. Those who were best informed were conscious that to these early struggles much of the dour and resolute nature which formed a distinguishing trait in his character was due. Thomsen, indeed, began life as a fighter, and a fighter he remained to the end of his four-score years."[13]

Although young Thomsen concentrated on his studies, these were not his only interest at the time. Attracted to poetry and art he wrote in the period from 1845 to 1847 a large number of poems, some of them very long, elegantly formulated and elaborately structured. These poems he kept for himself and they are only known through his extant notebooks kept at the Royal Library in Copenhagen. While most of Thomsen's poems were for occasional events or dealt with human emotions and nature's beauty, in a few of his aphorisms he mixed chemistry and poetry. One of them reads, in a free translation: "Chemistry tells us that heat, and flame, / is an indication of the love of matter. / Common folks say that this is the same, / for the former follows from the latter."

Immediately after graduation Thomsen was appointed an assistant to the Polytechnic College's chemical laboratory, first under Scharling and from 1847 under Johan Georg Forchhammer, the university professor of mineralogy. Between 1849 and 1851 he also

---

12. See the graph in Harnow (1998), p. 37.
13. Thorpe (1910), pp. 161-162, who acknowledged G. A. Hagemann for information about Thomsen's personal history.

did his compulsory military service, in an artillery regiment, but without being actively involved in the Three-Year's War raging at the time. Thomsen continued in this position as an assistant until 1853, since 1850 supplying his modest income by serving as teacher in a course on agricultural chemistry offered by the Polytechnic College. The course had started in 1849, but attracted very few students and still in 1855 there was no formal exam in the subject. With the official approval of the Royal Veterinary and Agricultural College in 1856 and its inauguration two years later this branch of applied chemistry was transferred to the new College, where it was taught by Christen T. Barfoed and Bendt S. Jørgensen. The result was that Thomsen lost his job. Yet it was in this early period that he laid the foundation for his later career and financial wealth. Not only did he invent and develop the method of soda manufacture based on the mineral cryolite, he also pioneered the study of theoretical thermochemistry by submitting in 1852 an important treatise on the subject to the Royal Danish Academy.

The treatise and what followed from it will be considered in Section 4.2. It was evaluated by a committee under the Academy consisting of Forchhammer, Scharling, and J. C. Hoffmann. Impressed, the three scientists concluded that it was a valuable contribution to science:

> Thomsen ... has consistently followed his ideas which he has supported and further developed by means of calculations. The committee is of the opinion that it [the treatise] opens up for important scientific progress. It therefore recommends to publish the memoirs in the proceedings and also to award the author the Academy's silver medal; and, to encourage him continuing the investigations, to offer him 50 Rbd. [rixdollars] for the purpose of acquiring accurately manufactured instruments.[14]

---

14. *Kgl. Da.Vid. Selsk. Forhandl., Oversigt* (1852): 238. Thomsen received further sums of money from the Academy in 1855 and 1872, in the first case to acquire an "electromagnetic measuring apparatus" and in the second to acquire a calorimeter. See Lomholt (1942-1973), vol. 2, p. 356. At the time 1 Rigsbankdaler (Rbd.) equaled 96 Skilling and a worker's daily salary was approximately 50 Skilling.

Figure 1.2. Julius Thomsen in the mid-1850s. Wikimedia Commons.

The money from the Academy, which was supplemented by a travelling scholarship from the Reiersen Foundation, allowed Thomsen to go abroad on an extended study tour to Germany and France in 1853-1854.[15] He first went to Hamburg and from there to Berlin,

15. The Reiersen Foundation was established in 1796 in order to assist young

from where he continued to Paris over Leipzig. After having studied the chemical life in the French capital – and enjoyed the famous opera – he proceeded to Frankfurt, Strassbourg, Dresden, and Munich. On his way back home in 1854 he also visited Prague. On the tour Thomsen visited many of the period's best laboratories in Paris, Berlin, Munich and elsewhere, and he became acquainted with some of Europe's distinguished scientists. In Berlin he met the physicist J. C. Poggendorff who in 1824 had founded the journal *Annalen der Physik und Chemie* in which Thomsen would publish many of his later articles on thermochemistry and other subjects. He also met Heinrich Rose, a brilliant inorganic chemist and former student of Berzelius, whom he described in his travel diary as "ultra-German."[16]

Shortly after Thomsen's return to Denmark his position at the Polytechnic College ceased, and he was forced to seek another job. From 15 May 1856 to 1859 he served as adjuster of weights and measures to the Municipality of Copenhagen, a purely practical and administrative position which required technical knowledge but did not include access to research facilities. The previous adjuster had been another polytechnic candidate, Peter Faber, who later became the first director of the Danish State Telegraph. Incidentally, also Julius' brother August served three years as an adjuster, in his case 1860-1863 in Schleswig.

In 1859 Thomsen applied for a position as chemistry teacher at the Royal Military High School, but with no success. The position was instead occupied by Jesper Bahnson, a graduate from the High School and Thomsen's junior by one year.[17] While Bahnson had neither chemical training nor publications to his credit, he had the advantage over Thomsen that he was an army officer who had

---

craftsmen and industrial entrepreneurs. Some of Ørsted's travels and lectures were supported by the same foundation.

16. Bjerrum (1909), p. 4977. On Thomsen and Poggendorff, see Section 6.4, and for Rose Section 2.4.

17. On the Military High School and its role in Danish chemistry, see Nielsen (2000), pp. 56-58. As Nielsen says about the positions in chemistry and physics at the High School, "it was the sort of position scientists accepted as long as nothing else was offered – as a preparation for an academic position or as a supplementary income."

served with distinction under the Three-Year's War 1848-1850 against the insurgents from Schleswig-Holstein. He later followed a military and political career which culminated in the period 1884-1894, when he served as the country's Minister of War. While Thomsen failed in obtaining the position in chemistry at the Military High School, in 1859 he was appointed teacher at the same institution, but in physics. He stayed in this position until 1866, but already two years earlier he began part-time work at the University of Copenhagen, first as a docent and lecturer. This was not quite his first experience with the university, for in 1859-1860 he had temporarily substituted for Scharling during one of the professor's periods of illness. Then, on 22 December 1866 the 40-year-old chemist was appointed full university professor and director of the university's chemical laboratory (more in Section 1.5). Now Thomsen could finally devote his time to research in the area that primarily interested him, experimental thermochemistry.

Although Thomsen was a private person and something of a workaholic, of course he also had a life outside the laboratory. On 25 October 1857 he married the six years younger Elmine Hansen, the daughter of a farmer tenant from the island of Langeland. We do not know the circumstances of the marriage except that the couple were married in Stoense Church on the northern part of Langeland. Elmine was a quiet and religious woman, with spiritual values quite different from those of her domineering husband, the patriarch of the family. She was worried over her husband's lack of religious feelings and fought through her life, unsuccessfully it seems, to convert him into a good and faithful Christian.[18] Elmine passed away in 1890, after having given birth to five children of whom four survived the *pater familias* (Figure 1.3). To the grief of Julius and Elmine, the couple's only son died of a brain infection in May 1883.[19]

---

18. This is indicated in some of the poems in Drechsel (1926) to which is referred in Section 8.3.

19. Julius Thomsen to his brother Sigismund, 14 May 1883. Royal Library, Copenhagen, box NKS 4306-4°. The son was named Julius after his father.

Johanne Drechsel, Ellen Richter, Anna Smith-Petersen og Marie Oppermann

Figure 1.3. The four daughters of Julius and Elmine Hansen at about 1930.
http://home1.stofanet.dk/l.h/slaegt/Anna-mfl.jpg

After Thomsen had become a professor, the family moved to a
spacious apartment in central Copenhagen, Ny Vestergade 11 just
opposite the present National Museum. To keep the home the family had two domestic servants. The apartment was located in the
same building which since 1859 housed the University's chemical
laboratory and where Thomsen would work for nearly thirty years.[20]
According to the obituary written by the chemist Niels Bjerrum, "In
their family life, [Thomsen] maintained patriarchal principles, and
the children's relations to him were more rooted in awe and solidar-

---

20. For a detailed description of the laboratory in Ny Vestergade, see Kjølsen (1965),
pp. 158-164.

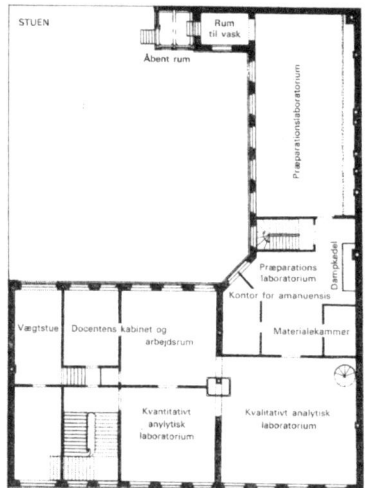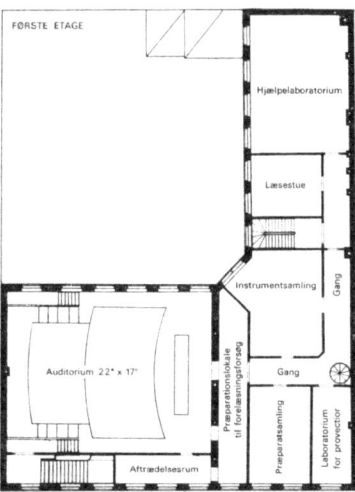

Figure 1.4. Copenhagen University's chemistry laboratory in Ny Vester-gade. Source: Kjølsen (1965).

ity than in friendship and confidence."[21] Whether this is true or not, it fits with the general impression of Thomsen's personality. How-ever, as we shall see in Section 8.3, the impression is in need of some revision. At some time after having become a widower Thomsen moved to a villa on Lindevej 13 in Frederiksberg, not far from the Agricultural College.

---

21. Bjerrum (1909). See also http://www.danishfamilysearch.dk/cid11367720. The daughters were Anna Sophie Frederikke Thomsen (1859-1950), Marie Franziska Thomsen (1862-1939), Ellen Thomsen (1865-1958), and Johanne Thomsen (1867-1963). The son who died in 1883 was named Julius. Johanne Thomsen became a recognised sculptress and author, first exhibiting her art in 1889. After her marriage with Fredrik L. V. Drechsel in 1892 (and change of name to Johanne Drechsel) she wrote a couple of books, including *En Fader og En Moder* (1926) in which she recollected her parents (see Section 8.3). I 1894 the Danish painter August Jerndorff painted her portrait on the order of her father. Her older sister Anna Sophie Frederikke, whose surname since 1884 was Smith-Petersen, was portrayed by Jerndorff three years earlier. Marie Franzisca married in 1895 the prominent art historian Carl R. Theodor Oppermann. For Ellen, see Section 7.5.

1.2. Science in a small country

Denmark's unfortunate intervention in the Napoleonic wars had serious consequences for the country's science, culture and economy.[22] In 1807 the British navy infamously bombarded Copenhagen, causing severe damages and the destruction of many buildings, books and instruments belonging to the university. Nonetheless, the king and his government kept to the alliance with France and even sought to expand the country's university system. One result was the establishment in 1811 of the Royal Frederiks University in the city of Christiania, or what since 1925 was renamed Oslo, the capital of Norway.[23] The new university in Christiania remained Danish for only a couple of years, namely until 1814 when Denmark lost Norway to the Swedish crown. To add to the calamities, the previous year the government was forced to declare the country bankrupt. Not until 1818, when an independent National Bank was founded on the ruins of the previous financial system, did the economic situation ease.

Even without the Norwegian university, until the loss of Schleswig-Holstein in 1864 Denmark had two universities as the University of Kiel, which was located in Holstein, had been under Danish administration since 1773. However, the significance of this institution to science in Denmark was limited, except that it provided temporary positions for promising Danish scientists unable to find positions in Copenhagen. One example is Forchhammer, who worked in Kiel 1815-1818, and another example is the medical doctor Peter L. Panum, later a distinguished professor of physiology in Copenhagen. Also several German chemists worked for periods in Kiel. One of them was Rose, who attained his doctoral degree at the University of Kiel in 1821.

---

22. This section relies on Kragh (2015b), Kragh (1998) and Nielsen (2000). References and further information can be found in these sources. See also Veibel (1939) and Kragh et al. (2008). For a general introduction to Danish history, see Jespersen (2004).

23. In the following I shall refer to Christiania rather than Oslo. In 1877 the official spelling of the city was changed from Christiania to Kristiania, but the new spelling was not generally adopted.

In spite of the country's economic and political troubles, the status of the natural sciences did improve, if only slowly and generally without the consent of the conservative university in Copenhagen. A number of new chairs were established, most of them provisional and funded by the State rather than the University itself. Chairs founded in the 1820s included botany, chemistry, and geology. As a consequence of the strengthened position of the sciences it became increasingly intolerable to be relegated to what was in effect a foreign faculty, the Faculty of Philosophy, in which the interests of the scientists were given low priority. The natural sciences were also represented in the Medical Faculty but only as subsidiary subjects. A prolonged battle over university politics, and the attempts to create a new institutional framework for the scientific fields, came to a culmination in the late 1840s. In 1850 a decision was made to split the Faculty of Philosophy into two, thereby creating an independent Faculty of Science, or a mathematical-scientific faculty. At its establishment it consisted of seven full professorial chairs, including physics (H. C. Ørsted), chemistry (E. A. Scharling), and mineralogy or geology (J. G. Forchhammer). The other chairs were in mathematics, astronomy, botany and zoology.

The formation of a Faculty of Science was primarily due to Ørsted, who was also responsible, to a large extent, for establishing Danish chemistry as an autonomous science independent of medicine. Ørsted saw no essential difference between physics and chemistry, but argued that both of these fields ought to have their own chairs and curricula. In 1820 he managed to obtain a new chemical university laboratory and two years later one of his pupils, William Christopher Zeise, was appointed professor with chemistry as his only subject. Zeise introduced laboratory work for advanced students from the very beginning. He even made it possible for students to do research work, which at the time was most unusual. "Those students," he wrote in 1822, "who have reached such a level of chemical competence that they are capable, in an able and orderly manner, to make new investigations by themselves, can use the laboratory and its apparatus if they pay for the additional expenses."[24] From that time on-

---

24. Quoted in Kragh (1998), p. 239.

ward, chemistry had obtained an autonomous scientific status and was taught by teachers with chemistry as their sole or main vocation.

In the period up to about 1850 it was usual to send young Danish chemists (and other scientists) abroad for one or two years of *Wanderjahre*. The purpose of these extended trips was primarily to gain new knowledge, but in some cases also to bring home secrets of chemical manufacture by more or less legitimate means. This element of industrial espionage ceased to play a role in later study travels. Ørsted, Zeise and Forchhammer all went abroad on such travels, which provided Danish chemistry with the international contacts necessary to counteract the provincialism which was always a danger in a small country on the periphery of the centres of European science. When Thomsen went on his study tour in the early 1850s espionage was not on the agenda but otherwise it was a continuation of the older tradition.

In addition to his many other activities, the indefatigable Ørsted also initiated one of the most important popular-science institutions of the century, the still existing society called Selskabet for Naturlærens Udbredelse (Danish Society for the Dissemination of Natural Science). He apparently got the idea for such a society during a trip to Britain in 1823, finding himself greatly impressed with the Royal Institution and some other institutions for the promotion and dissemination of science and technology that he saw in the large cities. Its Danish counterpart, founded in 1824, soon gained a solid membership base that included prominent citizens, university people, public servants, and industrialists. It was involved in a wide range of outreach activities of which its public lectures were the most important and the most visible. Although based in Copenhagen, from the very outset the society had specifically committed to spreading out its activities to include the larger towns around the country. A trained corps of "regional lecturers" was established for this purpose. Over a period of about twenty years, the Danish Society for the Dissemination of Natural Science held a total of 36 lectures programmes in nineteen different towns around the country, reaching an estimated audience of three to four thousand.

Thomsen was one of the young lecturers who contributed to the Society in the form of public lectures and also demonstration lectures to school classes. In 1855-1856 he lectured on the chemistry of non-metals, in 1863-1864 on optics, and in 1868-1869 – at a time he had become full professor – on chemical aspects of light. Moreover, between 1851 and 1875 he gave a long series of lectures to the paying members of the Society. The subjects ranged widely, including such topics as "climate zones of the Earth," "spectral analysis," "liquefaction of gases" and "the concept of combustion."[25] In 1886 Thomsen was elected chairman of the Society, but the busy scientist stayed in this position for only one year.

Important as the Society for the Dissemination of Science was in a national and popular context, another institution from the same period was even more important to Denmark's small scientific community. Inspired by the Gesellschaft Deutscher Naturforscher und Ärtzte founded in 1822, and also by the British Association for the Advancement of Science, which dates from 1831, scientists from the Nordic countries thought of creating a Scandinavian organisation along a similar line. The idea received stimulus from the "Scandinavism," a romantic, pan-Nordic movement that emerged as a strong political and cultural ideology in the 1830s. After much discussion and planning, the first convention of Scandinavian scientists, or the Society of Scandinavian Natural Scientists, was finally organised in Gothenburg in 1839, including a large section of the Danish scientific elite.

The Gothenburg convention and subsequent conventions in Copenhagen (1840) and Stockholm (1842) succeeded in building up a viable organisation that for decades continued to gather at regular intervals in the Nordic capitals. For several decades the inter-Scandinavian meetings were important for the exchange of scientific knowledge, but from about 1870 they began degenerating into formal gatherings with limited scientific substance. In the early period the leading figures of the society were H. C. Ørsted from Denmark, J. J. Berzelius from Sweden, and C. Hansteen from Norway. At the Copenhagen convention in 1847, the number of participants had swelled to 472 in all, of whom 338 were Danes, 91 Swedes, and 34

25. See the list of Thomsen's lectures in Harding (1924). Also Thomsen's two polytechnic brothers August and Thomas were active in the Society.

Figure 1.5. Celebration for Scandinavian scientists in front of Roskilde Cathedral on 12 July 1847. Painting by E. L. Henningsen. The speaker is H. C. Ørsted and other figures include J. J. Berzelius, J. G. Forchhammer, E. A. Scharling, C. Hansteen, and J. Steenstrup.

Norwegians. It was a great event. Heavily supported by the king and the City Council in Copenhagen the participants could listen to talks given at the Polytechnic College, in the Botanical Gardens, and in the lecture halls of the University. After the talks they could enjoy the new Tivoli Gardens and, even more interestingly, join a rail excursion to Roskilde on the recently opened railway line from Copenhagen, the country's first (Figure 1.5).

Twenty-one-year-old Julius Thomsen, a recent graduate from the Polytechnic College, was one of the participants in the Copenhagen conference, but of course he was quite anonymous compared to people such as Ørsted, Forchhammer and Scharling, not to mention Berzelius.[26] He was from an early date aware of the opportunity offered by the Scandinavian meetings and frequently participated in them, first giving a communication at the 1855 Stockholm meeting and subsequently at the 1856 Christiania meeting. Other meetings he is known to have attended included Stockholm in 1863, 1880 and 1898, Christiania in 1868 and 1886, and Copenhagen in 1892. At the latter meeting he gave the opening speech in which he somewhat controversially pointed out that the era of the polymath belonged to the past and that science inevitably became more and more specialised. Although he was just making an observation, indirectly he suggested that the whole idea on which the Society of Scandinavian Natural Scientists was founded might be obsolete.[27]

In addition to the University and the Polytechnic College, teaching and training in chemistry was also offered at the Military High School, the Royal Veterinary and Agricultural College, the Pharmaceutical College, the larger hospitals, and a few industrially oriented laboratories. Among these institutions, the agricultural university and the Pharmaceutical College inaugurated in 1892 were far the most important. However, none of them were oriented towards research, and chemistry was taught only as an auxiliary subject. During C. T. Barfoed's period as director of the chemical laboratory of the Agricultural College (1858-1887), he had to give all the lectures himself. To the list should also be added a non-educational institution, namely the chemical department of the Carlsberg Labo-

26. Thomsen's presence is confirmed in Erslew (1868), p. 385. On the early Scandinavian science conferences, see Eriksson (1991) and Christensen (2013), pp. 517-533.

27. Thomsen (1892a). See Eriksson (1991), p. 345, which includes details on the Scandinavian meetings. Thomsen made the same critical point at the 1880 Stockholm meeting and some other occasions. See Thomsen (1880f).

ratory founded in 1875 (see also Section 8.1). The Carlsberg Laboratory, devoted to basic research with an eye on subjects related to the fermentation industry in a broad sense, was an important factor in the Danish *fin-de-siècle* chemical landscape, but Thomsen had little to do with it and its researchers. The first director of the Laboratory's chemical department was Johan Kjeldahl, a polytechnic-educated chemist who specialised in amino acids and enzymatic processes. These were topics far away from Thomsen's research interests.

Yet another institution played an important if somewhat indirect role in Danish chemistry, namely the prestigious Royal Academy of Sciences and Letters founded in 1742. Although it was initially conceived as a humanist academy, within a few years it came to include also mathematicians and scientists. Only in 1796 was a chemist elected a member, the pharmaceutically trained Nicolai Tychsen who was the author of *Chemisk Haandbog* (Chemical Handbook). The few chemists who were members of the Academy in the nineteenth century – Thomsen among them – often published in its periodicals, which were however largely restricted to the Danish language. Only in 1902 did the Academy agree to publish in other languages than Danish and thus to make its publications accessible to readers outside Scandinavia.[28] In later sections we shall return to the Royal Academy of Sciences, the role it played for Thomson and the role that Thomsen played for it.

By the 1870s Danish chemists had definitely left the shadow of medicine and emerged as a distinct group of scientists. Indicating the growing professionalisation, a Danish Chemical Society was founded in 1879, the first such society in Scandinavia.[29] The chemis-

---

28. The rules for the doctoral dissertation at the University of Copenhagen were even more archaic. Traditionally they were in Latin, but from about 1840 it became common to write in Danish. Scharling's chemical dissertation of 1839 was in Latin. Only in 1921 was it allowed to write a doctoral dissertation in German, French or English! For details, see Smith (1950).

29. In Sweden a national chemical society was founded in 1883 and a Norwegian society was only established in 1893. For a detailed account of the formation and development of the Danish Chemical Society, see Nielsen (2008). See also Section 8.1.

try professor Sophus M. Jørgensen was elected the society's first president, a position he kept for more than twenty years. Remarkably, his distinguished colleague Thomsen was not among the group of fifty-nine "men with an interest in chemistry" who initially joined the society, nor was he later to become a member. On the other hand, his younger brother Thomas was actively involved in the formation of the society, which at the time of Thomsen's death in 1909 had grown to nearly 140 members.

At the time the Danish chemical community was still of a very modest size, comprising only five regular academic positions: two full professors (Thomsen and Jørgensen) and three associate professors. However, in addition to the few academic chemists there were also a much larger number of chemists working outside the University and the Polytechnic, including chemical engineers employed in industry and chemists working at hospitals, cooperative dairies, private laboratories and pharmacies. The teaching of chemistry at the University was overwhelmingly oriented towards non-chemists, and in particular to medical and pharmaceutical students. In the years 1865-1866 lectures to these two groups attracted an audience of approximately 100 students. And of the 79 students who on average trained in the chemical laboratory, 39 were medical students and 26 pharmaceutical students; seven were polytechnic students, four were people who had completed an education, and three were "other students."[30]

It is a striking fact that very few chemists graduated from the University of Copenhagen in the nineteenth century. In the latter half of the century only twenty students graduated with a master's degree. A graduate armed with such a degree could qualify to write a doctoral dissertation, but that happened only very rarely. The first doctoral dissertation in chemistry was completed in 1869 by S. M. Jørgensen and until 1899 it was followed by only six more doctorates. A list of chemical authors in the period 1846-1890 shows that there were very few university-trained chemists compared with the polytechnic chemists and even fewer if compared with the pharma-

---

30. From a list given by J. Thomsen and reported in *Aarbog for Kjøbenhavns Universitet 1864-1873*, p. 48.

cists. Generally the university chemists were more productive in writing papers – with the exception of the prolific polytechnic author Julius Thomsen. Here are the numbers:[31]

Professional background	Authors	Publications
Pharmacy	43	280
Polytechnic	26	540
Medicine, physiology	22	99
Agriculture	12	22
Military	3	4
University, chemistry	3	153
University, other fields	3	11
Unknown	6	10
Total	122	1127

Table 1. Danish chemical publications 1846-1890.

1.3. The Polytechnic College

The first advanced technical school or college in the kingdom of Denmark was established 1829 with Ørsted as its director.[32] However, already two years earlier a comprehensive plan for a school of this kind was presented by the twenty years younger mathematician, astronomer and technologist Georg Frederik Ursin, who at the time served as professor at the Academy of Fine Arts. This original plan was conceived as a school for artisans and craftsmen modelled after the German *Gewerbeschulen*, technical schools where science was taught at only a modest level. According to Ursin, the school was to be fundamentally practical, and the instruction aimed at skilled craftsmen and similar groups who would benefit from both scientific and technical knowledge, and thus in turn benefit Denmark's

---

31. Adapted from Kragh (1998), p. 247, which gives a more extensive table covering the period from 1801 to 1935. The large number of polytechnic publications is largely due to J. Thomsen, who wrote more than 210 papers and memoirs in the period. See Appendix A.

32. For the early history of the Polytechnic College, see Lundbye (1929) and Wagner (1999).

trade and industry. The idea of a technical college was eventually brought to fruition, but in a form quite different from the one Ursin had in mind. When the proposal was sent to the University for comment, Ørsted took the opportunity to transform it into a substantially different sort of institution inspired by the École Polytechnique in France rather than the *Gewerbeschulen* in Prussia and elsewhere in Germany. His proposal was much more scientifically oriented than Ursin's, offering only a minimum of training in the practical skills of engineering and craftsmanship.

Ørsted envisioned the institution to be closely linked with the University, and as a way of creating scientific-technical graduate competences outside the University, but still associated with it. He had unsuccessfully argued for a science faculty at the University, and the College was in a sense to serve as a substitute for such a faculty. As he phrased it in a provisional plan of 1828, he wanted "a higher educational institution in close connection with the University." This, he claimed, "would obviously be far more beneficial to the state." Moreover, the general goal of the new institution was "to instil, in young people possessing the necessary previous knowledge, such insight in mathematics and experimental natural science, and such skill at making use of these insights, that they may thereby become formidably useful in certain branches of State service, and in overseeing industrial enterprises."[33] Ørsted's proposal met with royal approval, and on 5 November 1829 the Polyteknisk Læreanstalt (Polytechnic College) was officially inaugurated in the presence of the king, Frederik VI. The first lectures were given less than two weeks later. Julius Thomsen was two years old.

The intensive courses were to run for two years, with the students divided into two classes of candidates, one specialising in "mechanics" and the other in "applied science." The latter mainly referred to technical chemistry or, slightly anachronistically, chemical engineering; whereas the former focused on mathematics, mechanical physics, machine construction, and similar subjects. It soon turned out that a study period of two years was unrealistic, and in 1831 it was extended with half a year. A completed study

---

33. Ørsted's provisional plan of 1828 is reproduced in Lundbye (1929), pp. 39-42.

	Mandag	Tirsdag	Onsdag	Torsdag	Fredag	Lørdag
8-9	Naturlærens mekaniske Del. *(Ørsted).*					
9-10	De vanskeligste Kapitler af Algebra og Trigonometri. *(v. Schmidten).*					Opgave‹ gennemgang *(v. Schmidten).*
10-11	Almindelig Kemi. *(Zeise).*					
11-12	Kemiske Indledningsfore‹ drag. *(Forchhammer).*		Krystallo‹ grafi. *(Forchhammer)*	Kem. Indled‹ ningsforedr. *(Forchhammer).*	Krystallo‹ grafi. *(Forchhammer).*	
12-1	Kemiske Øvelser. *(Zeise).* *(Forchhammer).*			Fysiske Øvelser. *(Ørsted).*		
1-2			Geometrisk Tegnelære. *(Hetsch).*			
2-3						
3-4	Arbejde i Værkstederne. *(Winstrup).*					
4-5						
5-6						
6-7	Natur‹		lærens kemiske *(Ørsted).*			Del.

Desuden daglig Kl. 12—2 Tegneøvelser *(Hetsch).*

Figure 1.6. Timetable and workload for students at the Polytechnic College during its first year 1829-1830. Source: Lundbye (1929), p. 69.

would result in a degree as either mechanical engineer or applied-science engineer, with the latter being largely equivalent to a chemical engineer. However, until 1857 the graduates were not, strictly speaking, engineers but just polytechnic candidates. During the first couple of decades civil engineering was not part of the Polytechnic College.

The close links to the University were ensured not only by teachers serving on both staffs, but also by the new institution taking over two of the University's buildings in central Copenhagen. Ørsted emphasised that the teachers at the College should never treat their offices as secondary to or "mere supplements to the [University] professorships."[34] The initial staff consisted of seven teachers, four of whom were associated with the University as either professors or lecturers. Ørsted himself taught physics and also served as director of the new technical college, retaining that position until his death in 1851. The courses in chemistry and applied science

---

34. Letter of 1828, quoted in Lundbye (1929), p. 36.

were given by Zeise and Forchhammer, and the young mathematician Henrik Schmidten, who died already in 1831, took care of the lectures in mathematics. As to Ursin, he was assigned the modest position as teacher in machine construction. See Figure 1.6.

Despite Ørsted's inspiration from the École Polytechnique, the new college of applied science and technology in Copenhagen was by no means a school of engineering during its early years. Later, from 1857 onwards, the college did develop to educate civil engineers, but under Ørsted's leadership it was foreign to engineering subjects. It was basically an academic institution with the purpose of educating graduates in technology and the useful sciences. The Polytechnic College had a positive impact on Danish science, amongst other reasons because it admitted talented students without a high school exam. Julius Thomsen was one of them, and there were many more. On the other hand, scientifically valuable as the College was, for a while it did not stimulate the country's industrial development to the extent that its creators had intended.

The purpose of the Polytechnic College and its relations to the country's small but growing business and industrial sector were matters of continual discussions. Not all agreed with Ørsted's academic ambitions and the low priority he gave to practical and directly useful technology. Another question of a more political nature emerged in the early 1850s when Forchhammer had replaced Ørsted as director. It was widely felt that the college was in need of drastic reform and there were even voices which recommended closing it down entirely.[35] One of the main questions concerned the relationship between the Polytechnic College and the Military High School, two institutions with a considerable overlap in structure and subjects. Could a small and poor nation afford two new institutions of higher technical education?

In 1851 the ministry of the interior established a committee to consider the future of the Polytechnic College. Among the members of the committee were Forchhammer, the mathematics professor Christian Ramus, and the military officer Johann C. Hoffmann who taught physics and chemistry at the Military High School. The

---

35. Lundbye (1929), pp. 100-118; Harnow (1998), pp. 48-58.

committee was divided, with one fraction of it suggesting an expansion and to replace the "mechanics" class with a new class of civil engineering. The other fraction, dominated by military officers, proposed to close down the College. Nothing concrete came out of the committee's work or other proposals made at the time. While some of the proposals recommended a reorganisation of the Polytechnic College as part of the University, others proposed to separate it completely from the University. Only in the late 1850s was a major reform introduced. As part of the reform a new chemical laboratory under Scharling was inaugurated in early 1859. Thomsen was not directly involved in the lengthy debate, but indirectly he was and his views probably played some role (see Section 3.1).

In its proposal of 1851 one of the fractions of the committee did not speak of an "applied science" line but instead referred to "education of chemists." In other words, the line was conceived as chemical engineering, an indication of the very prominent position chemistry held in the courses of this line or class. Of the fifty lessons originally offered per week, twenty were devoted to lectures and laboratory work in chemistry.[36] The educational schedule was ambitious and strict, leaving the students little time for other activities. The first semester they were given nine lectures a week in general chemistry and so-called chemical physics (heat, electricity and magnetism), and in addition six lectures a week in mineralogy and mineralogical chemistry.[37] The second semester meant more lectures on general chemistry and nine hours per week on organic and inorganic chemistry; the students also had to listen to two weekly hours

---

36. Veibel (1939), pp. 177-179. With the changed curriculum of 1831 also the courses in chemistry changed. Now chemical laboratory exercises only started in the second semester and continued until the fifth and last semester.

37. What at the time was called "chemical physics" was entirely different from the modern meaning of the term. In the nineteenth century it was common to divide the study of chemistry in two classes, where one of them, chemical physics, was mainly concerned with heat, light, electricity, and magnetism as chemical agents. Modern chemical physics (as distinct from physical chemistry) emerged in the 1920s and was largely synonymous with molecular spectroscopy and quantum chemistry. The *Journal of Chemical Physics* was founded in 1932. For aspects of the development, see Nye (1993).

of descriptive mineralogy (or "oryctognosy" as it was called, derived from a Greek name for mineral or fossil). Finally, in the third semester the students were introduced to aspects of technical chemistry through ten weekly lectures, whereas there were no chemical lectures in the fourth semester. In addition to the lectures, there were laboratory classes covering six hours per week. All this was more than enough to keep the students busy, indeed so busy that many of them dropped out of the courses and never completed their education.

1.4. Looking for a career

In one of his very last letters, Ørsted wrote to his close friend, the Norwegian physicist and astronomer Christopher Hansteen, concerning the future occupant of the chemistry chair at the University of Christiania. The chair had been held by Julius Thaulow, a young Danish-born and German-trained chemist, who had died in 1850. Hansteen asked a few prominent scientists, including Ørsted in Denmark, Carl Gustaf Mosander in Schweden, and Justus von Liebig in Germany, about suitable candidates; he also consulted the geologist Balthazar Keilhau, professor in Christiania. In his response to Hansteen, Ørsted used the occasion to lobby on behalf of young Danish chemists. On the other hand, he warned against hiring a chemist from Germany as the self-confident Germans "hardly will be able to identify themselves with the Norwegian people." A Dane would be much more suitable, he thought, and – not unimportantly – he would come more cheaply.

Ørsted consequently recommended three Danes, emphasising that he did it confidentially and without having consulted them. Among his candidates were Johannes Johnstrup and Christen Barfoed, who both had studied at the Polytechnic College. The same was the case with the third candidate, Julius Thomsen, whom Ørsted described as follows:

> Thomsen, a polytechnic candidate but with no exam as a university student. From he was 16 to about 18 years old he trained as an assistant to Scharling's chemical work at a time when Scharling only had

a small laboratory. He later studied at the Polytechnic College, graduating with honour. He subsequently became an assistant of Forchhammer, who is very pleased with him. Not only is he competent in
practical chemistry, indeed with a singular talent for it, he also has an
extensive knowledge of chemistry. He has given good lectures on agricultural chemistry and fine presentations on various subjects to the
Society of Industry. He has written a small popular chemistry, which
is much to his credit considering that he is a young man who contin-
ues making progress. Compared with the other two, he is perhaps the
one from which most can be expected. However, although he writes
very well, outside his field he is a man of less culture than the other
two. He is young and unmarried, so I guess it should be easy to get
him.[38]

None of Ørsted's candidates had at the time a record of research
publications, something which the ageing natural philosopher in
Copenhagen seems to have considered of no relevance. Hansteen
and the search committee in Christiania sensibly chose to disregard
Ørsted's recommendations. Instead they chose the 29-year-old
Adolph Strecker, a pupil of Liebig, who received strong support
from Mosander, Keilhau, Hansteen, and Friedrich Wöhler. Strecker
was appointed associate professor in the summer of 1851. His stay
in Norway was important but brief, as he left Christiania in 1860 to
become professor at the University of Tübingen in his native country.[39] After all, from a chemist's point of view Germany was much
more interesting than peripheral Norway. Young Thomsen was undoubtedly aware of the position in Christiania, but there is no indication that he thought of applying for it or that he knew about
Ørsted's recommendation of him.

With regard to issues to be discussed later in this book (especially Chapter 7), it is worth noting that Strecker had an interest in
atomic weights and, while still in Christiania, attended the famous
1860 Karlsruhe meeting that will be considered in Section 5.3. In

38. Ørsted to Hansteen, 13 February 1851, as reproduced in Harding (1920), vol. 1,
pp. 248-249.
39. On Strecker and Norwegian chemistry in the period, see Pedersen (2007). See
also Wöhler to Liebig, 21 May 1851, in Lewicki (1982), Part 1, p. 366.

1859, shortly before the meeting, he published a book on the atomic weights of the elements based on an atomistic conception of matter. Among other things, Strecker dealt with the relationship between the various atomic weights and the possibility that they expressed some deeper-lying regularity such as Prout's law of atomic weights being multiples of hydrogen's. These were questions that Thomsen later took up, but it is unknown if he was aware of Strecker's book. On the other hand, D. I. Mendeleev, another of the attendees in Karlsruhe, was aware of it. In his 1889 Faraday lecture, the Russian chemist called attention to and quoted from Strecker's book: "I now see clearly that Strecker, de Chancourtois and Newlands stood foremost in the way toward the discovery of the periodic law, and that they merely wanted the boldness necessary to place the whole question at such a height that its reflection on the facts could be clearly seen."[40]

In the late 1850s the ambitious Thomsen had not yet been able to ensure for himself a firm position in the small world of Danish science. He realised that one of the ways to increase his recognition and eventually to obtain an academic chair would be through the Royal Academy of Sciences and Letters, very much an institution for the elite. By 1855 members of the Academy counted only 42 Danish scientists and scholars of which less than half belonged to the natural sciences; in addition there were 65 foreign members. It was possible for non-members to have manuscripts accepted for publication in the proceedings (*Skrifter*) of the Academy's mathematical-scientific class, which happened quite regularly. Of course, submissions of this kind were evaluated by some of the members. During the period 1847-1880 twenty-nine memoirs were published by twenty authors who were not members of the Academy (but in many cases would become members).[41] Thomsen's thermochemical memoir of 1852 was one example and another is provided by his polytechnic collaborator August Colding, who published his first work on en-

---

40. Mendeleev (1889), pp. 637-638; reprinted in Jensen (2002), pp. 162-188. Strecker (1859).
41. See Lomholt (1942-1973), vol. 2, p. 37. On the Royal Danish Academy of Sciences, see also Pedersen (1992).

ergy conservation in *Skrifter* in 1850 and six years later was elected a member (see Section 3.2).

In 1858 Thomsen published another memoir in *Skrifter*, this time on the thermal effects of electrochemical processes. The memoir was based on a communication to the meeting of Scandinavian scientists held in Christiania two years earlier. Referring to his earlier work on the theory of thermochemistry Thomsen wrote that his intention had been to develop this work by means of an extensive series of experiments. However, "the circumstances have prevented me from following this route and I have been forced to seek new methods to determine the quantities [of chemical forces]."[42] In other words, in lack of a properly equipped chemical laboratory he was unable to determine chemical affinities by means of calorimetric experiments; he consequently focused on a temporary substitute, namely the relationship between electricity and the chemical forces operating in a battery. We shall return to the 1858 memoir and other of Thomsen's contributions to electrochemistry in Section 6.4.

Thomsen's serious preparations for entering the pantheon of Danish learnedness were crowned with success on 7 December 1860, when he was elected a member of the prestigious society. He was nominated by his first chemistry teacher, Scharling, who pointed out that he was already known to the members of the Academy as the author of two weighty memoirs in *Skrifter*. Scharling further referred to other of Thomsen's publications, including his work with L. A. Colding on the cause of the Copenhagen cholera epidemic (see Section 3.5). "Thomsen has thus enriched science by making careful and interesting experiments, and he has proved in many ways his extraordinary competence as a writer," Scharling wrote. "I am convinced that as his acceptance as a member of the Academy of Sciences will be an honour for him, so will the Academy, if he is elected, benefit from a valuable increase of its scientific strength."[43]

At the time the secretary of the Royal Academy of Sciences was J. G. Forchhammer, who in 1851 had followed another of Thomsen's teachers, H. C. Ørsted, and would continue on the post until

---

42. Thomsen (1858).
43. Letter of 19 October 1860. Royal Academy of Sciences, Main Archive 1859-1863.

his death in 1865. The Academy was without a president as no-one had been found to replace Anders Sandøe Ørsted, the brother of H. C. Ørsted, who had passed away in May 1860. Only 33 years old, Thomsen was one of the youngest members of the Academy and one of the very few without an academic degree in the form of a doctorate or a magister degree. He was not the only one though. The influential zoologist Japetus Steenstrup had also not written a doctoral dissertation and he was elected a member in 1842 at the age of 29. Another exception was the polytechnic engineer and physicist Carl Valentin Holten, who was elected a member the same day as Thomsen. Holten was Ørsted's successor as university professor in physics and had since 1849 served as a lecturer. Like Thomsen he had not passed the high school exam and it was only after he passed it that he was accepted as a university teacher.

The election was a personal triumph and an important step in Thomsen's career plan as it secured him recognition among the professors in Copenhagen. And yet it was only a step on the road, for he would have to wait another six years until he was appointed full professor of chemistry. For the next half-century the Royal Danish Academy of Sciences would become an important part of Thomsen's life; and conversely he became an important part of the Academy's life, serving as its president from 1888 until his death in 1909. Probably more than any other Danish scientist of his time, he made frequent use of the Academy's publications. Between 1852 and 1873 he wrote no less than 15 memoirs in *Skrifter* (proceedings) and from 1861 to 1905 he wrote 23 papers in the *Oversigter* (transactions) published by the Academy.[44] More on this subject follows in Section 8.2.

In 1861 Thomsen was elected a permanent member of the Royal Danish Agricultural Society (Det kgl. Danske Landhusholdnings-selskab), an institution founded in 1769 and one associated with high social prestige. Most of the country's prominent scientists and landowners were members. Although the Society no longer played an important scientific role, and although Thomsen had no particular interest in agriculture, membership of the institution added one more brick to his social capital.

---

44. According to Thomsen's own bibliography in Thomsen (1905a).

During the period 1860-1866 Thomsen's primary occupation was as a physics teacher at the Military High School. Although the job kept him busy, he found time for other activities as well, both scientific research and work related to the cryolite industry. To lay idle was not Thomsen's nature. Only a couple of articles from this period related to his teaching position, one of them being a paper on optical demonstration experiments published in the fateful year of 1864.[45] His successor at the High School was the three years younger Ludvig Valentin Lorenz, whom Thomsen undoubtedly knew at the time. Lorenz had in common with Thomsen that he was a graduate from the Polytechnic College in applied science (from 1852) and that he never wrote a doctoral dissertation. And yet he came to be recognised as the country's leading theoretical physicist and one of the few with an international reputation.[46] Lorenz was elected a member of the Royal Danish Academy in 1866, proposed by the mathematician Adolph Steen and supported by Thomsen and other Danish scientists.

What little research work Thomsen could do as a physics teacher was mostly concerned with electrical apparatuses and the chemical action of electricity and light. Two of his works from this period deserve brief mention (see further Chapter 6). In 1863 he published an important paper on the mechanical work exerted by rays of light, and two years later he constructed a special kind of battery known as a polarisation battery. As far as thermochemistry is concerned, in 1861 he presented a much revised version of his thermochemical system in which he formulated the basic chemical heat theorem sometimes known as "Thomsen's principle." His interests were not restricted to either pure research or industrially oriented work, for they also included dissemination of science to the educated public.

As mentioned above, in 1862 Julius Thomsen and his brother August founded one of the period's most important Danish science

---

45. Thomsen (1864).

46. On Lorenz's life and scientific work, see Kragh (1991). As a teacher at the Military High School he published textbooks in optics and heat theory for the army cadets. In his book on heat he dealt competently with Thomsen's thermochemical measurements and his theory of affinity. See Lorenz (1877), pp. 144-151.

journals. Its full title was *Tidsskrift for Physik og Chemie samt disse Videnskabers Anvendelse*, meaning Journal of Physics and Chemistry and the Application of these Sciences.[47] Although August seems to have been the driving force behind this private project, Julius also engaged whole-heartedly in it. The monthly journal was primarily oriented towards engineers and employees at the Polytechnic College and secondarily to school teachers and the Danish industrial sector. A large part of the papers were partial translations of articles in foreign periodicals, but there were also a fair amount of original papers written by the editors and other Danish scientists and engineers. To get an impression of the journal, consider its first issue of 1862 in which the two editor-brothers motivated its title and focus as follows:

> The steady progress of the sciences has brought physics and chemistry in increasingly closer contact, indicating that in the near future these two branches of science will merge into a single one. Numerous investigations have demonstrated that the physical and chemical properties of matter are intimately connected... For example, the atomic theory is founded on knowledge of the shape, density and heat capacity of matter, and the type theory and the theory of affinity have found support in other branches of the physical sciences.

Without further ado the journal started with an account of one of the period's most remarkable discoveries involving both physics and chemistry, an article on the sensational spectral analysis pioneered by Robert Kirchhoff and Robert Bunsen in Germany.[48] The first volume of *Tidsskrift*, covering 382 pages, was dominated by translated extracts from foreign journals. It only included a few ar-

---

47. On this and other Danish journals related to the chemical sciences, see Nielsen (2000), pp. 83-91 and Nielsen (2001).

48. The article was composed by August Thomsen but closely based on two of the German scientists' papers in *Annalen der Physik und Chemie*. The Thomsen brothers had good reason to introduce their journal with this subject, which was of equal importance to physics and chemistry. Moreover, it grew out of an exemplary collaboration between a physicist (Kirchhoff) and a chemist (Bunsen) at the University of Heidelberg.

ticles written by Danish authors, namely L. A. Colding (on heat diffusion from water pipes), L. Lorenz (on the theory of light), and J. Thomsen (on thermochemistry). August and Julius Thomsen continued as editors until 1879, after which time August Thomsen functioned as chief editor. With his death in 1894 the journal came to an end, although it was followed 1896-1898 by the short-lived and more physics-oriented *Nyt Tidsskrift for Fysik og Kemi* (New Journal of Physics and Chemistry).

Subject area	1862 (%)	1870 (%)	1880 (%)	1890 (%)
Original papers	20	21	10	33
Extracts from other journals	57	35	56	25
Short notices	8	4	6	3
Literature survey	0	0	1	11
Number of issues per year	12	12	12	12

Table 2. Content analysis of *Tidsskrift*. Adapted from Nielsen (2007).

1.5. Professor Thomsen

While still employed at the Military High School, in 1862 Thomsen was appointed titular professor, which was however purely an honorary title with no academic implications whatever.[49] The title meant a raised status in the social hierarchy but not that he came closer to an academic position. Then, in the autumn of 1863, he requested to become a member of the University's Faculty of Science and to be allowed to give lectures in chemistry and physics. The latter part of his request was accepted but not the first part. As stated in the University's yearbook, "Since Prof. Thomsen has no student's exam and assumedly is less acquainted with the conditions at the University he should probably be temporarily employed without entering

---

49. During the nineteenth century it was common to appoint honorable citizens titular professors. They could be scientists but in most cases they were artists, architects or authors who were in this way placed in the lower ranks of the official rank order. For example, Hans Christian Andersen was a titular professor.

Figure 1.7. Although best known as a geologist and mineralogist J. Georg Forchhhammer (1794-1865) was also an accomplished chemist. He significantly influenced young Thomsen's career. Image credit: DTU History of Technology.

the Faculty."[50] In April 1864 the Senate approved the proposal and Thomsen was hired, to be appointed extraordinary docent in chemistry on 22 August 1865 and now with a seat in the Faculty (but only after he had repeated his application and provided further arguments for it). Thomsen's obligations as a docent was the result of skilful negotiations with the Ministry in which he firmly denied to give separate lectures to the polytechnic students in subjects he cov-

50. *Aarbog for Kjøbenhavns Universitet 1864-1871*, p. 42. Accessible online at http://publikationer.ku.dk/aarlige_udgivelser/aarbog/

ered for the university students; he also would not lecture at the Polytechnic College in technical-organic chemistry. This was accepted by the Ministry.

In December 1865 Forchhammer died and as a result it fell on Thomsen's shoulders to give the ordinary lectures in chemistry. When Scharling passed away the following year, Thomsen was the natural successor and in a strong bargaining position. During the years 1859-1860 he had substituted for Scharling, who was frequently ill, so he was well prepared. Thomsen used the situation to propose a major revision of the chemistry teaching at the University and the Polytechnic College, which was accepted by the Faculty. Among other things he argued that chemical technology was too important a subject to be taught by the university professor of chemistry; it needed to be taken care of by a new docent with particular expertise in the field. Moreover, Thomsen proposed a new lectureship to be established at the University. The lecturer or docent should belong to the University but be responsible for the chemical laboratory at the Polytechnic College and give lectures there. Thomsen thus opened up for two new positions in Danish chemistry.

On 22 December 1866 he was appointed full university professor in chemistry. In addition to his duties at the University Thomsen also had to take care of parts of the chemical education at the Polytechnic College, his alma mater. The whole process from 1863 to 1866 was slow, tedious and bureaucratic as it concerned not only chemistry at the University but also at the Polytechnic College. By following the account in the university yearbooks one gets a detailed impression of the state of art of Danish chemistry at the two sister institutions in the 1860s.[51]

Shortly before Thomsen's professorship was confirmed, Sophus Georg Drewsen, a recent polytechnic-chemical graduate, wrote a letter to his friend Gustav A. Hagemann who was then staying in the United States. Drewsen had just become an assistant at the laboratory of the Polytechnic College, "so now I am together with Ju-

---

51. The negotiations can be followed in *Aarbog for Kjøbenhavns Universitet 1864-1871*, pp. 42-51.

lius ... with whom I get along quite well; there is much to do, though, for there are 49 people in the laboratory." He went on to tell about Julius Thomsen and his brothers: "Thomas Thomsen has been assigned assistant to Barfoed as August Thomsen has gone to Paris on some business of technical chemistry; he is to be docent in technical chemistry, a job that J. Thomsen refuses to take on. ... Thomsen rarely turns up in the university laboratory although he is supposed to be in charge of the teaching. At the Polytechnic College he works day and night in Forchhammer's laboratory and rarely shows up among the students."[52]

The period in the mid-1860s witnessed a change of guard in Danish chemistry, with new internationally oriented chemists replacing the older and more provincial generation of Ørsted's era. The change was to a large extent the result of Thomsen's negotiations. In 1867 Sophus Mads Jørgensen – always referred to as S. M. Jørgensen – was appointed docent and head of the chemical laboratory at the Polytechnic College and thus became a colleague of Thomsen. Contrary to Thomsen the eleven-year younger Jørgensen was university-trained and had passed the magister exam; in 1869 he wrote a doctoral dissertation on the iodine compounds of alkaloids. He became full professor at the University in 1887 and continued in this position until 1908. As to the proposed position in chemical technology, the first of its kind, after some delay it was occupied by August Thomsen, Julius' brother.

When Julius Thomsen was appointed professor at the university it caused raised eyebrows in parts of the academic establishment. Could it be true that a full professor at the University of Copenhagen did not master Latin and had not written a doctoral dissertation, that he did not even have a high school exam qualifying him for university studies? Although this was most unusual, Thomsen was not the only prominent scientist of his generation without the proper academic credentials. His chemical colleague C. T. Barfoed

---

52. The letter is quoted in extenso in Vinding (1942), pp. 40-43. Drewsen, who graduated in 1866 together with Thomas Thomsen, later worked as an engineer at Øresund Chemical Factories (cryolite) and the Tuborg Factories (beer). For Hagemann, see chapters 3 and 8.

also had not written a doctoral dissertation and the same was the case with the highly regarded zoologist J. Steenstrup who became a professor in 1845, only 32 years old. Thomsen was a self-made man who considered a scientist's credentials to lie with his scientific work and not with his academic titles. He never seems to have contemplated writing a formal dissertation and then become a "proper" member of the academic brotherhood.

In an age when old traditions slowly gave way to new norms and ways of thinking, Thomsen was not the only one who nourished a certain degree of disrespect for the traditional academic system and its rigid hierarchy. The botanist and fermentation physiologist Emil Christian Hansen, who in the 1880s would become famous as an innovator of yeast cultures at the Carlsberg Laboratories, had to carve his own path to the top of Danish science. Like Thomsen he came from a modest social background. He eventually wrote a doctoral dissertation, but only after having received dispensation from the ministry because of his lack of the academic magister degree. The Faculty of Science at the University of Copenhagen had decided against the dispensation, although the decision was resisted by Thomsen and a few other professors. It appears from Hansen's diary that he felt himself and Thomsen to be kindred spirits. In 1877 he had conversations with both of the professors of chemistry, Thomsen and S. M. Jørgensen. He felt that Jørgensen was too traditional and that he cared too much about formal qualifications and too little about scientific qualifications. Thomsen was different and more to his like:

> Thomsen, on the other hand, has gone his own way as I have done, and he understands my language. Although we have only talked together a few times, we seem to go along better; we are of the same origin and this is what matters. Also his lectures appeal more to me, they are remarkable by being the words of a living chemist and not of a book. There is something strangely indistinct, something pale and bloodless about most of the "genuine academics" with whom I am acquainted ... and which makes me want to slap them in their face.[53]

---

53. Quoted in Glamann and Glamann (2004), p. 99.

Thomsen never became what Hansen called a "genuine academic," but he nonetheless rose to an almost unrivalled position in the country's scientific and academic elite.

CHAPTER 2

Thomsen's white gold

Cryolite is a glassy white or white-reddish mineral which is largely limited to a small area in southern Greenland (Figures 2.1 and 2.2). At rare occasions is appears in other colours, such as gray or brownish or even black. Its density is 2.95 g cm^{-3} and its hardness 2.5 on Mohs' scale. The refraction index is 1.33, very close to water's, and it forms monoclinic crystals. The pretty but rather rare mineral has a unique position in Danish-Greenlandic scientific and commercial history, and is closely associated with Julius Thomsen's career and reputation. In brief, in the early 1850s Thomsen devised a chemical method which allowed him to produce soda from cryolite as raw material. After many problems had been overcome, the method was transformed into industrial soda works which for a limited period of time played an important role in the early phase of Danish industrialisation. As to Thomsen, it made him a wealthy man and established lasting relations between him and the elite of Danish trade and industry.

Remarkably, Thomsen was deeply engaged in this difficult and time-consuming project on industrial chemistry at the same time that he developed the thermochemical system for which he is best known. Whereas the first was a successful contribution to technical chemistry, the second was a no less successful contribution to pure or theoretical chemistry.[1] In the very same period, Thomsen wrote two elementary textbooks in chemistry and engaged in a major study of the Copenhagen cholera epidemic. These activities are covered in Chapter 3.

2.1. The icy stone from Greenland

The mineral cryolite is unusual in several respects, not least because its geographical distribution is essentially limited to a single depos-

1. Much of this chapter relies on Kragh (1995) and on Kragh and Petersen (1995).

Figure 2.1. Cryolite with siderite from Ivittuut, Greenland. http://www.johnbetts-fineminerals.com/jhbnyc/mineralmuseum/18478.jpg

Figure 2.2. Thomsenolite with pachnolite from Ivittuut, Greenland. http://www.johnbetts-fineminerals.com/jhbnyc/mineralmuseum/34858.jpg

it in the southern part of Greenland. When Thomsen got interested in the mineral it had been known by Danish and other scientists for more than fifty years. However, although cryolite attracted a good deal of attention among chemists and mineral collectors, no one thought that it might be exploited on an industrial scale. After all, why should they think so? But thanks to Thomsen's inventiveness and entreneneurial spirit, for a period of time the mineral became "white gold" not only to Thomsen but also to Danish industry and indeed for the Danish nation as a whole.

Since 1776, trade on Greenland was a monopoly under the state-owned Royal Greenland Trading Company (Den Kongelige Grøn-landske Handel, abbreviated KGH). In addition to furs from seals and polar bears, ships regularly brought to Copenhagen specimens of stones and minerals that were supposed to be rarities, hence of interest to museums, or were seen as objects of possible commercial exploitation.[2] At the end of the eighteenth century, one of the ship-ments from the ice-covered island included a white, glossy mineral. In 1795 it was examined by Heinrich C. F. Schumacher, professor at the surgical school Academia Chirurgorum Regia founded in 1787 and a prominent anatomist, botanist and mineralogist. Schumacher

2. See Gad (1976) for the important role of KGH in Greenland's history.

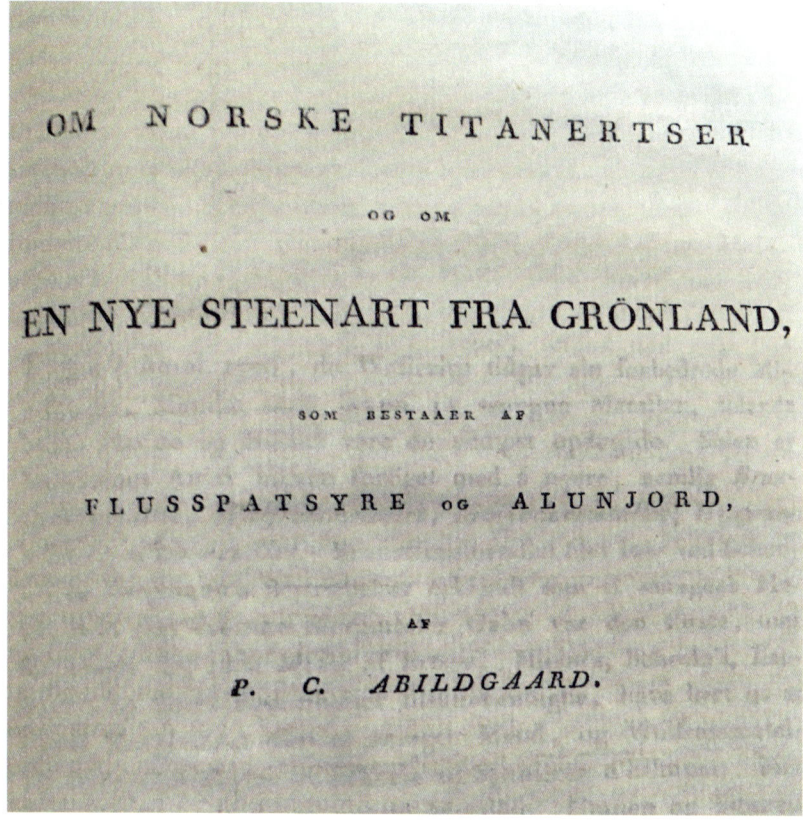

OM NORSKE TITANERTSER

OG OM

EN NYE STEENART FRA GRÖNLAND,

SOM BESTAAER AF

FLUSSPATSYRE OG ALUNJORD,

AF

P.   C.   ABILDGAARD.

Figure 2.3. P. C. Abildgaard's communication of 1800 on the chemical composition of cryolite.

was in 1789 a co-founder of the Society of Natural History (Naturhistorie-Selskabet), and he described the new mineral in a report to the society's journal.[3]

Schumacher's report caught the attention of another of the co-founders of the Society of Natural History, the no less prominent

---

3. Schumacher (1795). The journal of the Society of Natural History (*Skrivter*) appeared between 1790 and 1810 in a series of 11 half-volumes each containing about 200 pages. Although dominated by botany and zoology, *Skrivter* also included a fair number of articles on geology, topography, and chemistry. The short-lived journal was also published in German language, in a version entitled *Schriften der Naturforschenden Gesellschaft zur Kopenhagen.*

Copenhagen scientist Peter Christian Abildgaard.[4] Best known as a pioneer of Danish veterinary science, Abildgaard was also an able chemist and mineralogist. He was in contact with the Portuguese-Brazilian mineralogist Jozé Bonifacius de Andrada e Silva, whom he provided with specimens of the Greenlandic mineral for further examination. Andrada, who had described several new minerals in Scandinavia and elsewhere, mainly examined the physical and mineralogical characteristics. He found that the mineral was semi-transparent, with a wet gloss, and that it easily melted under the blowpipe, almost like ice. The refractive index turned out to be 1.33, very close to water's. More precisely, whereas the refractive index of water is 1.330, the one of pure cryolite is 1.338.

In letters to Abildgaard, Andrada reported his results and suggested to his Danish colleague that he undertook a proper chemical analysis. According to Abildgaard, the mineral could be dissolved in strong sulphuric acid, leaving fluorspar acid (hydrogen fluoride, HF) and alumina.[5] However, the volatility of the fluorspar acid prevented Abildgaard from determining the stoichiometric relationship between the mineral's two components. While a formula was still missing, Andrada and Abildgaard agreed on a name: they suggested calling it *cryolite*, meaning ice-stone or frost-stone in Greek.[6]

A quantitative analysis was first made by two of Europe's foremost analytical chemists, Louis Nicholas Vauquelin in Paris and Martin Heinrich Klaproth in Berlin.[7] Contrary to what had previously been believed, they found the alkali to be "mineral" or "marine" soda (natron), meaning that cryolite was composed by alumina, soda and fluorspar acid. At the time the element sodium had not yet been isolated (the element was first produced in pure form by Davy in 1807 by means of galvanic electrolysis). While Klaproth reported the weight percentages 24-36-40, Vauquelin obtained 21-

---

4. On Abildgaard's life and career, see Andersen (1985).

5. Abildgaard (1800); Andrada (1800).

6. The name first appeared in a brief report by Andrada in *Scherer's Allgemeine Journal der Chemie* **2** (1798): 502.

7. See Vauqelin (1800) and Klaproth (1801).

33-46. From a report of 1812 to the Royal Society of Edinburgh, we get an impression of this early phase of cryolite's chemical history:

> Abildgaard ... detected the presence of fluoric acid and alumina in its composition, the former of which had till then only been met with in combination with lime. The subsequent investigations of Klaproth and Vauquelin have given us the analysis of this stone in a more perfect form. Besides fluoric acid and alumina, they found soda, forming about one-third of the whole; thus adding to the catalogue of minerals the most interesting compound the kingdom affords. The name which was given to it by Abildgaard, was suggested by its wonderful fusibility, in which respect it surpasses every other mineral. ...This substance obtained very high estimation in the mineral market. I have in my possession a small specimen for which a friend of mine paid four pounds.[8]

It was only in 1823, when the great Swedish chemist Jöns Jacob Berzelius took up the matter, that the precise composition became known and cryolite was associated with a definite chemical formula. Berzelius concluded that the three components (alumina, soda, and fluorspar acid) appeared in the weight ratio 24.4-31.4-44.2.[9] Translated into values for aluminium, sodium and fluorine – an element only isolated in 1886 – the ratio corresponds to 12.91% Al, 32.83% Na, and 54.26% F (the latter percentage determined as rest). Given that today's values are 12.85% Al, 32.85% Na and 54.30% F, Berzelius' determination was remarkably precise. In modern nomenclature and ignoring crystal water, the formula is thus $3NaF, AlF_3$ or Na_3AlF_6.

During the first two decades of the nineteenth century, cryolite was considered a rarity of interest mainly to chemists, geologists, and collectors of minerals. The exploration of the cryolite deposit in Greenland, and hence the wider dissemination of the mineral, was due to the German-born polymath Karl Ludwig Giesecke, one of the more colourful figures in the history of early nineteenth-cen-

8. Allan (1813), p. 101. Thomas Allan was a Scottish banker and amateur mineralogist who had come into possession of some of the cryolite collected by K. Giesecke.
9. Berzelius (1824). The report on cryolite first appeared in *Svenska Vetenskaps Akademiens Handlingar* (1823): 315.

tury science. While staying in Vienna as a young man, Giesecke was primarily occupied with music, poetry and the theatre. He participated as a stage manager and actor in the first performance ever of *The Magic Flute*, Mozart's last and possibly most famous opera, which took place at the Freiherr Theater in Vienna on 30 September 1791. It is believed that he was involved in writing the libretto and may have been responsible for part of it.[10]

Only after having moved to Berlin in 1801 did Giesecke become seriously interested in science. Among those he met and learned from in the Prussian capital was Klaproth, the mineralogist Dietrich Karsten, and the famous geologist Abraham Werner known as the father of the "neptunist" conception of the Earth. He also got acquainted with the Danish pharmacist Johann Manthey, who taught chemistry in Copenhagen and was a mentor for the young H. C. Ørsted. As a result of his mineralogical travels to Scandinavia and the Faroe Islands, Giesecke was elected a foreign member of the Royal Swedish Academy of Science and in 1817 also of the Royal Danish Academy. In 1805 Giesecke settled in Copenhagen and the following year, on the recommendation of Manthey, he was hired by the KGH as a consultant. He arrived in Greenland in 1806, only to return to Denmark seven years later. His long but fruitful stay in Greenland was involuntary, owing to the Napoleonic wars that prevented an earlier return.

Giesecke subsequently settled in Dublin, where he morphed to the highly respected Sir Charles Lewis Giesecke, Professor of mineralogy at the Royal Dublin Society. In 1822 he published a brief account of the rich cryolite deposit that he had first located at Ivittuut (formerly Ivigtut) near the Arsuk bay in 1809. As Giesecke noted, he had found the deposit "by a mere accident" and was in no way to be regarded the discoverer of cryolite. "We owe the first discovery of cryolite to the Greenlanders," he insisted, "who, finding it to be a soft substance, employed water-worn rounded fragments as weights on their angling lines."[11] Indeed, the Greenlanders had for

---

10. On Giesecke's career, see Whittaker (2007). See also Whittaker (1991) for the Giesecke-Mozart connection.

11. Giesecke (1822). His detailed travel diary was published much later as Giesecke

long had a name for the mineral, which they called *Orsuksiksæt*, or "the stone looking like the seal's blubber."

Despite Giesecke's important contributions to the geology of Greenland, cryolite remained a mineral that appealed more to museums and collectors than industrially inclined scientists. The situation only changed in 1847, when J. G. Forchhammer began experimenting with a shipment of cryolite from the Ivittuut deposit.[12] Forchhammer's idea was to use the mineral as a glaze for porcelain and other earthen ware, but it turned out that it was not suited for the purpose. Nonetheless, the large amount of cryolite left over from Forchhammer's experiments, was more than just a waste product. Enter his assistant, young Julius Thomsen.

2.2. The cryolite soda process

Rather than using cryolite for the manufacture of porcelain glaze, Thomsen thought that the mineral might be decomposed so as to produce the commercially valuable soda crystals. Small amounts of soda were at the time produced domestically from kelp, but in quantities and purity which were unable to satisfy the demands of the local manufacturers of soap and glass. In the Leblanc process, so named after the French chemist Nicolas Leblanc who invented it in 1789, soda was obtained from common salt, sulphuric acid, carbon, and lime. The important Le Blanc process can (anachronistically, of course) be summarised as three consecutive processes:

$$2\,NaCl + H_2SO_4 \rightarrow Na_2SO_4 + 2\,HCl$$

$$Na_2SO_4 + 4\,C \rightarrow Na_2S + 4\,CO, \quad 2\,CO + O_2 \rightarrow 2\,CO_2$$

$$Na_2S + CaCO_3 \rightarrow CaS + Na_2CO_3$$

(1910), edited and supplied with a biographical foreword by the Danish geologist Knud Steenstrup.

12. Forchhammer did not report his experiments, but they are known from his correspondence. See Topp (1990), p. 3.

British soda factories based on the Leblanc process dominated the market in the mid-nineteenth century, when the country's alkali plants produced about 140,000 tons per year.[13] No factory of this kind was ever established in Denmark. With a growing need for soda, and practically no domestic production, the small country had to import most of the commodity from foreign factories. This was the economic situation that made Thomsen consider cryolite a possible raw material for soda if it could only be provided from Greenland in sufficient quantities and at a competitive prize. He did not consider exploiting cryolite's content of aluminium, such as a few other scientists did.

Experiments made in 1849 in the laboratory of the Polytechnic College, most likely a continuation of Forchhammer's work, convinced Thomsen that soda could indeed be produced from cryolite. Although he did not publish his method or reveal how he had arrived at his invention, it is possible to reconstruct it from an unpublished report attached to his 1852 patent application.[14] Rather than decompose cryolite with sulphuric acid, which would produce the harmful fluorspar acid (HF), he tried an alkaline decomposition, first by using cryolite suspended in baryta water, or what in our nomenclature is $Ba(OH)_2$. The result was surprising:

> A priori one could hardly expect anything to occur, and yet, by boiling a suspension of cryolite powder in baryta water a reaction between the components took place; I found in the liquid a solution of clayey soil in soda, whereas the precipitate was made of barium fluoride. This opened up the road for me... .[15]

Well aware of the chemical similarity between the elements barium and calcium, Thomsen next used quicklime (CaO) as well as slaked

---

13. See, for example, Haber (1958).

14. I rely on the extracts of the report reproduced in Topp (1990). Thomsen's only published account of his cryolite soda process appeared in 1862 and was mostly a description of the industrial version rather than dealing with the original discovery. See Thomson (1862a) and the German article Thomson (1863a).

15. Quoted from Topp (1990), p. 7. Thomsen's name for barium fluoride was "Fluorbarium."

lime $(Ca(OH)_2)$ with the same result, namely a precipitation of fluorspar (CaF_2). This initial step seems to have been guided by Thomsen's scientific insight and interest in the problem of affinity, an interest that he was about to develop into his important research programme of thermochemistry. In his 1852 report he stressed that the various methods he proposed were all "modifications in the application of the same principle – the strong affinity of fluorine to calcium – which I have succeeded in establishing." This was undoubtedly a reference to his new thermochemical theory.

Thomsen's reference to fluorine was to an element as yet unknown. As Thomsen wrote in his textbook of 1850 to be considered in Section 3.1, "Fluorine is the only element about which we know that it exists and yet have been unable to isolate in its free form."[16] Fluorspar acid and various salts of it were well known at the time, when it was suspected that they were compounds including a new chemical element analogous to chlorine. However, this highly electronegative element (fluorine) was still unknown and it would take many years until chemists succeeded in isolating it from its compounds. The most difficult problem of isolating the element and determining its atomic weight was eventually solved by the French chemist Henri Moissan, but it took until 1886 before the job was done.[17] Twenty years later Moissan would receive the Nobel Prize in chemistry for his preparation of fluorine.

It is tempting to speculate that Thomsen used thermochemical reasoning to infer or predict the alkaline decomposition of cryolite. However, his invention was hardly scientifically grounded in this direct manner. The strong affinity between calcium and fluorine was commonly known at the time, and Thomsen did not perform thermochemical measurements with cryolite or other substances active in the process leading to soda. Thomsen's first publication on his new principle of thermochemistry appeared in 1852, the same year as the patent report, but it contained no data relevant to the cryolite soda process.[18] Conversely, his publications on cryolite soda con-

---

16. Thomsen (1850a), p. 35.

17. See Weeks and Leicester (1968) for the complex story of fluorine and its discovery.

18. In Thomsen (1861), p. 119, he referred to his early experiments with cryolite and

tained no direct references to thermochemical reasoning. Thomsen proceeded in a systematic and thoroughly scientific manner, but his approach was largely empirical rather than theoretical. In any case, Thomsen's experiments proved that if pulverised cryolite was boiled with lime a chemical reaction occurred, leaving a suspension of alumina (also known as "clay soil") and a precipitate of fluorspar. Conceiving the clayey soil to consist of a mixture of alumina and soda ash, which here can be translated as Na_2O, he figured that the ash might be turned into soda by adding carbon dioxide:

$$Na_2O + CO_2 \rightarrow Na_2CO_3$$

Further experiments confirmed the hypothesis.

In a few cases Thomson referred in his report to "atoms" of matter. For example, he wrote that from "one atom of cryolite" it would be theoretically possible to produce "two atoms of soda and one atom of alum." He obviously used the term "atom" in a different sense than today, yet it is of some interest that he referred to atoms at all.[19] As we shall see, contrary to his former teacher Ørsted (and also to Forchhammer), Thomsen never doubted that matter consisted of smallest units, or what contemporary chemists typically called atoms. On the contrary, he was an outspoken atomist who in later works would praise the atomic hypothesis as the only true foundation of chemistry – without excluding the possibility that the chemical atoms were composites of even smaller bodies. Still, there are indications that in the early 1850s Thomsen did not yet subscribe to the atomic hypothesis in the same sense that he would do later on (see Section 3.1).

---

diluted sulphuric acid, noting that apart from fluorspar acid they also resulted in an unexpected evolution of hydrogen sulphide. The latter phenomenon puzzled him and apparently inspired him to think of a connection between affinity and heat evolution.

19. The term "molecule" as distinct from "atom" had not yet entered the chemical vocabulary. It only became commonly used after the Karlsruhe congress in 1858 (see Section 5.3). In his pioneering work on chemical atomism, *A New System of Chemical Philosophy* from 1808, John Dalton referred to, for example, the "atoms" of water, lime, sugar, and alcohol.

From Thomsen's report it is possible to reconstruct, in modern chemical language, the experimental basis of his original cryolite soda process. The reaction between cryolite and quicklime is

$$Na_3AlF_6 + 3\ CaO \rightarrow Na_3AlO_3 + 3\ CaF_2$$

What Thomsen described as "soda clay soil" can be written as

$$Na_3AlO_3 \cdot 3H_2O + Al(OH)_3 \quad \text{or} \quad 2\ Na_3AlO_3 = Al_2O_3 + 3\ Na_2O$$

The natron or soda ash thus being part of the earth is transformed into anhydrous soda by the action of carbon dioxide:

$$2\ NaOH + CO_2 \rightarrow Na_2CO_3 + H_2O \quad \text{or} \quad Na_2O + CO_2 \rightarrow Na_2CO_3$$

As Thomsen noticed, because of the absence of sulphuric acid in the process the soda produced in this way would not be contaminated with sulphides. As he knew, this was a severe problem in the Leblanc process, where large quantities of calcium sulphide were produced together with the soda. The first reaction between cryolite and quicklime was the key to Thomsen's process, but it occurred by means of intermediate solid phases that were not known to him. An examination based on modern thermodynamic data indicates several plausible candidates, which all transfer the sodium content of the cryolite into the water-soluble NaF.[20] Two of the candidate reactions are

$$2\ Na_3AlF_6 + 6\ CaO \rightarrow 6\ NaF + Al_2O_3 + 3\ CaF_2 + 3\ CaO^*$$
$$(\Delta G = -260\ kJ/mole)$$

and

$$2\ Na_3AlF_6 + 6\ CaO \rightarrow 6\ NaF + CaAl_2O_4 + 3\ CaF_2 + 2\ CaO^*$$
$$(\Delta G = -283\ kJ/mole)$$

---

20. See Jewess (2010). The symbol ΔG refers to the difference in Gibbs's free energy.

The symbol CaO* denotes unused quicklime and pressure and temperature are assumed to be 1 bar and 298 K, respectively.

It may be assumed that Thomson had produced small amounts of soda from cryolite by the end of 1849 and that he understood the chemistry of the involved processes. But his experiments were on a laboratory scale only, and he may not initially have considered an extension to an industrial scale. After all, the entire amount of available cryolite left over from Forchhammer was a few hundred kilograms. However, in 1852 the geologist and explorer Hinrich Johannes Rink was sent to Greenland by the Danish Government in order to, amongst other things, evaluate the mineral resources. Rink had originally studied physics and chemistry at the Polytechnic College and for a time served as chemistry teacher at the University of Copenhagen.[21] On the recommendation of Ørsted, he participated in the Galathea expedition around the globe between 1845 and 1847. Latest by the summer of 1852 Rink was aware of Thomsen's project and its potential commercial importance. When he returned to Copenhagen in the fall of 1852, he reported that the amount of cryolite was large and that the mineral was easily accessible.[22] With Rink's information Thomsen realised that his discovery might be transformed into an industry *if* he could be supplied with large quantities of the raw material. It was a great "if."

The first hurdle to pass would be to be granted a patent or royal privilege for the process, for which Thomsen applied on 2 November 1852. At that time, a Danish patent was only granted if the authorities judged the invention to be useful and able to be developed commercially in a national context. Novelty was less important. Furthermore, a patent – meaning the exclusive right to manufacture a product or exploit a method – was usually granted for five or ten years, but with the proviso that the invention had to be put into productive application within a year after the granting of the patent. If not so, the patent protection could be cancelled. It is no won-

---

21. See Brown (1894).
22. Topp (1990) and Garboe (1959-1961), vol. 2, pp. 213-221. Rink also mentioned that the Ivittuut deposit contained minerals based on iron, lead, copper and tin, but in much too small amounts to allow commercial production of the metals.

der, then, that Thomsen was eager to convince the authorities that his invention was indeed commercially useful and would mean an economic benefit for the country. He stressed in his report that the cryolite soda process "is no empty speculation, but the method is really practically feasible." Thomsen estimated that an annual extraction of 200 tons of cryolite would result in a profit of 7,800 Rbd (rix-dollar), a result obtained from the assumed sale of 350 tons of soda (21,000 Rbd) minus the total costs, which he estimated to 13,200 Rbd. At the time the annual income for a fully employed worker in Copenhagen was 180 Rbd and the rate of exchange about 9 Rbd = £1.

Thomson's figures in his report of 1852 show that he had a clear grasp of the chemical mechanisms in the described process. Using the above reactions, the figures can be reconstructed as follows. Theoretically, one equivalent of cryolite will result in 1.5 equivalent of anhydrous soda and then also 1.5 equivalent of the commercial product crystalline soda ($Na_2CO_3 \cdot 10H_2O$). In terms of weights, 200 tons will yield an amount of 392 tons of crystal soda.[23] Realising that the industrial process would not follow the theoretical scheme, Thomsen reduced the figure to 350 tons. That is, he presumed a yield of 90% of what was theoretically possible, which was still a somewhat optimistic estimate. With regard to the amount of lime, a stoichiometric calculation shows the theoretical cryolite-to-lime weight ratio to be 1.37, as compared to Thomsen's 1.31. However, his experiments had shown that a complete decomposition of the cryolite required an amount of lime about 1.3 times as large as the theoretical value. That is, the two figures approximately cancel, and the needed mass of lime will equal the mass of cryolite.

Although Thomsen emphasised the manufacture of soda, he also called attention to the commercial aspects of the alum, which could easily be produced by the sodium alumina not otherwise used. Alum – usually potassium alum in the form $KAl(SO_4)_2 \cdot 12H_2O$ – was widely used in paper manufacture, tanning of leather, manufacture of mordants, and several other industries. Although

---

23. Eleven years later Thomsen, using more accurate atomic weights, corrected the figure to 408 tons. See Thomsen (1863a), p. 366.

the alum resulting from cryolite was sodium alum – roughly $NaAl(SO_4)_2 \cdot 12H_2O$ – Thomsen believed it would find equally good industrial application. The fact that alum manufacture would require large quantities of sulphuric acid was, Thomson argued, only an advantage. He maintained it would stimulate a great increase in domestic production of the important acid and thereby would pave the way for a chemical industry in Denmark centred on the two chemicals, soda and sulphuric acid, that already had revolutionised the industry elsewhere in Europe – and all would be based on local raw materials.

2.3. Towards a Danish chemical industry

Thomsen was granted his patent on the manufacture of cryolite soda on 23 January 1853. Although it was valid for a period of ten years, it included the standard clause that the invention had to be put into practical use within a year, meaning by early 1854. Since the entire scheme rested on large supplies of cryolite, and there still was no organised extraction of the mineral in Greenland, Thomsen was in a dilemma. A subsequent application for the free right of mining of cryolite was turned down by the Ministry of the Interior which on 29 March 1853 decided that the rights of mining belonged exclusively to the state-owned KGH. A concession for mining in Greenland had earlier been applied for by Jacob Henrik Lundt, a wealthy merchant from Aalborg, who was primarily interested in the deposits of graphite, lead and tin minerals but also had an eye for the possible use of cryolite. In fact, as early as 1850 Lundt sold about 250 kg of cryolite in England, which was the first commercial transaction involving the mineral. The small amount was not used to manufacture soda, but boiled with slaked lime it was used as a substitute for soda by British soap makers in the Staffordshire county.[24] Lundt's applications for mining rights 1851-1852 were declined

24. In the years 1847-1869 the Danish-Jewish novelist and publisher Meïr Aron Goldschmidt ran a political magazine titled *Nord og Syd* (North and South). Quoting the trade magazine *Mining Journal* (4 October 1856): p. 670, in a detailed article of 1857 *Nord og Syd* referred very positively to the emerging industrial use of cryolite. The

for largely the same reasons that Thomsen's application was declined. The Government wanted to keep mining in Greenland a state monopoly.

Not only had Thomsen no financial backing for his project, in much of the period 1853-1854 he was abroad on a chemical study tour and for this reason unable to communicate easily with authorities and potential investors in Copenhagen. He was on the verge of giving up the whole scheme when he joined forces with Johan Christian Georg Howitz, a 32-year old engineer with good connections to industrialists and financiers. Howitz was an old friend of Thomsen with whom he had studied chemistry and physics at the Polytechnic College. He graduated in 1846, the same year as Thomsen, and for a short period in 1848-1849 he served as assistant to Forchhammer. Howitz was subsequently employed as technical assistant at the Copenhagen Water Works and in 1854, when the city decided to establish its own gas works, he was put in charge of its construction. A pioneer in the establishment of coal gas supply in Denmark, he became the first managing director of the Copenhagen Gas Works (Vestre Gasværk) founded in 1857.[25] His interest in coal gas technology led him the following year to apply local bog iron as an efficient purifier to remove hydrogen cyanide and hydrogen sulphide from coal gas. Bog iron is cheap and rich in Fe_2O_3. The processes in Howitz's method to get rid of the toxic gases H_2S and HCN can be reconstructed as

$$Fe_2O_3 + 3\,H_2S \rightarrow Fe_2S_3 + 3\,H_2O$$

and

---

article was unsigned but possibly written by Goldschmidt, for which reason I have placed it in the bibliography as Goldschmidt (1857). See also Sveistrup (1956) and Topp (1990), pp. 13-22.

25. The first Danish coal gas system was established in Odense in 1853. A detailed historical account of gas works in Denmark is given in Hyldtoft (1994), which includes an extended summary in English. As we shall see in Chapter 8, Thomsen took a keen interest in the Copenhagen gas system both from a scientific and an economic point of view.

$$2 \, Fe_2O_3 + 3 \, FeO + 18 \, HCN \rightarrow Fe_4[Fe(CN)_6]_3 + 9 \, H_2O.$$

The method soon spread to other Danish gas works and also to other countries. It resulted in a considerable export of Danish bog iron for this purpose.[26]

After their graduation in 1846 Thomsen and Howitz had remained in contact, and in the summer of 1853 the latter offered his assistance in the still more problematic cryolite case. The result was a company formed by the two polytechnic chemists. In Thomsen's absence it was Howitz who in January 1854 applied, this time successfully, for a renewal of the royal privilege concerning cryolite soda manufacture. Although Howitz tried hard to raise capital among Danis industrialists and investors, at first his efforts were in vain. Apart from lack of capital, the company formed by Thomsen and Howitz also had to fight with the bottleneck of insufficient supplies of cryolite from the KGH. While they received 56 barrels in 1854, the following year the shipment from Greenland had dwindled to only 3 barrels, and a supply of about 30 barrels from KGH in 1856 did not help much (1 barrel is approximately 10 kg). It was much too little for the cryolite soda industry of which they dreamed. Parts of the small amounts were used by Thomson in experimental work on a larger scale than in his earlier experiments. Most of the rest was sold to local soap makers.

Still in 1856 Thomsen had not published his invention in either the scientific or popular literature. As cryolite and its possible use in the manufacture of either soda or aluminium became more known, it led to various claims that the relevant processes were due to other people. In this case, as in several later cases, Thomsen felt obliged to defend his priority. For example, in an article in the Copenhagen journal *Dagbladet* from the summer of 1856 it was suggested that Lundt was the originator of the industrial use of cryolite. Thomsen at once objected: "It is the writer of this letter and no one else who first demonstrated the importance of cryolite for industry and trade. When I made my investigations concerning cryo-

---

26. See Kragh and Petersen (1995), pp. 285-287.

lite and established the methods of production that would create a new line of industry in Denmark, ... there was still no technical application of cryolite."[27]

The fortunes of the frailing cryolite adventure changed to the better in the fall of 1855, when Thomsen made contact with the young, up-and-coming businessman Carl Frederik Tietgen, who was on his way to becoming the country's leading industrialist, banker and entrepreneur.[28] Denmark's first telegraph line, connecting Copenhagen and Hamburg, had opened in early 1854 and Tietgen was quick to realise the commercial importance of the new technology. He had just begun his involvement with the telegraph business, which soon would lead to the establishment of the Great Northern Telegraph Company, but he was on the lookout for any profitable investment and considered cryolite a promising one. Tietgen had come to know about the mineral while staying in Manchester, where one of his associates had mentioned the possible use of cryolite in connection with the production of aluminium (and not soda). His interest was not so much concerned with Thomsen's patent as with the mining of cryolite for the purpose of selling it to manufacturers of aluminium in France.

In late 1855 Tietgen arranged with Thomsen and Howitz that if they could obtain the right to extract cryolite, he would finance and organise an expedition to Greenland with the object of shipping large quantities to Copenhagen. The application of Thomsen and Howitz of 28 January 1856 was granted and shortly thereafter the two chemist-entrepreneurs made an agreement with Tietgen, according to which he took over the company and the right to use Thomsen's patent. In September the schooner *Sønderjylland*, acquired by Tietgen and a consortium of five business partners, returned to Copenhagen loaded with 134 tons of cryolite. However, contrary to Tietgen's expectations, it proved impossible to find

---

27. *Dagbladet*, 6 June 1856. Thomsen's full letter is reproduced in Topp (1990), pp. 64-65. Thomsen also highlighted his priority in Thomson (1862a) and, aimed at an international audience, in Thomsen (1863a).

28. A full biography is given in Lange (2006), which deals with Tietgens' early involvement in the cryolite industry on pp. 66-69.

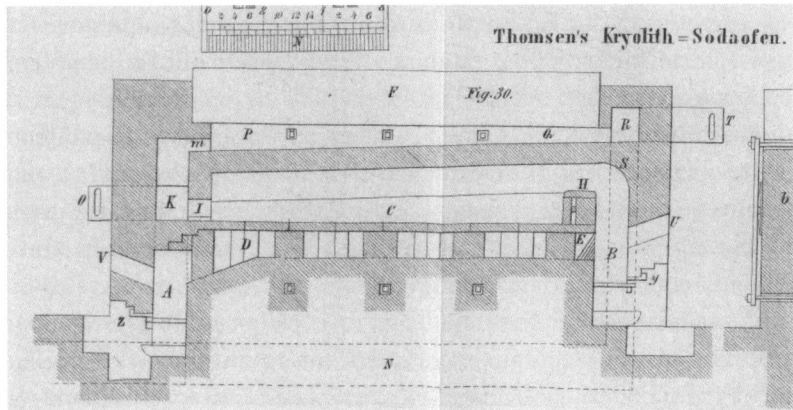

Figure 2.4. Thomsen's kiln for the manufacture of soda from cryolite. Source: Thomsen (1863a).

enough customers for the commodity in the small European aluminium industry.

As a result of the disappointing investment adventure, in May 1857 Tietgen sold the rights to use Thomsen's patent to the Danish trading company Theobald Weber & Co., which also took over approximately 120 tons of cryolite left over from the shipment of the previous year. According to the contract, Thomsen and Howitz were each secured one third of the profit of the cryolite business. Weber & Co. wanted to use cryolite soda in their new manufacture of mirror glass, the first production of such glass in Scandinavia. For this purpose, where chemically pure soda was required, a factory at Christiansdal near Haderslev in Southern Jutland was established. In the autumn a small cryolite soda industry began to operate, proving on an industrial scale that Thomsen's method actually worked. On the other hand, the Christiansdal factory was not really successful, as it was plagued by technical and economic problems. After about a year's operation it was decided to move the production to Copenhagen, where a new and much bigger factory was established.

The new factory, named "Kryolithfabriken Øresund" (so named after the nearby belt between Denmark and Sweden), was inaugurated on Thomsen's birthday, 16 February 1859, an indication that the factory would not have existed had it not been for Thomsen's research. It was owned by Weber & Co. and had Thomsen as technical director and joint owner. Whereas the Christiansdal works had purchased only about 200 tons of cryolite and manufactured some 150 tons of soda, the Øresund factory was designed to an annual production of 2500 tons. The actual production of soda during the early years is not known, but with a consumption of 1500 tons of cryolite in 1859 it must have been about 2200 tons. In the five-year period 1860-1864, a total amount of 13,100 tons of raw cryolite was shipped from the Ivittuut deposit, peaking in 1862 with 4700 tons.

The main features in the production process at the Øresund factory in the early period 1857-1865 were as follows.[29] The raw materials were limestone and cryolite, which were pulverised and mixed in the ratio 3:2. The powder was transferred to a specially designed kiln where it was heated for about three hours to complete the decomposition into fluorspar and soda clay soil (Figure 2.4). After cooling, the mixture was treated with water, resulting in a precipitate of fluorspar and a concentrated solution of soda clay soil, roughly of the form Na_3AlO_3.

This first phase of the process differed in some respects from Thomsen's original proposal where the decomposition was performed by boiling aqueous suspensions of cryolite and quicklime. This was also the process initially used at the Christiansdal works in 1857, but Thomsen soon realised that a solid-state reaction, or what was known as the "dry way," was preferable if the heating could be carefully controlled so as to heat the mixture without melting it. Comparative tests showed that boiling with quicklime gave a somewhat higher soda yield than heating with solid limestone, a result of the formation of Glauber salt (Na_2SO_4 , 10 H_2O) due to impurities of sulphur in the cryolite. However, it also turned out that the heating method resulted in a more concentrated solution of soda after lixiviation. It therefore required less heat for the final evaporation and was, in effect, considerably more economic. After many difficulties, in 1861 Thomsen designed a new type of kiln, specifically to meet the demands of solid cryolite-lime heating. Apart from giving

29. For further details, see Thomsen (1862a), Jarl (1909a), and Halland (1911).

a satisfactory soda yield (150-160%), Thomsen's kiln also had a better fuel economy, requiring 800 kg of pit coal for a daily processing of 2.5 tons of cryolite into soda.[30]

According to a later description, due to the American chemist and metallurgist Joseph Richards at Lehigh University and referring to the larger production at the Øresund works:

> Thomson constructed a furnace in which the flame from the fire first went under the bed of the furnace, then over the charge spread out on the bed, and finally in a flue over the roof of the hearth. The hearth ... is charged twelve times each day, each time with 500 kilos of mixture, thus roasting 6000 kilos daily, with a consumption of 800 kilos of coal. ... In this furnace the mass is ignited thoroughly without a bit of it melting, so that the residue can be fully washed with water. The reaction commences at a gentle heat, but is not completed until a red heat is reached. Here is the critical point of the whole process, since a very little raising of the temperature above a red heat causes it to melt.[31]

In the second phase of the process, the solution of clay soil was treated with carbon dioxide, coming entirely from the heating of the limestone whereby soda was formed. After evaporation and precipitation of impurities, soda crystals were obtained as the end product. Tests of the new kiln fed with pure cryolite showed a yield of 197% or very close to the theoretical maximum of 204%. However, in the ordinary production (where the cryolite contained impurities), the factory only managed to get a yield of about 160%.

Apart from soda, another valuable product was obtained from the precipitate of clayey soil that, according to Thomsen's analyses, consisted of bauxite (45%), water (35%), and soda ash (20%), where the latter product refers to sodium oxide. In the early 1860s Thomsen developed a titrimetic method to determine the composition of

---

30. Thomsen (1862a) and (1863a).

31. Richards (1887), pp. 146-149. Richards referred to "Julius Thomson," without given any information about the inventor of the process. The misspelling was and still is common among English-speaking scientists.

clay soil products with the aim of standardising the sulphuric clayey soil produced by the cryolite factory.[32] The clever method was based on titration with a dilute solution of sodium hydrogen carbonate ($NaHCO_3$) and comparison with analysis based on weight and volume proved it to be very accurate. Other specially developed titrimetic methods were used to determine the strength of the soda lye, which could not be reliably determined by the standard titration with sulphuric acid and litmus as colour indicator. At the Christiansdal works, the soil was either sold directly to consumers or sent to Germany where it was turned into the marketable "sulphuric clayey soil" by treatment with 60% sulphuric acid. This clay consisted mostly of hydrated aluminium sulphate. The large quantities of sulphuric acid needed for this part of the process were primarily bought from the domestic works established by the British-Danish industrialist Joseph Owen, but supplemented with imported acid. During the years 1870-1886 the Øresund factory had its own production of sulphuric acid. The clay soil from cryolite attracted much interest and was a major product of the industry, both in Copenhagen and elsewhere.[33]

2.4. On aluminium and cryolite

Thomsen's interest in cryolite focused on its use for the manufacture of soda, while he paid little attention to the mineral's content of aluminium. All the same, the histories of the mineral cryolite and the metal aluminium are tightly interwoven, and in a national context Thomsen played a part in both. In the long run it was not the soda manufacture that made cryolite a valuable commodity but rather its use in the aluminium industry, which only took off at the time of Thomsen's death. A brief account of aluminium's early history and its connections to cryolite will be appropriate.[34]

---

32. Thomsen (1863b). The paper was first given at the ninth meeting of Scandinavian scientists held in Stockholm in July 1863 and also appeared in the proceedings of the meeting only published 1865 (*Beretninger*, pp. 331-340).

33. A detailed description is given in Wagner (1862). On Owen and his manufacture of sulphuric acid, see Kragh and Petersen (1995), pp. 137-142.

34. The literature on the history of aluminium is extensive. For the early phase, see

Although aluminium was only isolated in metallic form in the 1820s, its existence had been recognised since about 1780. For example, Lavoisier's list of chemical elements appearing in his seminal *Traité Élémentaire de Chimie* from 1789 included "alumine" as an earth, but as Lavoisier pointed out, the metal had not yet been separated from the clayey earth supposedly containing it. Some twenty years later Humphry Davy suggested the name "alumium" (and later "aluminum"), admitting that he had been unable to isolate the metal by means of electrolysis. The modern name "aluminium" dates from 1812, although in American spelling it is "aluminum." To put it in a nutshell, the problem was that in clay earths such as bauxite, which contains 52% of the oxide Al_2O_3, the metal is very tightly bound to the oxygen. Only in 1825 did Ørsted report a positive result based on a chemical method which he described in a communication to the Royal Danish Academy of Sciences in Copenhagen. Ørsted's method proceeded in two phases, the first one consisting in the production of anhydrous aluminium chloride. It can be written

$$Al_2O_3 + 3\ C + 3\ Cl_2 \rightarrow 3\ CO + 2\ AlCl_3$$

To separate out the metal from the chloride, Ørsted used in the second phase potassium in the form of an amalgam (the mercury alloy KHg, where Hg is chemically inactive):

$$AlCl_3 + 3\ K \rightarrow 3\ KCl + Al$$

The result of Ørsted's experiments was what he described as "a lump of metal which in its colour and lustre somewhat resembles pewter." He did not investigate this first sample of impure aluminium – or what he proposed to call "argillium" – any further. Ørsted wrote to a few of his colleagues about his discovery, which received a brief notice in Schweigger's *Journal für Physik und Chemie* and also in Poggendorff's *Annalen der Physik und Chemie*. Apart from this, "the dis-

---

for example Richards (1887), Hall (1976), Morel (1992), and Williams (1993).

covery was not trumpeted to any great degree," as Ørsted's biographer puts it with an understatement.[35]

It was only when the young German chemist Friedrich Wöhler visited Ørsted in 1827 that something happened. Ørsted generously passed over his method to Wöhler, who the following year succeeded in producing aluminium with pure potassium (instead of the KHg amalgam) and to determine the main properties of the new metal. During the nineteenth century Wöhler was generally recognised as the discoverer of aluminium, whereas Ørsted's pioneering work of 1825 was either ignored or not appreciated as a proper discovery.[36]

Wöhler's aluminium remained a rarity and luxury product, extremely expensive and of no practical use. The situation only changed in 1854, when Henri Sainte-Claire Deville, professor of chemistry at the École Normale in Paris, found a new method to produce the metal in pure form, namely to reduce aluminium chloride with sodium instead of potassium. In Deville's process, alumina (Al_2O_3) was mixed with charcoal and, after heating, it was treated with dry chlorine to form $AlCl_3$; a stream of chlorine vapour was passed over molten sodium to reduce the chloride to pure aluminium.[37] The pure metal produced by Deville was much less expensive than previously, if still as expensive as silver. It created a sensation, catching the imagination of people who hailed it as the metal of the future. As the Danish periodical *Nord og Syd* (North and South) noted in 1857, the light aluminium seemed an obvious metal to use for airships. A few years later, in his famous science fiction novel *From the Earth to the Moon*, Jules Verne described the metal as possessing "the whiteness of silver, the indestructibility of gold, the tenacity of iron, the fusibility of copper, the lightness of glass." With aluminium objects displayed at the Paris Exhibition in 1855, and with Deville's metal being enthusiastically received by the public, aluminium

35. Christensen (2013), p. 428.

36. Today Ørsted's priority has won general recognition. On the controversy concerning the discovery of aluminium and the relevant historical sources, see Kjølsen (1965), pp. 102-121.

37. Deville (1859). See also Foster (1859) for a contemporaneous account of aluminium and its potential uses.

was suddenly *en vogue*. No wonder, then, that the aluminium content of cryolite also caught the attention of a few chemists and industrialists who hoped to make the Greenlandic mineral a raw material of this metal of the future.

Thomsen was aware of the possible use of cryolite for aluminium manufacture, and in 1857 he experimented with electrolytic decomposition of cryolite, not in molten form but suspended in water. We know about Thomsen's experiments from a letter Tietgen wrote to a business associate: "Thomsen believed that in the next week he would succeed [in making aluminium] by using a hitherto unexplored method, namely the wet way in which it [the cryolite] is dissolved and aluminium is precipitated by means of a galvanic current."[38] However, Thomsen was forced to conclude that the electrolytic method was unsatisfactory and consequently he never published his results. During his journey to Germany and France in 1853-1854 Thomsen was in contact with the distinguished chemist at the University of Berlin, Heinrich Rose, whom he informed of his recent work and also of the possible manufacture of aluminium from cryolite as raw material. Perhaps as a consequence, Rose devised a method in which he placed, in an iron crucible, finely divided cryolite between thin layers of sodium. To this he added a layer of potassium chloride as a flux. By strong heating in a furnace, globules of aluminium were formed, due to the process

$$Na_3AlF_6 + 3\,Na \rightarrow 6\,NaF + Al$$

Rose was very optimistic with regard to the use of cryolite. "I am of the opinion," he wrote, "that cryolite is the best adapted of all compounds of aluminium for the preparation of this metal. It deserves the preference over aluminium-sodium chloride or aluminium chloride, and it might still be employed with great advantage even if its price were to increase considerably."[39] Four years later, at a meeting

---

38. Letter of 16 May 1857 quoted in its Danish original in Topp (1990), p. 91. Tietgen described his involvement with Thomsen and the manufacture of cryolite and aluminium in Tietgen (1904), pp. 41-45.

39. Rose (1855), p. 161. Without mentioning Ørsted, Rose ascribed the discovery of

of the British Society of Arts, the secretary of the Society repeated Rose's optimistic judgment: "I am led to believe that the cryolite process is the one that will ultimately be preferred to that of the chloride of aluminium."[40]

In his article of 1855, Rose mentioned that "powdered cryolite is completely decomposed by quicklime and water," which was an indirect reference to Thomsen's process and a main reason why this process became widely known despite Thomsen's decision not to publish on it. As it happened, Rose was often mentioned as the one who had discovered the decomposition of cryolite by lime. Thomsen of course new better, and in 1862 he protested: "I am the inventor of the entire cryolite industry; the process of obtaining soda from cryolite is fully and in detail my invention."[41] He did not refer to either Rose or other chemists.

There was at the time a great deal of interest in cryolite-based aluminium production, which was investigated not only by Rose but also by Wöhler in Germany and Allan Dick in England.[42] The latter, a metallurgist at the Government School of Mining in London, packed layers of cryolite and sodium in a platinum crucible, which he heated to redness over an air-blowpipe. At a meeting of the Royal Institution in March 1855 he demonstrated small pieces of aluminium prepared in this way. Attempts to apply the new methods on an industrial scale took place immediately after the works of Deville and Rose became known. In 1859 F. W. Gerhard opened the first manufactory in Britain, using powdered cryolite mixed with common salt to which was added pieces of sodium.[43] However, his plant in Battersea was short-lived and it was only in France that larger amounts of aluminium were produced in this early era.

---

aluminium to Wöhler. Although he did not refer to Thomsen, there are reasons to believe that Rose's process was indebted to his conversations with Thomsen. See Kragh (1995).

40. Foster (1859).

41. Thomsen (1862b), p. 443.

42. Wöhler (1856); Dick (1855). Wöhler, who obtained his cryolite from Forchhammer in Copenhagen, used cryolite mixed with sodium and potassium chloride. See the description in Richards (1887), p. 289.

43. On Gerhard and his works, see Foster (1859) and Williams (1993).

Between 1855 and 1857 factories were established in Amfreville near Rouen and in Nanterre, both of them originally managed by Paul Morin, a mining engineer and friend of Deville. In the Nanterre works an improved version of Deville's method was used and cryolite only added as a flux; the Rouen plant, on the other hand, was based on the Rose-Dick cryolite method. In 1859, the Rouen works produced about 960 kg of aluminium and the Nanterre works about 600 kg, but there were constant difficulties with finding a market for the still very expensive metal. The fact that both of the existing aluminium works used cryolite, if in two widely different ways, made it natural for Tietgen and his business partners involved in the cryolite trade to focus on this market. Although only small amounts of cryolite were used at the Nanterre works, at one stage Deville considered using the mineral as a raw material. Together with Morin, he experimented with cryolite, both by means of the Rose-Dick method and by means of electrolytic decomposition.[44]

Baruch Levy, a young Danish-Jewish chemist and pharmacist, stayed in Paris when Deville made his first experiments with aluminium. In April 1854 he sent a small sample of Deville's metal to Forchhammer, his former teacher, mentioning that it had been "manufactured by means of the galvanic current and thus in a way which is entirely different from the one applied by Wöhler." Levy acted as an intermediary between Forchhammer and Deville in the latter's attempt to obtain enough cryolite for a test production of aluminium. In another letter to Forchhammer, from the autumn of 1855, Levy wrote: "I had the pleasure of meeting Thomsen during his stay in Paris and he promised me that, after he had returned, he would hopefully be able to comply with my wish; he expected very soon a new shipment from Greenland."[45] Levy's meeting with Thomsen probably took place in 1854.

---

44. Deville (1856). Methods of electrolytic decomposition based on cryolite mixed with sodium chloride were developed by chemists as late as the 1880s, but the yield of aluminium was very low.

45. Both letters are quoted in Garboe (1966). Deville had wanted 200 kg cryolite, but Forchhammer was only able to send him about a tenth of the amount. It is unknown if Thomsen met with Deville or other French scientists during his stay in Paris.

Aluminium production based on cryolite as raw material never developed into a mature industry. Not only was the demand for the new metal very limited, production also depended on the availability of large amounts of the Greenlandic mineral at a competitive price. In 1856 a French corvette visited Ivittuut with the purpose, among other things, of evaluating the size of the cryolite deposit and the possibilities of turning cryolite into the basis of commercial aluminium production. The report of the involved French scientists concluded that cryolite would remain a costly mineral unsuited for industry at a larger scale. Deville agreed and added to this a patriotic argument: "It is fortunate that cryolite is not indispensable, for no one would wish to establish an industry based on the employment of a material which is of uncertain supply. ... The Société de Nanterre has judged that there is no advantage in establishing a manufacture of aluminium and soda based on cryolite."[46]

Tietgen had sold 10 tons of cryolite to the Rouen works in late 1855, but attempts to sell larger quantities failed. In 1857, the works at Nanterre ordered 5 tons, but this was all the cryolite that was sold for the manufacture of aluminium. Although suggestions of using cryolite for this purpose continued up to the 1890s, by the end of the 1850s it had become clear to most chemists and industrialists that an aluminium industry based on cryolite would not become a reality. The works at Amfreville, the only place where cryolite was used as a raw material, closed in 1864, unable to compete with the Deville process. Although aluminium production based on Deville's method continued, the earlier euphoria was replaced by a more realistic view of the metal's limited use. "There was a time," wrote August Thomsen, Julius' brother, in a book of 1879, "when people thought that this new metal faced a bright future ... but the high price of 30 kroner for 1 pound has prevented its use as a common commodity."[47]

It was only with the emergence of the Héroult-Hall process in 1886 that the fortune of the aluminium industry began to improve, first slowly but eventually drastically and irreversibly. In this impor-

---

46. Deville (1859), p. 110.
47. A. Thomsen (1879), p. 490.

tant industrial process, invented by the two 23-year-old chemists Paul Héroult in France and Charles Hall in the United States, cryolite only entered as a flux. Thomsen followed the new development with interest. In 1892, only four years after Hall had established the first aluminium plant in Pittsburgh, he offered a prophecy much more optimistic than his brother's. "It is likely," he said, "that we are entering an era when iron, steel and bronze in many cases will be replaced by another metal and its alloys; the metal is aluminium, which just forty years ago was such a rarity that it was not to be found in even the collections of chemical laboratories."[48] By that time the world production of aluminium was about 1,000 tons per year (today it is approximately 40 millions tons).

2.5. Expansion and decline

The soda cryolite works Øresund was granted the right to unlimited extraction of cryolite at its foundation in 1859. Some years later, after a period of economic difficulties, the trading company J. P. Suhr & Son entered the board of directors and bought the rights from Thomsen and Howitz. In 1865 the Cryolite Mining and Trading Company Ltd (Kryolit, Mine- og Handelsselskabet, KMHS) was formed as a company independent of, but with close contacts to, the Øresund works. J. P. Suhr & Son was the main shareholder, Tietgen owned approximately 20 per cent of the shares (a total of 500.000 rdl.), and Thomsen and Howitz were minor shareholders with about 2 per cent and 1.2 percent. Thomsen was appointed technical director of the new company and received a substantial bonus of 7.000 rdl. Whereas KMHS was responsible for all extraction and sale of cryolite, working up of raw cryolite and preparation of soda and other products took place at Øresund. The long-time collaboration between Thomsen and the Suhr trading company, until 1875 in the form of Ole Bernt Suhr, proceeded to the satisfaction of the both parties. When the company ceased in 1899 it left a large bequest of 100,000 kroner to the Royal Danish Academy in recognition of Thomsen's seminal contributions to the cryolite in-

48. Thomsen (1892a), p. 34.

Figure 2.5. Loading of cryolite from the Ivittuut mine. https://c3.static-flickr.com/4/3783/9912955186_7148427cd5_c.jpg

dustry and Danish trade in general. The bequest, which was named "J. P. Suhr and Sons's Grant in Memory of Professor, Dr. Med. & Dr. Phil. Julius Thomsen," was to be used for the general purposes of the Academy at the discretion of its president – which happened to be Thomsen.[49]

By the early 1860s, a large-scale production of cryolite soda and alum earth based on Thomsen's methods and designs was in operation in Copenhagen. The Øresund factory was successful in the sense that it produced large amounts of soda by a new and technically well-functioning process, and also in the sense that it provided the Danish State with a very substantial income in terms of the taxes put on the mined cryolite. It has been estimated that by 1910 the

49. See Lomholt (1942-1973), vol. 1, pp. 612-615. Although Thomsen was sometimes titled "Dr. Med. & Dr. Phil." (doctor of medicine and doctor of philosophy), in fact he never wrote a doctoral dissertation. The titles were honorary (doctor honoris causa) and dating from 1877 and 1879, respectively.

total income was 25-30 million kroner, far more than the State had used for scientific research in the same half-century period.[50] According to contemporary sources, the Øresund factory was technically advanced and a site bustling with industrial activity. Employing about 100 workers, it was one of the largest industries in the country (Figure 2.5). During the early 1860s, the purchased amount of cryolite was about 2000 tons per year, which by the standard soda-to-cryolite yield of 1.6 translates into an annual production of soda of roughly 3200 tons.

The production of soda in Copenhagen was only part of the cryolite business in the 1860s, when the Thomsen process was transferred also to a few other countries. In 1861, Harburger Alaun-Werk started production of soda and alum earth from cryolite, and during the next three years other factories based on Thomsen's process were established in Ludwigshafen, Warsaw, Breslau, Prague, and Uithoorn (near Amsterdam). The production of the latter factory peaked in 1866, when the Dutch company employed 38 workers and produced about 800 tons of soda. Of much greater importance was the contract that Pennsalt (Pennsylvania Salt Manufacturing Company) signed with Thomsen, Howitz and the Øresund company in early 1865.

Realising the value of cryolite for its consumer products, in 1865 the General Superintendent of Pennsalt, Henry Pemberton, travelled to Copenhagen to negotiate the contract with the involved Danish parties, not least the Ministry of the Interior.[51] Thomsen was deeply involved in the negotiations, and one gets from his involvement a good impression of his skills as a diplomat and a businessman.[52] Over a period of ten years, the American company agreed to purchase an annual amount of 6,000 tons of cryolite at a fixed price of £3 per ton. The factory at Natrona, Pennsylvania, produced soda and soda-based products according to Thomsen's method and also

---

50. See Jessen-Hansen (1910), p. 350, who used the figure to illustrate with "a concrete example" the usefulness of pure science.

51. Pennsalt was founded in 1850 and still exists, if now as the Pennwalt Corporation. For a brief company history, emphasising the importance of the cryolite business, see Grossman and Jennings (2002).

52. See Topp (1990), pp. 304-310.

used cryolite as a component in the production of aluminium, in-secticides, ceramics, and abrasives. From about 1870, the Americans sold part of their cryolite to glassware manufacturers who used it to make milk-white glassware marketed as "hot-cast porcelain" or, in Europe, "American porcelain." During the five-year period 1865-1869, KMHS exported 44,700 tons of cryolite, of which 63% went to Pennsalt and the other 37% to the five European factories.

The total output of cryolite soda in the late 1860s is unknown, but must have been very substantial. The output can be estimated by assuming that 90% of the cryolite was used for soda and to have had an average purity of 90%, which leads to an average annual output of cryolite soda of about 14,000 tons. About 53% was pro-duced in Pennsylvania, 15% in Copenhagen, and 32% in the other European works. In comparison, the annual German production of soda in the same period was about 6,000 tons, and that of France about 45,000 tons, almost all of it produced by means of the Leb-lanc process.[53]

While Thomsen acted as technical director and supervisor for Øresund, he was assisted by a series of young chemists who under-took the daily routine work. In the period 1864-1867 the assistants included some of the country's most promising chemists and engi-neers, including Vilhelm Storch, Gustav Hagemann, and Vilhelm Jørgensen.[54] In 1869 the Øresund factory was sold on an auction to the two last-mentioned polytechnic chemists, 25-year-old Vilhelm Jørgensen and the two years older Gustav A. Hagemann. Under their leadership the company continued as the Øresund Chemical Factories (Øresunds Chemiske Fabriker).

Shortly after graduation from the Polytechnic College in 1865, Hagemann was sent to Pennsylvania to assist Pennsalt with its new production of cryolite soda, and during his stay in the United States

53. Haber (1958), pp. 41-47.

54. V. Storch would later become a prominent dairy chemist and professor at the Agricultural College. V. Jørgensen was Tietgen's brother-in-law. In addition to the three mentioned chemists, also Peter T. H. Schiellerup, who became a pharmacist, and Thomas Thomsen, Julius' younger brother, served as assistants at Øresund. All of Thomsen's assistants were polytechnic candidates rather than university-trained chemists.

he also established chemical factories producing bromine from magnesium salts.[55] It was the profit he earned from selling the bromine factories that allowed him and Jørgensen to buy the Øresund cryolite factory. While in Natrona, Hagemann examined some of the cryolite shipped from Ivittuut and found several new fluorine minerals related to cryolite. By chemical analysis he established tentatively their composition and crystal forms. What is known as chiolite, thomsenolite and hagemannite are rare cryolite-like minerals first described by Hagemann.[56] In modern nomenclature, the compositions of the two first are $Na_3(AlF_4)_3$, $2NaF$ and $NaCaF_3 \cdot H_2O$, respectively, whereas hagemannite is an impure form of thomsenolite.

The industry based on Thomsen's experiments and chemical insight was widely and not unreasonably seen as one more proof of the amazing industrial power of applied science. August Thomsen, Julius' brother, published in 1879 a comprehensive and semi-popular book on the industrial applications of science entitled *Naturkræfterne i Menneskets Tjeneste* (The Forces of Nature in the Service of Man). Without naming the inventor, he included a section on cryolite soda, a product "based on a Danish invention." The new manufacture, he wrote, was "a beautiful example of the impact of science on industry." As he explained:

> Supported by the spirit of entrepreneurship, laboratory experiments on a limited scale have given rise to a relatively important industry. In this way a mineral that previously was only available in mineralogical collections has turned into a valuable product; it provides work and dignity to many people and secures the public finances with a significant income.[57]

---

55. See Vinding (1942), pp. 44-81. On Hagemann's later career see also Nielsen and Wistoft (1996). Bromine was at the time used for photography (daguerreotype) and, in the form of potassium bromide, also had medical uses as a sedative.

56. Hagemann (1866a) and Hagemann (1866b). Naturally, the name "hagemannite" was not suggested by Hagemann himself. It is due to the American geologist Charles Sheppard, who wrote a paper on various minerals in the *American Journal of Science and Arts* of 1866.

57. A. Thomsen (1879), p. 333.

August's older brother undoubtedly agreed.

In spite of the expansion of the cryolite business, the production of cryolite-based soda soon declined. The factory in Copenhagen was constantly in economic troubles and the European factories had difficulties in selling their products. One reason for the difficulties that faced the industry in the 1870s was that supplies of cryolite became of lower quality, hence more costly to work up, as the high-grade cryolite was used up. More important was the general decrease in prices for soda, which followed the competition between the established Leblanc process and the new, highly successful Solvay process. In this process, developed by the Belgian chemist and inventor Ernest Solvay in the 1860s, soda is produced from common salt and ammonia, which may be schematised as

$$NaCl + NH_3 + CO_2 + H_2O \rightarrow NaHCO_3 + NH_4Cl$$

$$2 NaHCO_3 \rightarrow Na_2CO_3 + CO_2 + H_2O$$

As a result of economic and other factors, cryolite was no longer competitive on the soda market. The European works ceased production between 1870 and 1884, and the Øresund factory was forced to change to other products in order to survive.[58] Production of cryolite soda peaked with 3,010 tons in 1871 and then declined rapidly. In 1894, soda production from cryolite ceased in Copenhagen; six years later, it also ceased at Pennsalt.

Although the Danish cryolite business continued and in fact prospered in the early decades of the twentieth century, when Thomsen died in 1909 his great invention belonged to the past. Cryolite was now increasingly used as flux in the electrolytic production of aluminium, but in the end the natural and still more low-grade Greenlandic mineral was replaced by other fluxes sometimes called "synthetic cryolites." These are today typically manufactured

---

58. Still in the early 1880s, students at Copenhagen's Polytechnic College were taught the production method of cryolite soda alongside the more important methods of the Leblanc and the Solvay processes. See A. Thomsen (1883), pp. 194-214, a textbook in technical chemistry written by Julius Thomsen's brother, August.

Figure 2.6. The Greenland vessel "Julius Thomsen" at its maiden voyage in 1927. http://7seasvessels.com/wp-content/uploads/2011/09/julius-Thomsen-klubben.jpg

by hydrofluoric acid, sodium carbonate, and aluminium. In 1962, after more than a century's operation, the extraction of Greenland's white gold ceased for good. Thomsen's crucial contribution to Greenland's economy was not forgotten. In 1927 the KHMS built a new steamship, which was named "Julius Thomsen" (Figure 2.6). The ship played an important role in the last years of World War II, when it was Greenland's only passenger ship.[59]

Thomsen was not only interested in cryolite as a source for commercial products, but also in the many related minerals found in the Greenland deposit. As late as 1904 he investigated a red-brown fluorite mineral, showing the presence of helium in it (see Section 7.6). His mineralogical interest seems to have been stimulated by the professor of mineralogy at the University of Copenhagen, Niels V.

---

59. Cryolite plays an important role in Peter Høeg's novel *Smilla's Sense of Snow* (1992), which mentions the ship a couple of times. Unfortunately the English edition gives its name as "Johannes Thomsen" rather than "Julius Thomsen."

Ussing. The same year, 1904, Ussing described a new lithium-rich mineral in the form of single-crystal inclusions found in cryolite blocks. Following a proposal by Thomsen, he named the mineral cryolithionite, a name derived from *cryo*lite and *lithi*um.[60] The chemical formula of cryolithionite can be written $Na_3Li_3(AlF_6)_2$, showing its relation to cryolite, Na_3AlF_6.

---

60. See Pauly (1986).

CHAPTER 3

Years of the cholera

During the 1850s Thomsen was not only occupied with developing his new method of cryolite soda manufacture; it was also in this period that he founded the theory of thermochemistry with which his name in the history of chemistry is primarily associated. Moreover, he found time to write popular books and articles on science, with some of them aimed at the popular and educational sector. At the time a recent graduate from the Polytechnic College he took an interest in the teaching of the College and how to disseminate science to the broader public. This part of Thomsen's life and work, which was largely limited to his younger days, is not well known but adds a dimension to the traditional picture of his personality and scientific work. Importantly, it was in a semi-popular context that he first encountered the idea of force or energy conservation and what in the 1860s became known as the two laws of thermodynamics. While the first law became the foundation of his thermochemical system, the second law remained foreign to it until the advent of chemical thermodynamics in the 1880s.

Thomsen had at the time close relations to the slightly older Danish physicist and engineer August Colding, whose views concerning the forces of nature may have inspired him. Quite independent of their shared interest in the natural philosophy of forces, in 1853 Colding and Thomsen collaborated in an innovative study of the causes of the cholera which that year haunted Copenhagen and other parts of Denmark. In this case Thomsen applied his chemical skills in the service of sanitary medicine, the first and only time he did so. The Thomsen-Colding work has been largely forgotten but deserves a place in the history of epidemic diseases as well as in the history of science more generally. While Thomsen would soon specialise in the particular research field of thermochemistry, in the 1850s he was a generalist with broad scientific and literary interests. We shall return to some of these interests in later chapters.

86

3.1. Science education

In his letter to Hansteen of 13 February 1851 cited in Section 1.4, Ørsted mentioned that young Thomsen had written "a small popular chemistry which does him much honour." The book he referred to was *Kortfattet Lærebog i den Uorganiske Chemie* (Brief Textbook in Inorganic Chemistry), which was prefaced December 1849 and thus the very first of his numerous publications.[1] In fact, Thomsen did not consider his book a popular introduction to chemistry but an account of the essence of chemistry aimed at serious students. Although written as a textbook it is unclear if it was actually used at the University, the Polytechnic College, the gymnasium schools, or some other of the country's institutions for higher education. The exposition differed from that of most other elementary textbooks by placing much emphasis on theoretical rules and concepts, and less on the purely descriptive parts of chemistry. As the 22-year-old author proudly stated in the preface, he had chosen to focus on "the general qualitative and quantitative properties of the elements," or "the true meaning of theoretical chemistry ... as a science." This was an unusual focus for an elementary chemistry textbook.

Thomsen accordingly started with a discussion of the force of affinity which supposedly kept the chemical elements together in composite bodies, the chemical compounds. This was a subject that would eventually form the nucleus of his thermochemical research programme, but at the time he did not suggest a definite measure of affinity. He preferred "quite abstractly" to characterise an element in a binary compound such as $NaCl$, Al_2O_3 or NH_3 by its relative degree of either a "positive" or "negative" property. Oxygen was the most negative of the elements and the others of the common elements followed a series ending with the strongly positive potassium:

---

1. Thomsen (1850a). In 1849 Thomsen wrote a critical review in *Berlingske Tidende* of Erik G. Silfverberg's *Lærebog i Chemien* (Textbook in Chemistry). Silfverberg, who had studied at the Polytechnic College, but without graduating as a candidate, worked as a school teacher in mathematics, physics, chemistry and geography.

O - Cl – S – H – N – P – C – Si – Au – Bi – Fe – Al – Mg – Ba – K

Contrary to Berzelius and his followers, Thomsen did not associate the property with the electrical forces of the elements. Affinity was simply considered a measure of how willingly two elements united or the ease at which a binary compound could be decomposed. It was not determined only by the elements themselves, for the affinity would also depend on external factors such as heat and electricity. A compound, he said, would usually decompose if acted upon by a substance with a stronger affinity to one of the compound's constituents. As an example he mentioned that if lead sulphide is melted together with iron, metallic lead will be produced according to

$$PbS + Fe \rightarrow FeS + Pb$$

Consequently, sulphur's affinity to iron is greater than its affinity to lead. At the end of his book Thomsen, adopting an empiricist attitude, warned against speculations concerning the force that manifested itself in the form of affinity: "We do not know the true nature of the forces," he wrote, "but only recognise their presence through their effects and from these we infer what they are." And yet Thomsen was convinced that affinity was not a new force peculiar to the chemical kingdom:

> We have good reasons to assume that the chemical forces are not new and unknown, but that they are previously observed actions which in this case appear in a different form. If we recall that other parts of the sciences have demonstrated a close association between the forces we call heat, light, electricity and magnetism, and when we reckon that nearly every chemical union or decomposition is followed by heat, light and electricity – when we consider that these forces are able to promote some chemical unions and also to decompose some compounds … then we have a well-founded reason to assume that the chemical properties of the substances are effects of these general forces of nature.[2]

---

2. Thomsen (1850a), pp. 90-91.

This may be seen as reflecting Ørsted's pet idea of a unity between the forces of nature, but nowhere in the book did Thomsen mention Ørsted by name or, for that matter, the names of other scientists. Again, in his section on metals he described Ørsted's method of isolating aluminium on the basis of potassium and aluminium chloride, but without revealing to his readers when the method was discovered or by whom. *Brief Textbook* was factual and completely ahistorical.

In regard of Thomsen's later work it is of interest to notice that he presented the empirical laws of chemistry purely as stoichiometric rules, that is, as weight relations between elements and their compounds. It was in this stoichiometric sense, and only this, he introduced an atomistic terminology: "The weight relations according to which substances unite can be expressed in numbers, which are called atom-numbers; by an atom of a substance we understand the weight given by its atom-number."[3] For example, on the scale O = 100, commonly used at the time, he assigned to iron the atom-number 350 corresponding to 56 on the H = 1 scale; similarly, for chlorine he stated the value 221.6 or 35.5. It is not quite clear why Thomsen spoke of atom-numbers rather than using the term atomic weights, which for long had been generally adopted. He further spoke of atoms of compounds and not only of elements. When he wrote that one atom of Fe_2O_3 consisted of two atoms of Fe and three atoms of O, he simply referred to the corresponding amounts of weights: "The atom of a compound is the weight expressed by its formula; FeO means a weight of 450, Fe_2O_3 on the other hand of 1000, and these numbers are consequently the atom-numbers of the compounds." Moreover, by expressing the density d and atom-number a in terms of oxygen as unity ($d_o = a_o = 1$), he found that

$$\frac{a}{d} = x,$$

---

3. Thomsen (1850a), p. 6. Thomsen's used the word "atomic number" (atomtal), but in order not to confuse it with the much later concept used today, designating the number of positive charges in the atomic nucleus, I find "atom-number" to be more appropriate.

where x is a small integer. The ratio a/d he referred to as the atomic volume or "the space occupied by an atom," a phrase which indicates that he may after all have conceived atoms to be real entities and not merely a calculational device.

What matters is that Thomsen did *not* refer to atoms as elementary and undivisible material particles in the sense of John Dalton. Although he knew of Daltonian atomism, apparently he preferred to follow Ørsted, Zeise and Forchhammer (and many chemists abroad) in their non-materialistic interpretations of the stoichiometric laws. Ever since reading as a young man Kant's *Metaphysische Anfangsgründe der Naturwissenshaft*, Ørsted had adopted the Kantian metaphysics of nature in which there was no room for atoms á la Dalton. Yet, slowly admitting the force of Dalton's theory, at about 1830 Ørsted sought to formulate a compromise between atomism and his favoured dynamism, suggesting that the atomic weights – or "chemical numbers" as he preferred to call them – expressed the intensity of localised chemical forces. He argued that Daltonian atomism was "metaphysical" and that the only acceptable form of atomism was one based directly on the phenomena of nature.[4]

Some of the concepts and phrases used by Thomsen in his textbook were reminiscent of Ørsted's, which possibly indicates an influence from his teacher and mentor. By the mid-nineteenth century scepticism with regard to Dalton's atoms was still widespread among chemists, and Thomsen's position in no way heterodox. There is little doubt that he received direct inspiration from Forchhammer, who in his widely used 1842 textbook in chemistry conceived atoms as weight ratios and referred to "chemical numbers" or "atom-numbers" as synonyms for atomic weights; his favoured word for atoms was "elemental parts" (grunddele).[5] Although he mentioned Dalton several times, it was only in connection with the gas laws and without referring to his atomic theory. Like Thomsen

---

4. On Ørsted's hostility to the atomic hypothesis, see Kragh et al. (2008), pp. 191-196. A full account of the complex story of atomism in eighteenth-century chemistry is given in Rocke (1984).

5. Forchammer (1842), p. 63. Forchammer's plan was to write two more volumes of the textbook, one on chemical mineralogy and the other on technical chemistry, but the two volumes never appeared.

did in 1850, Forchhammer referred to Avogadro's law, again without interpreting it in terms of atoms or molecules. Forchhammer also discussed the concept of affinity, but only briefly. "Two or more substances have a tendency to attract one another and enter into a chemical compound," he wrote. "We call this tendency *affinity*; many circumstances influence and modify the affinity, and heat is particularly important." Apart from stating that "the entire action [in a chemical process] is determined by the electrical forces," he did not comment on the nature of affinity.

With regard to atomism it is of interest to note that Forchammer discussed William Prout's hypothesis that the chemical elements might be composite bodies with hydrogen as their common denominator. In fact, he was quite sympathetic to the idea that "hydrogen should be a *Communis divisor* for all the other elements."[6] Forchhammer was aware that the "chemical numbers" of some elements deviated from Prout's hypothesis but thought that the deviations might be due to inaccurate atomic weight determinations.

This was a subject that Thomsen would later take up and develop in great details, but it did not appear in his book of 1850. The problem of the reality of the atoms continued to occupy chemists up to the end of the century. As late as 1884 the Danish chemist John Sebelien advocated that the term "atom" should be restricted to its stoichiometric sense, as Thomsen had done in his textbook.[7] Among the chemical rules highlighted by Thomsen was the well-known Dulong-Petit rule going back to Pierre-Louis Dulong and Alexis-Thérèse Petit in 1819. This rule or empirical law states that the specific heat capacity c of most solid substances varies approximately inversely with their atomic weights:

---

6. Forchhammer (1842), pp. 451-452, who did not mention Prout by name. Although Forchhammer became known primarily as a geologist, he also worked extensively in chemistry and was an important figure in the Danish chemical community. See Nielsen (1994).

7. Sebelien (1884a), who referred to "atom-number" or "equivalent number" as a definite amount of weight units. Thomsen's later work on Prout's hypothesis is described in Section 7.3. In Jørgensen (1860) the 23-year-old chemistry student Sophus Mads Jørgensen discussed Prout's hypothesis but without subscribing to it.

$$c \times A \cong 6 \text{ cal degree}^{-1}$$

Without mentioning the two French scientists or the history and previous use of the rule, Thomsen defined the "atomic heat" as "the product of the specific heat capacity and the atom-number, meaning the amount of heat an atom contains." As he pointed out, the rule was not only valid for the elements, with atomic heats varying in the range between 37.6 and 42.8, but also for compounds of similar constitution. As an example he referred to the sulphates of Ba, Sr and Ca for which he stated the atomic heats to be 164.0, 163.6 and 167.0, respectively. Thomsen made no attempt to explain the various rules he discussed, but merely presented them as empirical generalisations.

While *Brief Textbook* was the most interesting of Thomsen's early contributions to chemical education, it was not the only one. In 1853 he published a book on preparative chemistry and the same year an elementary exposition of chemistry aimed at a general readership and schoolteachers in particular. In the first book, a manual of practical laboratory chemistry, he described the most important chemical operations and the laboratory equipment needed for them, and it also contained recipes for preparing a wide range of simple inorganic chemicals (Figure 3.1). He chose not to include organic chemistry. Although the book contained no theory at all, Thomsen was convinced that experimental chemistry without a theoretical foundation would make sense only for those who regrettably were satisfied with chemistry as a mere craft. This was not the audience for whom the book was written. "Only students with knowledge of theoretical chemistry can advantageously participate in chemical exercises," he wrote.[8] For this purpose he recommended his textbook of 1850 as a necessary supplement.

Chemistry was not a separate subject in Danish public schools in the 1850s, when science of any kind was given very low priority. When Thomsen was approached by schoolteachers suggesting him to write an elementary textbook, he decided that a general introduction to chemistry and its many applications would be more appro-

8. Thomsen (1853a), p. 1, preface dated December 1852.

114

netop er ligesaa stor som Vægten af det anvendte Sukker; der
findes ligesaameget Kul, Brint og Ilt i Sukkeret, som i den
Mængde Kulsyre og Viinaand, der ved Gjæringen er dannet
af Sukkeret. 90 Vægtdele Sukker give 46 Dele Viinaand og
44 Dele Kulsyre; men

46 Dele Viinaand indeholde 24 Dele Kul, 6 Brint og 16 Ilt,
44 — Kulsyre — 12 — —   32 —

90 Dele Sukker indeholde 36 Dele Kul, 6 Brint og 48 Ilt,
hvilket netop er den Sammensætning, som ovenfor er angivet
for Sukkeret.

Vinens vigtigste Bestanddeel er Viinaanden. Betyde-
lige Mængder Viin blive aarligen destillerede for deraf at fremstille
Viinaand, der i en passende Blanding med Vand danner Bræn-
deviin. Destillation af Viin foretages ofte af Chemikeren
ved Undersøgelser af Vinens Styrke; han benytter dertil et Apparat,
der i sine Hoved-
træk er fremstillet
i hosstaaende Fi-
gur. Karret a,
som indeholder Vi-
nen, der skal under-
søges, kaldes en
Retort og er af
Glas; Karret b,

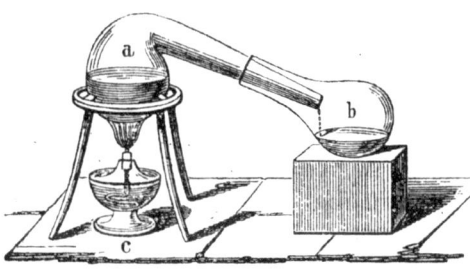

der ligeledes er af Glas, tjener til at opsamle den overdestillerede
Vædske og kaldes et Forlag. Naar man nu ved en lille Lampe,
c, opvarmer Retorten, begynder Vædsken i denne snart at koge.
Dampene fylde Retorten og strømme ud i dens Hals og der-
igjennem ned i Forlaget; men ved den Afkjøling, som de
lide deels i Retorthalsen deels i Forlaget, som man sætter i
koldt Vand, fortættes Dampene atter til Vædske og samle sig
saaledes i Forlaget b. Viinaanden koger ved lavere Varmegrad
end Vandet, nemlig ved 78 Grader, og derfor vil den største Deel
af hiin overdestillere, inden Varmen bliver saa høi, at Vandet
kommer i Kog, hvilken Varmegrad som bekjendt kaldes 100 Gra-

Figure 3.1. A page from Thomsen's 1853 textbook on preparative chemistry illustrating alcoholic distillation (Thomsen 1853a).

priate. The result was *Et Forsøg paa en Almeenfattelig Fremstilling af Chemiens Vigtigste Resultater* (An Attempt at a Popular Exposition of the Main Results of Chemistry), a comprehensive, pedagogical and highly informative introduction to the many facets of the chemical landscape.[9] Although the main part of the book described technical chemistry and its uses in industry and agriculture, its higher aim was to enlighten the reader about chemistry as a science (Figure 3.2). Like in his textbook of 1850, Thomsen disregarded the historical development and mentioned no names of either chemists or industrialists. His sections on plant, animal and nutritional chemistry were particularly detailed, not unlike Liebig's treatment in his popular and widely read *Chemical Letters* which had been translated into Danish a few years earlier.[10] Concerning the chemistry of plants, also known as phytochemistry, he wrote: "Man is unable to form from the chemical elements a single vegetable substance, be it sugar, yeast, oil, resin or some other organic substance." This was not quite right, as a total synthesis of acetic acid had been achieved by Hermann Kolbe in 1844. In the 1890s another German chemist, Emil Fischer, succeeded in synthesising sugar in the forms of glucose and fructose.

Popular Exposition demonstrated, as did his *Excursions* three years later (Section 3.3), that Thomsen was a gifted popular writer and in his younger days had a genuine interest in the genre. While the book was primarily a wide-ranging account of chemistry as a productive force and a key to understanding nature, Thomsen also found space to include the interrelationship between the forces of nature as revealed by the new principle of energy conservation:

> All known forces of nature are interconnected in the sense that one force can change into another. A piece of burning coal produces heat and light; when used in a steam engine it results in motion, a mechanical force. And in the form of friction between two bodies motion can be used to generate heat and electricity; or, if the heat formed by combustion radiates on soldered metals electricity will be pro-

9. Thomsen (1853b). A second edition was published 1854.
10. Liebig (1846). On Danish agricultural chemistry in the period, see Nielsen (2002).

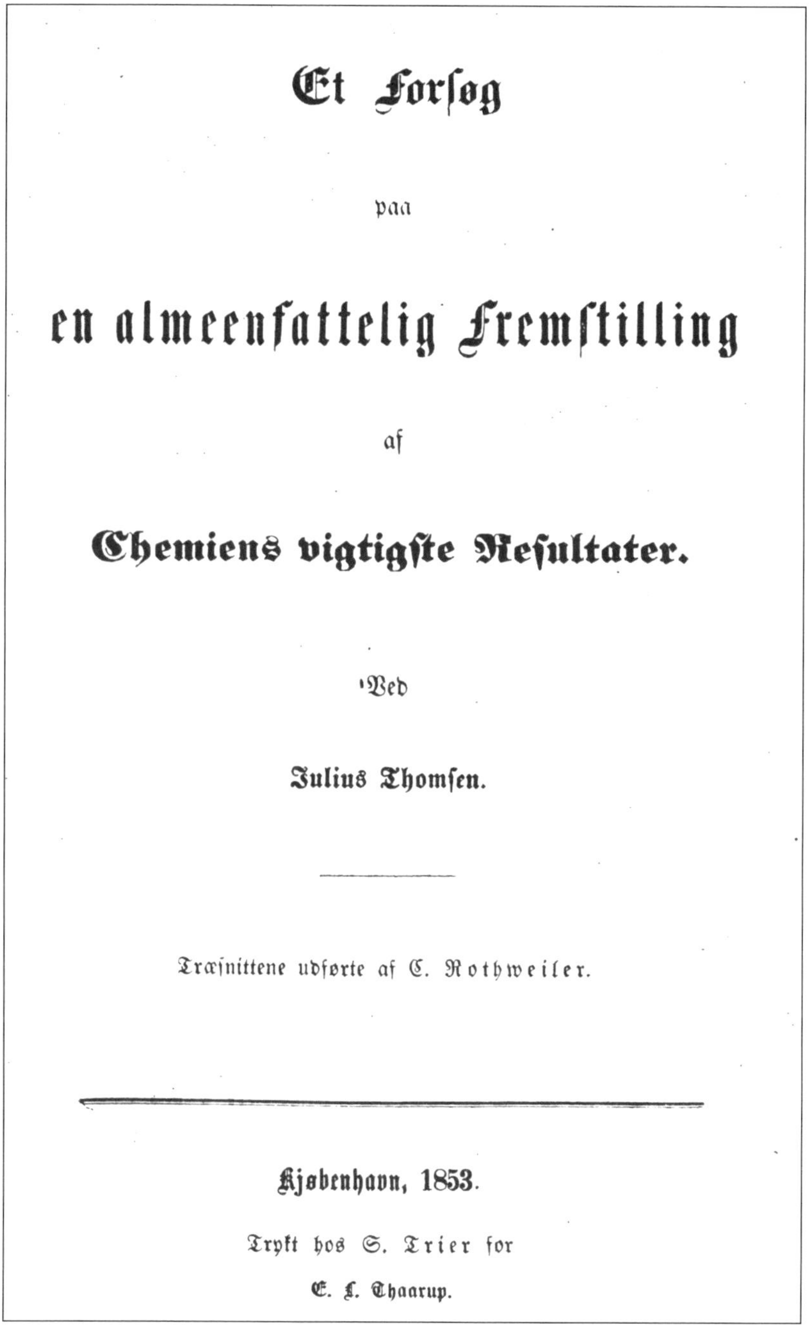

Et Forsøg

paa

en almeenfattelig Fremstilling

af

Chemiens vigtigste Resultater.

Ved

Julius Thomsen.

Træsnittene udførte af C. Rothweiler.

Kjøbenhavn, 1853.

Trykt hos S. Trier for

C. F. Thaarup.

Figure 3.2. Thomsen's popular book on chemistry (Thomsen 1853b).

duced which can again generate magnetism and chemical actions. Thus all the forces of nature stand in such an intimate relationship that one cannot help assuming that they all emerge from a common source.[11]

Thomsen was at the time a lecturer on agricultural chemistry at the Polytechnic College, which explains his interest and competence in the field. It also accounts for a booklet he published on his own account on plants and their circulations in nature, a much extended version of a lecture he gave to the Gardeners' Association in 1850.[12] In one of the chapters he outlined how first plants and later animals had emerged on the ancient Earth as a result of natural processes. Much of the content in this publication was of a chemical nature and it reappeared with little change in his later *Popular Exposition*. Similarly, in 1854 he published a work on the use of science in agriculture which was based on a talk he gave at a meeting of the Danish Farmers' Association.

Thomsen's books and booklets from the early 1850s were his only contributions to the teaching of elementary chemistry. They were not widely used, if at all, in the "learned schools" or high schools (gymnasia) which prepared students for a university education.[13] Realising the need for strengthening the scientific subjects compared to humanistic studies and classical languages, in 1871 an educational reform was introduced in the Danish school system. According to this reform, chemistry would be an obligatory part of the gymnasium curriculum, if only in modest doses and in combination with physics. As a result of this and other reforms, several textbooks in elementary chemistry were published, either by schoolteachers or academic chemists. The most widely used were Hannibal Jespersen's *Kortfattet Lærebog i Uorganisk Kemi* (Brief Textbook in Inorganic Chemistry) from 1874 and S. M. Jørgensen's *Kemiens Begyndelsesgrunde* (Introductory Chemistry) from 1876. By then Thom-

---

11. Thomsen (1853b), p. 56.

12. Thomsen (1850b).

13. On chemistry education in the Danish gymnasium schools, see Riis Larsen (1991) and Riis Larsen (1998).

sen was absorbed in his extensive research on thermochemistry and he may not have found it worthwhile to write yet another textbook for the gymnasium schools.

An employee at the Polytechnic College, Thomsen took a keen interest in the organisational and educational aspects of his school. At the time of Ørsted's death there was much discussion concerning the aim of the school and the needs to reform it.[14] First and foremost, should the candidates receive a higher scientific education or rather a more practical one corresponding to the needs of public services and Danish business interests? In a report of 1855 Thomsen entered the debate, emphasising that the existing conditions for education and research were highly inadequate. "Most likely, there is not a single lecture hall for chemistry abroad which is as poor as the one belonging to the Polytechnic College," he lamented. "Of the twenty or so of these auditoria I have seen in Germany and France, none comes close."[15] Moreover, young Thomsen pointed out that the hygienic conditions were most unsatisfactory, even tending to be harmful to the students' health. "When the auditorium is used for chemical lectures the audience is affected in a bad way; for all the vapours, which inevitably are produced by the many lecture demonstrations, are mixed with the air in the room and cause problems for the audience."

On a more constructive note Thomsen proposed that the students' exercises in physics should consist of real experiments and not only paper-and-pencil exercises. He listed a variety of physical instruments and experiments with which the students ought to be familiar by means of their own experience and not only that of their lecturer. It was a sensible and progressive proposal but ahead of its time. Only in the 1880s did Thomsen's ideas become a reality at the Polytechnic College when they were implanted by the physics professor Christian Christiansen.

Perhaps surprisingly, Thomsen advocated a reorientation of the

---

14. See Harnow (1998) and Wagner (1999).

15. Thomsen (1855b), pp. 10-12 and Harnow (1998), p. 68. Thomsen's report was widely known and reviewed in influential newspapers such as *Berlingske Tidende*, *Dagbladet*, and *Fædrelandet*. See Erslew (1868).

College from Ørsted's university-like institution to an engineering school responding to the needs of Danish society. He deplored that the Polytechnic College had "more the character of a scientific academy than a technical school," and that the students "learn too much and too little – too much theory and too little application." As far as lectures in pure science were concerned, Thomsen held that they belonged to the University, whereas the aim of the Polytechnic College should be "to give the students a particular technical education in agreement with their future employments."[16] It is unknown which effects Thomsen's proposal had, if any, but several years later a major reform turned the College into a school of civil engineers. As we shall see in Chapter 7, Thomsen would come to play a most important part in the further history of Ørsted's old institution.

Thomsen's memoir of 1855 was known to the students and graduates from the Polytechnic College, which had established a Polytechnic Association (Polyteknisk Forening), a union of students, younger candidates, and a few senior engineers. In 1856 Thomsen became a member of a committee under the Association which sought to reform the institution. He was for a period an active member of the Polytechnic Association and served as its president 1852-1853.[17]

3.2. August Colding and energy conservation

In his 1910 Thomsen Memorial Lecture the esteemed British chemist Edward Thorpe wrote that Thomsen, in his search for a thermochemical theory based on mechanical principles, was "doubtless under the influence of Ludwig Augustus Colding, an engineer in the service of the Municipality of Copenhagen, and a pioneer, like Mayer, in the development of that theory [of force or energy conservation.]."[18] While it is reasonable to assume that part of the inspiration for Thomsen's system came from the nine years older Colding, there is no direct evidence for it and I would consequently

16. Thomsen (1855b), p. 20.
17. Harnow (1998), p. 59 and p. 100. See also Section 8.1.
18. Thorpe (1910), p. 162. Schelar (1966), p. 107, repeats Thorpe's statement.

Figure 3.3.
Ludvig August
Colding. Wikimedia
Commons.

prefer to speak of "most likely" rather than "doubtless." In any case, Thomsen and Colding had close relations over a long period of time, and the history of the latter belongs to some extent to the history of the former.

Ludvig August Colding (who rarely used his first name) was born in 1815 in Holbæk, Zealand, the son of a farmer and former ship's captain.[19] The father happened to be acquainted with H. C. Ørsted and it was on Ørsted's recommendation that young August in 1836 became an apprentice to Anders Christian Olsen, a well-known instrument-maker and craftsman in Copenhagen. Ørsted, who recognised August Colding's talents as a mechanic and potential scientist, advised the young man to study for the entrance examination that was necessary to become a student at the Polytech-

19. English-language scholarly literature on Colding includes Dahl (1972) and Caneva (1997). Marstrand (1929) provides valuable details about Colding's life and career. Several of Colding's publications are translated into English in Dahl's book.

nic College. In 1837 Colding enrolled at the Polytechnic, where he received a solid training in mathematics, physics and engineering, and four years later he graduated as a candidate in the mechanics class. During the winter 1841-1842 he gave popular lectures on applied science in Nakskov under the auspices of another of Ørsted's inventions, the Society for the Dissemination of Science. Colding also assisted his teacher and mentor in some of the experiments that Ørsted made on the compressibility of water. After a period as private tutor, he became involved in the plans of providing Copenhagen with a modern system of water, gas and sanitary needs. In 1851 he was appointed acting engineer in charge of the ambitious project, to end up six years later as Copenhagen's first City Engineer (Stadsingeniør).

Colding owes his place in the international history of science to the work he did in the 1840s on what became the principle of energy conservation, or the first law of thermodynamics, and which makes him one of several co-founders of this most fundamental principle or law of nature.[20] Even before his graduation in 1841 he speculated that the various forces of nature were somehow interrelated and the totality of forces preserved. By Colding's own account, his original ideas on the subject were of a metaphysical and religious nature, as he believed that the forces of nature expressed God's essence and consequently could neither be created nor annihilated.[21] The religious element in the new concept of energy was particularly strong in Colding, but not exclusive to him. Other contributors to the concept, including Mayer, Joule, and the amateur scientist William

---

20. The literature on the discovery of the law of energy conservation is extensive. See, for example, Elkana (1974). Although Colding is sometimes mentioned as a co-discoverer, credit for the discovery is normally ascribed to J. R. Mayer and J. P. Joule, and to H. Helmholtz for a full understanding of the law. As convincingly argued by Caneva (1997), Colding's concept of "force" did not quite correspond to the modern concept of energy as established by Helmholtz and other physicists at about 1850. See also Christensen (2013), pp. 552-559.

21. It is quite wrong when Donald Cardwell, a leading historian of technology, wrote that "the Danish engineer L. A. Colding had been led to the dynamical theory of heat through his considerations of the working of steam-engines." Cardwell (1989), p. 229. It was God who served as Colding's inspiration, and not the steam engine.

Grove, stressed that energy conservation was a useful weapon against atheism and a sure sign of a divinely created universe. It is one of history's many ironies that later in the century the law of energy conservation was embraced by materialists and atheists.

In 1856 Colding was elected a member of the Royal Danish Academy, and on this occasion he gave a lecture summarising his ideas on force conservation. "My first thoughts," he said, were indebted to "the view that the forces of nature must be related to the spiritual in nature, to the eternal reason as well as to the human soul; thus it was the religious philosophy of life that led me to the concept of the imperishability of forces."[22] Colding further expressed his indebtedness to "the immortal H. C. Ørsted [who] first taught me to recognise and treasure, that even the forces of nature must be real, imperishable entities." He ended his lecture by formulating his credo in a manner that most likely reflected the influence of Ørsted: "We must never forget the law of nature which states that only that which is in accord with the soul in nature can persevere, while everything which offers resistance is perishable and must sooner or later be destroyed."

According to Colding, the law of force conservation included not only natural forces such as heat, chemical processes, electricity, and mechanical work, but intellectual activity as well. His early speculations concerning a link between the physical forces of nature and mankind's intellectual life were in part inspired by Ørsted, and probably also by the writings of Daniel F. Eschricht, a prominent Danish professor of physiology. In a book of 1850 Eschricht had suggested the permanence of things intellectual and spiritual, at the expense of the transience of things bodily.[23]

However, religious and philosophical speculations do not qualify as science and nor did they at the time. In the early 1840s Ørsted advised Colding to conduct experiments on the heat produced by frictional motion, and it was on the basis of such experiments that

---

22. Colding (1856). I cite from the English translation in Dahl (1972), pp. 105-128.
23. Colding (1856) admitted his debt to lectures given by Eschricht: "It was really at these lectures that I decided to present my favourite idea on the relationship between the forces of nature and the spiritual life." Dahl (1972), p. 120.

Colding, in 1843, established that heat and mechanical work were proportional. At the time he was unaware of the slightly earlier works of Julius Robert Mayer in Germany and James Prescott Joule in England. Improved and more elaborate experiments led him to conclude that "1 pound of water heated 1 degree Celsius can be expressed by 1185 pounds raised 1 foot," which in modern units corresponds to 3.7 joule as the mechanical equivalent of one calorie, which is 13 per cent short of the modern value. This is what Colding reported to the 1847 meeting of the Scandinavian Association of Natural Scientists held in Copenhagen. His detailed report was only published in 1850 and it took until 1871 before it appeared in an English translation in the *Philosophical Magazine*. For some twenty years and with one noteworthy exception Colding's work remained unknown outside the Scandinavian countries.[24]

Nor did his work create much interest in Denmark, where only few scientists commented on it or supported Colding's claim that he (rather than Mayer) was the true originator of the principle of energy conservation. One of the few was Julius Thomsen, who in a paper of 1855 dealt with the question. According to Thomsen, the law of energy conservation owed its existence to four scientists who, largely independently, had reached the insight each in their own way: "J. R. Mayer in Heilbronn, A. Colding in Copenhagen, and Joule in England started their investigations on this subject at about the same time; and a few years later, again apparently independently of the others, it was treated by Helmholtz in Königsberg."[25] Although Thomsen thus gave credit to Colding for his work, he did not highlight his contribution. Many years later, in a speech given in 1890, Thomsen paid tribute to his former and by then deceased collaborator by placing him before Mayer: "The basic features of the theory of the constancy of the amount of work were developed in the years 1842-43 by one of the candidates of the Polytechnic Col-

---

24. The exception was none other than Helmholtz, who in an influential address of 1854 stated that "a Dane named Colding" had found the same law as Mayer. The address appeared in English translation in *Philosophical Magazine* **11** (1856): 489-518 and is reprinted in Helmholtz (1995), pp. 18-45. See also Section 3.3.

25. Thomsen (1855a), p. 236. More about this essay follows in Section 3.3.

lege, the late professor Colding, and about the same time by Dr. Mayer in Heilbronn."[26]

While Thomsen was a chemist with a strong interest in chemical heat as a manifestation of the more general concept of energy, Colding was a physics-trained mechanical engineer. All the same, he was fully aware of the relevance of chemical phenomena for the law of energy conversation. As early as 1847, in his oral contribution to the meeting of Scandinavian scientists, he referred to galvanic processes and thermochemical measurements in support of his theory, much like Helmholtz did in greater detail in his seminal treatise of the same year. Colding's presentation of 1847 was published in the proceedings of the Scandinavian Society three years later. In the early 1850s Thomsen developed his new system of thermochemistry, and at the same time he and Colding collaborated in their analysis of the probable causes for the Copenhagen cholera epidemic to be considered below. I suspect that Thomsen's influence was in part responsible for an interesting but previously unnoticed section on "the chemical powers and those results which have emerged from chemical physics" that Colding included in his 1856 address to the Royal Academy of Sciences.

Concerning thermochemistry and the forces of affinity, Colding pointed out that "many excellent men have laboured, and still labour, to precisely determine the evolution and absorption of heat which occurs during various chemical combinations and separations." According to the view of Colding, the most excellent of these men was his younger friend Thomsen:

> No one has worked with greater success in this direction than our countryman Polytechnic Candidate Julius Thomsen, as he has not only determined the strength of the chemical power for many elements and their combinations with a high degree of precision ... but in addition has laid the groundwork for the mathematical treatment of chemistry. He has, purely with the aid of his mathematical computations, not only demonstrated that it is a simple consequence of the

26. Speech of 1 September 1890 on the occasion of the inauguration of the new Polytechnic College, as reproduced in Lundbye (1929), on p. 225. Colding died in 1888.

existing forces that the chemical combinations or separations ... must take place as long as the manipulations are carried out according to the teachings of chemistry, but on the basis of computations he has even predicted several other previously unknown actions. ... He has also investigated the matter in the laboratory and found his conclusions to be fully confirmed by nature.[27]

The dream of a mathematical chemistry obviously appealed to Colding, who thought that with the new theory of energy conservation the dream was about to become a reality. As he expressed it in his 1856 address:

I had not expected that within the span of a few years we would see chemistry placed on a mathematical foundation – to the extent that we even now can anticipate that it will not be long before we, with the aid of mathematics, will probe the smallest constituents of matter with the same clarity and confidence which enabled us to peer out and survey the conditions existing in the boundless universe. Soon the time will come when the chemist, by way of mathematical formulae and computations, will be able to predict in advance the results of experiments in the laboratory; yes, I believe it is not an overstatement to say that we have no idea to what height science will reach in this direction, and all this is based on the law of nature which states that the forces of nature are imperishable.

The optimistic hope expressed so eloquently by Colding was to some extent shared by Thomsen, whose reform of thermochemistry rested on mathematical calculations based on the principle of energy conservation (see further in Section 4.2). Similar beliefs continued to occupy the minds of some later researchers within the tradition of physical chemistry, but then it was the more general theory of thermodynamics and not only its first law that was thought to provide an axiomatic basis for the mathematical chemistry of the future.[28]

---

27. Colding (1856), in Dahl (1972), p. 116.
28. See, for example, van Laar (1901). A proper mathematization of chemistry only

Colding and Thomsen were both influenced by Ørsted, if in different ways and to different extents. Ørsted's impact on Colding was generally much stronger than on Thomsen, and although Colding shared much of his mentor's metaphysics he dismissed Ørsted's dynamic perception of matter and heat. Like Thomsen, Colding supported the new kinetic theory of heat based on the motion of material atoms and molecules. In unpublished notes he even speculated that atoms might be internally structured entities, an idea that Thomsen would entertain later in his career in connection with the periodic system of the elements.[29] In spite of some similarities between the scientific outlooks of Colding and Thomsen – compared to Ørsted they were both "moderns" – there were also marked dissimilarities. First and foremost, the religious-spiritual dimension that was so strong in Colding's thinking was completely absent in Thomsen's.

3.3. The forces of nature

During the busy decade of the 1850s Thomsen was not only occupied with establishing the new cryolite soda industry and developing his system of thermochemistry; he also found time to write broad expositions of science aimed at the general reader. Contrary to many contemporary Danish scientists Thomsen was not a "populariser," but in his younger days he recognised the need to contribute to the growing literature on popular science and in this way to make his name better known. For example, between 1856 and 1866 he wrote a couple of small pieces for the *Folkekalender for Danmark* (The Danish People's Calendar), a popular yearbook founded in 1852 and which included among its contributors the famous fairy-tale author and poet Hans Christian Andersen. On one occasion he contributed to the popular bourgeois magazine *Illustreret Tidende* (Il-

became a reality with the emergence in the 1970s of so-called computational chemistry based on computer simulations and methods of quantum chemistry.

29. In an unpublished note of 1854, Colding suggested that an atom was composed like "an infinitesimally small planetary system." See Dahl (1972), p. 177. For Thomsen's later speculations on the complexity of atoms, see Chapter 7.

lustrated Journal) with an article on Laskaris and Johann G. Böttger, two alchemists from the early eighteenth century.[30] Although Thomsen did not believe in the art of gold-making, he pointed out that a proof for the impossibility of transmutation of metals was actually missing. The only "proof" was that repeated attempts of turning ordinary metals into gold and silver had all failed. He used the occasion to ridicule unnamed Danish scientists for their interest in spiritualist séances and belief in spiritual phenomena. According to Thomsen, spiritualism was nothing but fraud and superstition.

Two of Thomsen's works from this period, published in Danish only, are of particular interest because they give an insight in how he perceived science in general, including the methods and aims of science. They also counter the traditional view of Thomsen as a scientist who focused narrowly on his research in experimental chemistry. In 1855 a new journal titled *Tidsskrift for Populære Fremstillinger af Naturvidenskaben* (Journal of Popular Expositions of Science) was launched in Copenhagen, adding to the already considerable body of Danish science journals and magazines addressing a general audience.[31] Most of the articles in the new journal were serious and comprehensive contributions written by the country's leading naturalists and with a strong emphasis on subjects of natural history. Papers on chemistry and physics filled little, and mathematics was absent. The subscribers were primarily priests, physicians, wealthy farmers, and the educated bourgeoisie.

Thomsen contributed to the first volume with a lengthy article in two sequels on the interaction of the forces of nature, a free translation of a famous lecture that Helmholtz gave in Königberg (the

---

30. Thomsen (1860a). As Thomsen pointed out, Böttger's competence in the art of alchemy was not in vain. In 1710 he used it to manufacture the first European porcelain in Meissen outside Dresden.

31. *Tidsskrift for Populære Fremstillinger af Naturvidenskaben* was founded by the zoologist Christian Lütken, the geologist Carl Fogt, and the botanist Christian Vaupell. In 1863 it included an extensive summary of Darwin's theory of evolution, appearing anonymously but in fact written by Lütken. The journal ceased to be published in 1883. See Hjermitslev (2004). Danish popular science periodicals in the eighteenth century related to chemistry are covered in detail in Nielsen (2000), pp. 83-90, and Nielsen (2007).

present Kaliningrad) on 7 February 1854. The translation was free indeed, for at places it left out parts of Helmholtz's address and at other places it added to it or changed its structure. Thus Thomsen's account of the simultaneous discovery of the energy principle, as quoted in Section 3.2, deviated significantly from Helmholtz's. While the latter stressed Mayer's priority,[32] according to Thomsen's version Colding and Joule shared the priority with the German physician. More conspicuously, readers of *Popular Expositions* would be unaware of Helmholtz's references to the Bible and his cautious analogy between the scientific scenario of the cosmos and "the Mosaic tradition." Thomsen simply left this part out. The Königsberg lecture was scientifically important because it it introduced a new theory of the generation of solar heat, which Helmholtz suggested was due to a slight contraction of the Sun (whereby potential gravitational energy is transformed into thermal energy). Thomsen referred to it as follows: "A diminution of the Sun's diameter by a ten-thousandth part would cause an evolution of heat sufficient to cover the loss of radiation over a period of more than 2000 years."

The Thomsen-Helmholtz article in *Popular Expositions* was probably the first time that the two fundamental laws of thermodynamics were presented to a Danish audience. However, Thomsen's account of the second law was more vague and incomplete than the one of his source. For example, Thomsen left out Helmholtz's references to the works of Sadi Carnot, William Thomson, and Rudolf Clausius. Yet he included the notorious heat death, an early and most destructive consequence of the second law of thermodynamics. There will come a time, Thomsen said in his free translation, when all the forces of nature have been transformed into one uniformly distributed soup of heat.

> Then any source of change will be extinguished and a complete cessation of all natural processes will have occurred. Of course, plants and animals can no longer exist; the Sun will have lost its higher heat

---

32. "The first who saw truly the general law here referred to, and expressed it correctly, was a German physician, J. R. Mayer of Heilbronn, in the year 1842. A little later, in 1843, a Dane named Colding presented a memoir to the Academy of Copenhagen … ." Helmholtz (1995), p. 27.

and then also its light. And all the constituents of the surfaces of the globes will have entered into those compounds that correspond to their nature.[33]

Stated differently, the heat death is the result of the irreversible increase of entropy in the universe – except that in 1855 the term "entropy" had not yet been coined. At the end of his Königsberg lecture, Helmholtz stated that the laws of thermodynamics threatened humanity with "a day of judgment" and he referred to the "higher moral problems" of the human race. These allusions to Christian faith may not have been to Thomsen's taste. His end was significantly different: "The Earth will most likely continue its evolution, and as one being has replaced the other since the earliest time to the present, so the time may come when the human race has fulfilled its destiny on Earth and must make way for organic beings of a higher order."[34] In Helmholtz's address there was no reference to these beings of a higher order.

Whereas the article in *Popular Expositions* was meant as a presentation of frontier science and aimed as much at Danish scientists as at general readers, in a book of 1856 Thomsen entered the field of popular science in the true sense of the term. Alexander von Humboldt's famous and highly successful *Kosmos* had just been translated into Danish by Christian Anders Schumacher, an army officer and writer, with the four volumes appearing 1847-1858. Humboldt's book may have inspired Thomsen to write a similar if less ambitious exposition of the natural world and the wonders of science. The result, *Vandringer paa Naturvidenskabens Gebeet* (Excursions into the Landscape of Science) was a fine example of science for the people, highly informative and written in an accessible, even elegant language. Thomsen invited the reader to travel with him to near as well as far corners of the world, even to the end of the universe, explaining carefully and patiently the natural phenomena experienced along the journey. The sciences he covered in this way were primarily geology, meteorology, physics, oceanography, and astronomy. The lat-

---

33. Thomsen (1855a), p. 240.
34. Thomsen (1855a), p. 374.

ter subject alone took up about half of the book's 287 pages. Somewhat surprisingly, given the author's own field of science, *Excursions* contained very little chemistry.

In the introductory chapter Thomsen asserted that in modern science there was no way around specialisation and that the dream of the past, to encompass all the fields of science, was unrealistic and counterproductive. And yet he maintained that for all their divisions and subdivisions, the sciences worked towards a common goal. The Romantic belief in a unity of the sciences was gone, but not the belief itself:

> The great task that natural science has to solve is to obtain a true understanding of nature itself: In each individual branch of science work is being done to solve this task, and gradually as the separate parts of science move closer to this common goal, they also enter into a more intimate contact with one another. There will come a time when all individual branches of science will once again have merged into one strong current which, flowing quickly, seeks to reach the goal, infinitely distant even now.[35]

The later development only confirmed Thomsen in his view of a dialectic process between unity and disunity in the sciences. In a speech in the Royal Danish Academy of 24 March 1899 he said that the first half of the nineteenth century had been characterised by generalists – people who were "natural philosophers in the wider sense of the term." But in the second half of the century progress depended on increased specialisation. While individual scientists were necessarily specialists, collectively this healthy tendency towards specialisation opened up for a no less healthy interdisciplinarity and cross-fertilisation of the apparently fragmented knowledge:

---

35. Thomsen (1856), p. 4. The book originally appeared in two parts in 1855-1856. It was positively reviewed in leading newspapers such as *Fædrelandet* and *Berlingske Tidende*.

The points of connection between the various branches of science continued to increase and the borders between them became more diffuse; the many branches of science showed ever more clearly that they had a common origin. Even those sciences that were apparently quite far apart from the natural sciences established connections to the latter and adopted their exact and analytical methods. In this way the different parts of science approach increasingly a common and higher unity.[36]

In Thomsen's view the tendency towards a methodological unity showed itself in the humanities and the social sciences taking over the mathematical and experimental methods of physics and the other exact sciences. It was a common view at the time, an integral theme of the predominating spirit of positivism. Like most of his contemporaries, Thomsen was unable to imagine that the natural sciences could benefit from adopting some of the methods of the humanist branches of knowledge.

Seven years earlier, in an address to the fourteenth meeting of Scandinavian scientists, Thomsen dealt with some of the same issues. Apart from praising the age of specialised science, he prophecied that psychical and paranormal phenomena – or what he called "the mysticism of natural science" – were on the verge of being conquered by the scientific mind. "Those phenomena known under names such as somnambulism, hypnotism, spiritualism, telepathy, etc. are covered by a veil; but it surely will be lifted a as they have now become subjects of scientific investigation."[37] Thomsen presumably had in mind the Danish experimental psychologist Alfred Lehmann whose work he seems to have valued.[38] Lehmann did pioneer research on paranormal experiences.

---

36. Quoted in Lomholt (1942-1973), vol. 1, p. 476. The speech was given in front of the king, Christian IX, and his son prince Frederik.

37. Thomsen (1892b), p. 34.

38. Lehmann graduated as a chemical engineer from the Polytechnic College and for a few years, until he changed to psychology, worked as assistant in chemistry at the Agricultural College. He published two papers in *Tidsskrift*. In the mid-1880s he created an important school of experimental psychology in Copenhagen. Lehmann was elected a member of the Royal Danish Academy in 1902. In 1890 Thomsen

At yet another occasion Thomsen connected his view of the development of science with the more popular presentations that were an important part of the Scandinavian meetings since the age of Ørsted. Commenting on the Stockholm meeting in 1880 he said, not without a grain of sarcasm, that "the meeting has definitely satisfied those who place particular weight on the social character of such meetings."[39] He was less certain about its scientific value, suggesting that in the future the Scandinavian meetings should be limited to "those branches which are narrowly scientific." He presumably had in mind the exact natural sciences and not fields such as medicine, botany and ethnography which he considered less scientific. Moreover, Thomsen argued that the trend toward specialisation and precise experiments – a trend he welcomed – stood in unavoidable conflict with the older tradition of popular science:

> The development of the basic sciences – physics and chemistry – during the last decades has fundamentally changed the purely empirical nature that characterized them previously and brought them closer to the mathematical and philosophical sciences. Theorems or laws derived from philosophical considerations about force and matter now form the foundation of these sciences; phenomena are often developed from existing theories with mathematical stringency. And precisely for this reason those sciences loose part of their earlier popular nature; many results of great scientific importance cannot be justified and exposed clearly in a brief lecture for an audience in lack of the necessary qualifications.

To return to *Excursions*, if there were a main message in Thomsen's book, it was this: Science was an unqualified gift to humanity which not only annihilated the superstitious beliefs of the dark past but also showed the way to a bright future with continual epistemic, material, and social progress. The book was throughout a trium-

---

helped him to move his private psycho-physical laboratory to the old buildings of the Polytechnic College. Lehmann later expressed his gratitude to Thomsen's positive attitude. See Lehmann (1920).

39. Quotations from Thomsen (1880f).

phalist account of the marvels of science, which the author painted in colours corresponding to the new empirical positivism. Distancing himself from the previous generation of romantically inspired natural philosophers, he wrote that now observations and experiments had "changed belief into knowledge, destroyed the confused natural philosophy [Naturphilosophie] and established science on the sure ground of experience." While this may have been an implicit critical reference to Ørsted and his age, at one place – and only one – the heritage from his former teacher did shine through. Recent progress in science, Thomsen said, has indicated a unity of the natural forces, but we still do not know the unitary force which is ultimately responsible for all phenomena of nature. "The true nature [of this force] is the great enigma – it is the soul in nature."[40]

For Ørsted and Colding the soul in nature was divine, and science without religion impossible. Still in the 1850s elements of natural theology and religion generally were commonly found in popular books and addresses on science, but the winds were changing. In a lecture of 1864 Forchhammer felt it necessary to defend the view, now under attack by materialists and atheists, that science was in agreement with and even a testimony to "our highest religious truths."[41] Thomsen's *Excursions* was not explicitly materialist in outlook and there was no trace of atheism in it, but then there was no trace of theism either. In fact, it was silent about religion, systematically avoiding words such as "God" and "the creator." It did refer to the church, though, and then critically and in connection with cases from the history of science. The readers were told that it was only the death of Copernicus in 1543 that saved him from the cruel fate of Giordano Bruno, the martyr of science who in 1600 was burned at the stake as a heretic. The Copernican world view, "now accepted by even the less educated man," contradicted the dogmas of the church which consequently "fought by all means the new system and brought its adherents to the stake."[42] He wrote broadly about

---

40. Thomsen (1856), p. 2 and p. 101. The title of Ørsted's last book published 1849-1850 was *Aanden i Naturen*, translated into English as *The Soul in Nature*.
41. See Kragh et al. (2008), pp. 135-136.
42. Thomsen (1856), pp. 107-108. Of course, today it is known that the story about

"the church," without distinguishing between Catholicism and protestant-evangelical beliefs.

Although Thomsen was well acquainted with the history of science, he rarely referred to it in his writings. On the few occasions when Thomsen did outline the historical development of chemistry, he presented it as a progressive march towards the truth beginning with the chemical revolution in the last quarter of the eighteenth century.[43] One of those occasions was the centenary of Lavoisier's death, when Thomsen gave a talk to the Royal Danish Academy in which he dealt not only with Lavoisier but also with his contemporary Luigi Galvani. He presented both as heroic pioneers of science but in very different ways. While Lavoisier was the systematic investigator building on a wealth of earlier experiments, Galvani's discovery of animal electricity was purely fortuitous yet no less important. Thomsen, evidently a great admirer of Lavoisier, compared the French chemist to Newton – not an uncommon parallel at the time. Contrary to his Danish colleague S. M. Jørgensen (not to mention his French adversary Marcellin Berthelot), Thomsen was not seriously interested in the history of science and never studied the sources of past science.[44] He probably considered it a waste of time.

To return to the book of 1856, in a section dealing with meteorites – those "pieces of stones or metals which fall from the unlimited space in the heavens" – Thomsen pointed out that according to chemical and mineralogical analyses the elements contained in the

Copernicus is wrong and that Bruno was not burned because of his belief in the Earth being a planet circumventing the Sun. Incidentally, as late as 1837 the priest, philosopher and author N.F.S. Grundtvig, more than just a "less educated man," wrote a paper in *Nordisk Kirke-Tidende* in which he denied the validity of the Copernican world system.

43. E.g., Thomsen (1887a), pp. 1-12, and Thomsen (1894e). On the other hand, Thomsen recognised the value of earlier and less scientific theories of chemistry, such as alchemy and the phlogiston theory. See Section 7.1.

44. Thomsen (1894e). In 1907 Jørgensen examined in great detail and with historical insight the discovery of oxygen and later the development of the concept of acidity. The first work was translated into German and the latter published posthumously by the Royal Danish Academy. Jørgensen's interest and competence in the history of science went back to his youth, as exemplified by Jørgensen (1860), an informative account of the atomic theory and its historical development since the Greeks.

meteorites were the very same as those found on Earth. At the time some twenty chemical elements had been identified in stone and iron meteorites, a finding which was widely seen as confirmation of the material unity of the universe. Thomsen suggested that our planetary neighbours Venus and Mars were not only similar to the Earth, but "we cannot reasonably doubt that these planets, like our Earth, are populated with organic beings."[45] So-called pluralism, the belief that life is abundant throughout the universe, was commonly accepted by the mid-nineteenth century. Ørsted, for one, subscribed to the idea and so did Colding.[46] Although Thomsen too believed in extra-terrestrial life, he did not elaborate and cautiously avoided to refer to intelligent life or otherwise speculate about the properties of the organic being on other planets.

At the end of his book Thomsen dealt in some detail with the mysterious nebulae studied by the astronomers, which attracted great interest at the time. Were they gaseous clouds slowly evolving into "new worlds" or just conglomerates of very close and distant stars unresolvable by even the most powerful telescopes? The first answer was widely accepted and considered support for the popular nebular hypothesis also referred to as the Kant-Laplace hypothesis.[47] It became associated with the fashionable view of nature being in a state of continual evolution, with the result that it was highly rated by evolutionists many years before Darwin gave a new meaning to evolution. The question of the nebulae was however controversial for both scientific and ideological reasons. The nebular hypothesis ostensibly offered a naturalistic explanation of cosmic evolution that left little room for divine agency. Indeed, to the new

45. Thomsen (1856), p. 177.

46. On Ørsted's pluralism and the history of extra-terrestrial life in the nineteenth century, see Crowe (1999), pp. 256-257.

47. See Brush (1987). The term refers to a cosmological scenario suggested by Immanuel Kant in 1755 and a theory of the origin of the solar system that Pierre-Simon Laplace proposed in 1796. However, the two theories had little in common and for this reason the name is somewhat unfortunate. Helmholtz dealt with the Kant-Laplace nebular hypothesis in his 1854 Königsberg lecture, which Thomsen included in his free translation of 1855. There is little doubt that Helmholtz was Thomsen's main source regarding the nebular hypothesis.

breed of materialists, atheists and socialists the nebular hypothesis was attractive precisely because it made the creator superfluous or at leat unemployed. On the other hand, there was no unanimity regarding the question, for the nebular hypothesis and the associated evolutionary world view could also be conceived as arguments in favour of religion. This is what Colding and several other authors did.

Thomsen was clearly a protagonist of the nebular hypothesis, which he knew primarily from Helmholtz's address but also, presumably, from Humboldt's *Kosmos*. Whatever the sources, he enthusiastically embraced the hypothesis in both of its forms, as a theory of the universe at large and a theory of the origin of the solar system. Young Thomsen was a cosmic evolutionist but not yet an atomic evolutionist. As he described the universe, it was continuously evolving but not coming to an end in an absolute sense. In the far future the Sun and the other stars would become extinguished, disappear from the heavens, but new stars would be formed to replace them, the process going on eternally in grand cosmic cycles. Apparently Thomsen did not realise that a cyclic cosmology of this kind disagrees with the irreversibility of the second law of thermodynamics; or perhaps he did and just chose to disregard the problem.

Another feature of Thomsen's picture of the universe is worth noticing, namely that he conceived it to be infinite in both its spatial and temporal dimensions. "Space is unlimited, just as time is," he said, as were it a scientifically established matter of fact.[48] Thomsen undoubtedly realised that the claim of an infinitely old universe was controversial as it contradicted a creation in the past and then one of Christianity's key doctrines. But he refrained from commenting on the wider implications of his stated view. Nonetheless, his description of the universe and the nebular hypothesis in *Excursions* played an indirect role in the emerging struggle between the new positivist-materialist world view and the Christian belief with which science had been traditionally associated.

---

48. Thomsen (1856), p. 286. He further wrote that "the universe as a whole is infinite in time and space."

August Colding was no less in favour of cosmic evolution than Thomsen, but his reasons were quite different. In the same year that Thomsen published his popular book, Colding gave his lecture on the forces of nature to the Royal Danish Academy, and in this lecture he fully accepted the nebular hypothesis and cosmic evolution. Colding suggested that even the atoms of the chemical elements had come into being through an evolutionary process: "None of the chemical elements could have been formed, had not the general force of attraction existed from the ... time when the material itself was created in the form of an unbelievably diffuse mass of vapours occupying all of the infinite void." Unlike Thomsen he concluded: "Thus nature compels us to acknowledge that 'God has created the world seemingly out of nothing'."[49] Although Colding's universe was infinite in space, it was not eternal. Not only had God created the primordial nebulous matter, he had also created the laws of nature governing all matter and forces.

Colding's religious metaphysics of forces caught the critical attention of Rudolph Varberg, a journalist and writer known for his sympathy for materialism, atheism, and Darwinism. In a paper of 1857 he vehemently objected to Colding's claim of having demonstrated the immortality of the soul on a scientific basis.[50] Varberg also criticized Colding's interpretation of the law of force conservation and his understanding of the nebular hypothesis. In both cases he referred to Thomsen's publications which he seems to have considered authoritative expositions of the subjects. Apparently unaware that Thomsen's article of 1855 was in fact a free translation of Helmholtz's address, he described it as "instructive and interesting" and valued Thomsen's book of 1856 no less highly. Varberg may have thought that the young chemist was a potential ally in the fight against spiritualism and theism, but if so he was mistaken. Thomsen had no wish to be drawn into a dispute which was essentially ideo-

---

49. Colding (1856), in Dahl (1872), p. 125.

50. R. Varberg, "Naturkræfterne og Sjælens Udødelighed," *Dansk Maanedsskrift*, October 1858, reprinted in Varberg (1868), pp. 17-58. Varberg (1828-1869) became known in particular as an enthusiastic supporter of Darwinian evolution theory. In 1851 he defended Ludwig Feuerbach's controversial view of the Bible, which at the time was regarded as pure atheism.

logical and political. He wanted to follow his scientific interests. As to Colding, he chose not to respond to Varberg's attack.

Thomsen was also invisible in the heated discussion that from about 1870 took place in Danish intellectual circles concerning materialism, science, idealism, and Christian faith. In this discussion, which primarily involved philosophers, theologians and literary critics, Thomsen's thermochemistry must have appeared irrelevant and too technically demanding. Apart from Varberg, I am only aware of a single case in which a prominent participant in the debate, the professor of philosophy Rasmus Nielsen, referred to Thomsen's work.[51]

Finally, it is relevant to mention that another August, Julius Thomsen's brother, also contributed to the popular literature on the unity of the natural forces. In an article of 1860 on the Sun's energy and its terrestrial effects he repeated much of what his older brother had written in *Excursions*. As to what became known as the "solar constant" he informed the readers that in one year the Earth received enough solar heat to melt a 100-feet thick layer of ice (at 0 °C) surrounding the entire Earth. The illustrative analogy was not due to August Thomsen, who merely took it over from contemporary physicists and astronomers.[52]

3.4. Water analysis and public health

In a joint publication of 1853, Colding and Thomsen reported their detailed investigations of the causes of the cholera epidemic which hit Copenhagen in June the same year. Part of this remarkable work, to be described in the next section, relied on chemical analyses of the quality of the drinking water coming from the many wells and springs in the city. By the mid-nineteenth century investigations of

---

51. See Nielsen (1873), p. 305, who in a discussion of atoms and matter referred to Thomsen's early treatise on thermochemistry published in the proceedings of the Royal Danish Academy of Sciences.

52. A. Thomsen (1860). The solar "constant" or irradiance is the solar power that the Earth receives per square metre. It was first approximately determined in 1838 by the French physicist Claude Pouillet, who made use of the ice analogy. J. Thomsen (1855a) referred to Pouillet's result in his free translation of Helmholtz's 1854 lecture.

this kind – what may be called early environmental or hygienic chemistry – were still uncommon in Denmark if not quite without precedence. Colding and Thomsen relied to some extent on earlier investigations on the purity of the waters in the Copenhagen area.[53]

The first attempt to evaluate, in part by chemical means, the quality of the Copenhagen drinking waters goes back to 1756, when the medical doctor Johannes Christian Lange published a book on the subject. However, written at a time when the new analytic chemistry associated with the so-called chemical revolution was still in the future, Lange's work had no impact on later investigations concerning the health effects related to the poor quality of the drinking water. It soon came to be seen as a curiosity, a document of the past. In 1807 another medical doctor, Heinrich Callisen, an influential professor at the Academy of Surgery, published in two volumes a careful study of the hygienic state of affairs in Copenhagen. His *Physico-Medical Considerations Concerning Copenhagen* (Physisk Medizinske Betragtninger over Kjöbenhavn) included a chapter in which various water sources in the capital were examined chemically by Johan Gottlieb Blau, one of the city's apothecaries. For example, Blau used ammonia water to determine qualitatively the presence of clay earth in the water and similarly silver nitrate to determine chloride compounds. Ørsted, who on Callisen's instigation commented on Blau's results, was not impressed. According to him, chemical analysis was of limited use in determining the quality of drinking water. "In this matter," he wrote, "the witnesses of our senses are of greater importance, as they tell us more than the chemical means of testing."[54]

Little more occurred in this area of applied medical chemistry until the 1840s, when the population of Copenhagen had grown considerably and the pollution problems in the filthy and over-

---

53. Sources for this section can be found in Vestergaard (1999). Aspects of environmental chemistry in Denmark until about 1860 are summarised in Kragh (2010a). See also Bostrup (1996) for the development until the early nineteenth century.

54. Callisen (1807), p. 362. Two years earlier Ørsted had critically discussed the use of eudiometry as a means of measuring the pollution of air and its influence on health. See Bostrup (1996), p. 184.

crowded city even more so. At the time the methods of chemical analysis had improved drastically, leaving Ørsted's judgment obsolete. With the explosive growth of organic chemistry much of the attention shifted from inorganic pollution to the health hazards caused by organic compounds in water and air. Environmental chemistry was emerging elsewhere in Europe, where the health problems related to impure water were more serious than in Copenhagen. In London and other of the large British cities there were several chemical investigations of drinking water and attempts to improve its purity. Aware of what happened abroad Danish authorities were inspired to take up similar initiatives.[55]

In 1846 a royal commission belatedly decided that an examination of the water quality of the wells and springs of Copenhagen was needed. The task was assigned to Johannes Frederik Johnstrup, a young mineralogist who would eventually, after Forchhammer's death in 1865, become university professor of geology. Johnstrup had studied "applied science" at the Polytechnic College from where he graduated in 1844, just two years before Thomsen. In 1851, when the University of Christiania looked for a professor in chemistry, Forchhammer mentioned the possibility to Johnstrup – while Ørsted spoke warmly of Thomsen as a possible candidate.[56] In his water analysis of 1846, Johnstrup paid particular attention to the amount of nitrates dissolved in the water, which he measured by intense heating of the solid substance left after evaporation. If reddish vapours were observed, he concluded that the sample contained alkali nitrates. In modern notation the reddish vapours were N_2O_3 formed by

$$NO_2 + NO \rightarrow N_2O_3$$

In the case of potassium nitrate the reaction can be written as

---

55. See Hamlin (1990) for a detailed account of water analysis in Great Britain in the nineteenth century.

56. On Forchhammer, Johnstrup and the position in Christiania, see Garboe (1959-1961), vol. 2, p. 290. The chemistry chair is mentioned in Section 1.4.

$$2 \, KNO_3 \, (s) \rightarrow 2 \, KNO_2 \, (s) + O_2 \, (g),$$

followed by

$$2 \, KNO_2 \, (s) \rightarrow K_2O \, (s) + NO_2 \, (g) + N \, O \, (g)$$

Also Forchhammer took up the study of water analysis in the late 1840s, in his case with a focus on the organic compounds present in Copenhagen's lake waters. As he wrote in his report communicated to the Royal Academy: "Clearly, the influence of the [drinking] water on people's health depends much more on the content of organic compounds than the inorganic compounds, for the first determine to which degree the water will ferment and putrefy."[57] Forchhammer's method to determine the amount of non-nitrous organic compounds was new as it relied on oxidation by means of potassium permanganate ($KMnO_4$), a method which became widely used also internationally. He further recognised the importance of organic nitrogen compounds and developed means of detecting them in water.

The need for pure water was not only a concern for physicians and the public, but also for manufacturers of beverages and other goods based on water. For example, when the brewer Jacob C. Jacobsen planned the location of his new Carlsberg Brewery in about 1845, he realised that plentiful supplies of pure water were essential for commercial success. He consequently had one of the country's chemists, Christen T. Barfoed at the Military High School, to analyse the quality of the water near Valby, the site outside Copenhagen where the brewery was to be built.[58]

In 1846 the Danish Medical Society established a commission which under the authorship of the medical professor Andreas Sommer four years later published a report on the quality of Copenhagen's drinking water. *Om Kjøbenhavns Drikkevand* (On Copenhagen's

---

57. Translated from quotation in Vestergaard (1999), p. 28.
58. Christen Thomsen Barfoed (1815-1889) was yet another of the polytechnically trained chemists of Thomsen's generation. He graduated from the Polytechnic College in 1835 and later worked at the Royal Veterinary and Agricultural College.

Drinking Water) was highly critical, leaving no doubt that the city's water supply was inadequate and the water much unhealthier than in other of Europe's major cities. The country's first chemistry professor, William C. Zeise, assigned his pharmaceutically trained assistant Frederik Zedeler the task of analysing organic substances in the lakes around the central city. Some of the analyses were performed by Thomsen, who may have alternated with Zedeler.[59] Zeise, recognised as an expert in organic chemistry, died in late 1847 and seems not to have participated in the work. The analyses performed by Zedeler and Thomsen did not result in a publication and are only known from the data given in Sommer's book from 1850.

It is not obvious why Thomsen in 1853 decided to join forces with Colding in an investigation of the causes of the cholera and its dissemination by means of water. While Colding served at the time as inspector of Copenhagen's water works and was engaged in reforming the city's supply of water, Thomsen had no experience with either public health or public water systems. On the other hand, he was a promising and versatile young chemist who had previously substituted for Zedeler in water analyses. Moreover, it is likely that he obtained further insight in analytical methods through his work as an assistant for Forchhammer.

Two years after his collaborative work with Colding related to the cholera epidemic, Thomsen again performed a series of water analyses of wells, this time in the Frederiksberg area of Copenhagen. The physician Emil Hornemann, an advocate for sanitary reforms and a leader of the hygienic movement in Denmark, suspected that health problems in the area might in part be caused by water polluted from a cemetery. After consultations with Colding, the two men requested Thomsen to investigate the matter by means of chemical analysis. In his report to the Health Commission, Thom-

59. Vestergaard (1999), p. 24. Apart from his mentioned work under Zeise, Zedeler seems to be an unknown figure in Danish history of science. According to the national census of 1845, Frederik Adolph Zedeler was born in Sæby, Northern Jutland, in 1822. I have found no publications from his hand.

sen found strong evidence that the water from one of the wells was seriously polluted by putrefied organic substances coming from the Frederiksberg churchyard.[60]

3.5. On the causes of the cholera

The feared Asiatic cholera came to Copenhagen on 11 June 1853, causing 7,219 people out of the city's approximately 130,000 inhabitants to fall ill within the next few months. The number of deaths was 4,743, meaning a mortality rate of 66 per cent. A total of 6,688 Danes lost their life to the cholera. Even before the epidemic had passed, officially by 1 October, Colding and Thomsen had finished their scientific investigation on the causes of the spread of the disease. Their detailed 112-page report, titled *Om de Sandsynlige Aarsager til Choleraens Ulige Styrke i de Forskjellige Dele af Kjöbenhavn* (On the Probable Causes for Variations of Cholera Outbreaks in Different Regions of Copenhagen), was published in late September or early October.[61] Most likely Colding and Thomsen wrote the treatise on their own initiative and not on the request of the Copenhagen Municipality or the Health Commission. What interested the two authors was primarily the uneven distribution of cholera attacks in various districts of the city, which they assumed was mainly due to local differences in the composition of the soil on which the houses were built. As the other main factor for the variation they took the local population density.

In medical circles there were at the time two alternative theories of how infectious substances were transmitted, the "miasmatic" and the "contagious" theory. According to the first theory, on which Colding

---

60. The main content of Thomsen's report appeared in the first volume of a new journal on public hygiene founded by Hornemann, *Hygiejniske Meddelelser* (Hygienic Communications). See Hornemann (1857), pp. 3-8. Together with Thomas Segelcke, a 23-year-old graduate from the Polytechnic College, Thomsen also analysed the water from a well in Copenhagen, as described in Hornemann (1856), pp. 14-17. Segelcke later became professor at the Agricultural College and the country's leading expert in dairy chemistry.

61. Colding and Thomsen (1853). See Vestergaard (1999) for a detailed analysis. The study of Colding and Thomsen is generally ignored in the international literature. A brief summary appears in Raestad (2000).

Figure 3.4. The cholera comes to Copenhagen. Satirical drawing in the popular journal *Folkets Nisse* of 2 July 1853 commenting on the inefficiency of the Copenhagen city council regarding the cholera epidemic.

and Thomsen implicitly based their work, diseases such as cholera were caused by a noxious form of "foul air" (called miasma) emanating from organic matter fermenting and putrefying in the soil. Contagionists, on the other hand, believed that the carrier of diseases was direct or indirect physical contact between people and they consequently recommended quarantine as a means to stop the spreading of cholera. The lack of success was a main reason why medical experts generally favoured the miasma theory in one version or other.

Referring to remnants of manure and renovation from the past, Colding and Thomsen stated that the purpose of their work was to show "that those substances, which have been in the soil for 200 or even 500 years, are still in a state of putrefaction and they emanate their products to the wells of the city through air and soil."[62] They

---

62. Colding and Thomsen (1853), p. 24.

reasoned that the assumed noxious emanations from the bad soil would in part be absorbed in the underground water, but not that the water itself was the carrier of the cholera. The carrier was principally the air. When they nevertheless focused on the water, it was because they conceived it as a reservoir of the gases produced by rotten organic matter. The prime indicator was methane, CH_4, a gas which was harmless but served as an indication of the "badness" of the soil. Thomsen, who was undoubtedly responsible for the chemical analyses, could have chosen to determine methane in the air but found it more practical and precise to measure the content of the gas absorbed in water. For this purpose he (or he and Colding) developed a new analytical method.[63]

The essence of the Thomsen method was the following. From a sample of well water Thomsen boiled out the gases and removed H_2O, H_2S and CO_2 by standard methods. He next transformed the remaining gas, in the form of CH_4 and other hydrocarbons, to CO_2 by passing it over red-heated copper oxide:

$$CH_4 \text{ (g)} + 2\,CuO \text{ (s)} \rightarrow CO_2 \text{ (g)} + 2\,Cu \text{ (s)} + H_2O \text{ (l)}$$

The carbon dioxide was precipitated with baryte water as $BaCO_3$:

$$CO_2 \text{ (g)} + Ba(OH)_2 \text{ (aq)} \rightarrow BaCO_3 \text{ (s)} + H_2O \text{ (l)}$$

After drying the precipitate the amount of CH_4 could be determined gravimetrically. Colding and Thomsen confirmed by various control measurements that the amount of precipitated $BaCO_3$ was indeed a valid expression of methane in the water, hence an expression of the state of fermentation in the soil. They examined 32 of the wells in Copenhagen and found methane in 19 of them, of which wells near the churchyards had a particularly high concentration. The next step was to relate the fermentations, or state of the soils, to the frequency of cholera attacks.

63. A later version of the method was described in Fleury (1875), written by the Danish chemist August Fleury who at the time served as Thomsen's assistant at the chemical laboratory of the University of Copenhagen.

For this purpose Colding and Thomsen constructed a measure of the soil's health quality B in terms of the local population density T and the frequency of cholera incidents A, namely

$$B = c\frac{A}{T},$$

where c is an empirical coefficient. The claimed proportionality between A and BT needed to be confirmed empirically, which the two scientists did by means of a statistical correlation. Although their use of statistics was in fact problematic and not nearly as convincing as they claimed, it confirmed the hypothesis to their own satisfaction. It led them to the following conclusion:

> Based on statistical analysis, we conclude that cholera is closely linked to two factors: infected drinking water and high population density. … We must therefore consider it beyond any doubt that it will be perfectly possible to diminish the power of the disease considerably, although not preventing it altogether.[64]

The Colding-Thomsen investigation was scientific in the sense that it was based on chemical and statistical methods, but its purpose was entirely practical, namely, to identify the causes of the cholera in order to prevent a future epidemic. More generally the publication was an argument for improving the living conditions for a large part of the population in Copenhagen. It was in part motivated by the authors' social indignation that building had become an object of profitable speculation. They deplored that "large, beautiful and airy quarters have been almost destroyed as a result of unrestrained building" and recommended a firmer public control of new buildings and their environments. "In order to take care of the necessary hygiene," they wrote, "it should be an indispensable rule that all new streets are provided with lines for gas, water and sewage."[65] Thomsen and Colding further recommended a more even and regulated distribution of the citizens to avoid overcrowding. Although

64. Colding and Thomsen (1853), p. 107.
65. Colding and Thomsen (1853), p. 65.

Figure 3.5. In 1853 most of Copenhagen's inhabitants lived in small and unhealthy apartments, which meant that the spread of cholera had almost ideal conditions. Photograph of the crowded Vognmagergade quarter in central Copenhagen from the late nineteenth century. Copenhagen City Museum.

their recommendations did not lead to immediate action, in the long run they did have an effect. Hornemann and other advocates of the hygienic movement used the analysis of Colding and Thomsen as ammunition in their fight for a cleaner and healthier city. Thomsen kept an interest in problems concerning sanitation and hygiene. For example, he was active in the preparations which in 1880 led to the Society for Hygiene in Denmark (Selskabet for Sundhedsplejen i Danmark), an organisation with Hornemann as its first president.[66] For another example, see Section 8.2.

The English physician John Snow, one of the fathers of modern epidemiology, is famous for his investigation of the 1854 London cholera. Although Snow did not believe in the miasma theory, there are close parallels between his work and the one made the year before by Colding and Thomsen in Copenhagen.[67] Snow made use of chemical water analysis, if not of methane but of chloride, and he plotted all registered cholera cases on maps to show how poor living conditions were correlated with the frequency of the cholera. Contrary to Colding and Thomsen, he also used microscope examination of the water to demonstrate its poor quality. Snow was unaware of the work by Colding and Thomsen, which was published in Danish only. Whatever the share of the two Danish researchers in the history of epidemiology, in 2010 Colding (but not Thomsen) was appointed an honorary member of the Scandinavian branch of the John Snow Society – very much post mortem.

66. See Lindegaard (2001), p. 183. Online as http://dendigitalebyport.byhistorie. dk/bibliografi/dokumenter/phdafhandling_lindegaard.pdf.
67. Vestergaard (1999), pp. 75-76; Raestad (2000).

CHAPTER 4

Pioneering thermochemistry

If known at all among modern chemists, Julius Thomsen will be vaguely remembered as one of the nineteenth-century pioneers of classical thermochemistry and especially for his systematic and very precise measurements of thermal data over a wide range of chemical compounds. Indeed, this was the basis for the great reputation he enjoyed during his own time. But although Thomsen was the first to suggest a definite theory of thermochemical reactions, he did not invent thermochemistry. This branch of chemical research has roots farther back in time. The roots can be found in calorimetric measurements made in the late eighteenth century, and in the 1840s thermochemistry was developed on an experimental basis by a small but gowing number of chemists. The main product of Thomsen's extensive work on the subject was an impressive four-volume book published between 1882 and 1886, the *Thermochemische Untersuchungen* (Figure 4.1). Thomsen based his system on the principle that reactions of a proper chemical nature always evolve heat and that the amount of heat is a measure of the affinity of the involved substances. However, at the time of publication of Thomsen's book several other chemists had entered the field of thermochemistry, and in some cases they challenged his data and interpretations. The most prominent of Thomsen's rivals was the brilliant and versatile Frenchman Berthelot, who based his theory on what he called a "principle of maximum work."

While the theory of thermochemistry proposed by Thomsen and Berthelot was solidly founded on the first law of thermodynamics, it disregarded the second law and generally the new chemical thermodynamics formulated in the 1880s by Helmholtz and others. It turned out that the classical theory did not give a correct picture of the role of heat in chemical processes. As became clear latest in the 1890s, the Thomsen-Berthelot principle is of limited validity and

THERMOCHEMISCHE

UNTERSUCHUNGEN

VON

JULIUS THOMSEN,

DR. PHIL. ET MED.

PROFESSOR DER CHEMIE AN DER UNIVERSITÄT UND DIRECTOR DER
POLYTECHNISCHEN LEHRANSTALT ZU KOPENHAGEN.

VIERTER BAND.

ORGANISCHE VERBINDUNGEN.

MIT EINER TAFEL.

LEIPZIG,

VERLAG VON JOHANN AMBROSIUS BARTH.

1886.

Figure 4.1. The fourth and last volume of Thomsen's great work on thermochemistry.

only true at absolute zero temperature.[1] Thomsen as well as Berthelot were unwilling to admit that their thermal theories were at best approximately true, but by the turn of the century this was the verdict of the new generation of physical chemists – and it is still the verdict.

4.1. Affinity, calorimetry, and thermochemistry

Why do certain chemical reactions occur while others do not? As seen from the standpoint of early nineteenth-century natural philosophy the question could be answered in terms of the forces binding together the constituents of matter or what was generally known as the "chemical affinities" of substances. In the previous century several tables of relative affinities had been proposed, first in 1718 by the Frenchman Étienne François Geoffroy. Later in the century more elaborate tables were presented by Pierre Joseph Macquer in an influential textbook of 1749 and by the Swedish chemist Torbern Bergman in 1775. The framework of these tables was the idea of certain "elective affinities" which depended solely on the nature of the substances involved in reactions. Within this framework reactions were supposed to proceed in one way only, which ruled out reversible processes. However, tables of this kind were of limited use, for other reasons because they failed to take into account the physical conditions under which chemical reactions occur.

Chemical affinity was that something which caused substances to enter into or resist decomposition, but what was that something? The question was often conceived within the framework of Newtonian mechanics, with chemical force being largely synonymous with chemical affinity. The causes of chemical processes were seen as residing in the attractive forces supposed to act on a microscopic level and somehow being of a nature similar to that of Newtonian gravity

---

1. I use the term "Thomsen-Berthelot principle" as a convenient name for the basic idea of classical thermochemistry as developed by Thomsen and Berthelot. The name was rarely used at the time, van't Hoff (1905) being one of the few exceptions. Neither Thomsen nor Berthelot, rivals as they were, would have liked the name.

on the macroscopic level.[2] With the rise of galvanism and electro-chemistry in the first decades of the nineteenth century it became more common to think of affinity as a manifestation of electrical rather than gravitational forces. Inspired by Berzelius' widely accepted dualistic theory of composition some chemists sought for an electrical explanation of affinity, but not very much came out of their endeavours. Whatever the nature of the elusive concept, it was realised that chemical affinity, in order to be of scientific use, needed to be associated with experiments. One way of providing affinity with a quantitative and operational measure was by means of the heats evolved – or sometimes absorbed – in chemical processes.

Among the first attempts to determine the amount of heat that evolved in chemical reactions with the object of obtaining a measure of chemical affinity were a famous series of experiments conducted by Antoine-Laurent Lavoisier and Pierre Simon de Laplace in about 1780. Their interpretation of experiments with, for example, sulphuric acid and water in different proportions did not rest on any particular theory of heat except that it presupposed heat to be a conserved quantity. The ice calorimeter that Lavoisier and Laplace used was instrumental in establishing calorimetry as a science, but it was a delicate and expensive instrument which was limited to a small temperature interval. Moreover, the results obtained by means of the ice calorimeter were not very consistent and therefore not very accurate. Incidentally, the word "calorimeter" was coined by Lavoisier in 1789 to refer to the instrument he and Laplace had developed about a decade earlier. As the name indicates, it was originally associated with the belief in *caloric*, the material but imponderable and self-repulsive substance that Lavoisier assumed to be the cause or even the nature of heat. In the list of chemical elements stated in his famous *Traité Élementaire de Chimie* of 1789 caloric figured as a "simple substance."

Another pioneer of calorimetry was the Irish chemist and physician Adair Crawford, who in books from the same period discussed various designs of calorimeters to measure the heats of combustion

---

2. See Levere (1971) and Duncan (1996) for detailed accounts of the early history of affinity.

of a few chemical substances.[3] Like Lavoisier, he was particularly interested in the "animal heat" associated with physiological chemistry. Crawford reported in 1788 that he had exploded a mixture of oxygen and hydrogen and estimated the heat of combustion. Not being a convert to Lavoisier's new chemical system he referred to the gases as "dephlogisticated air" and "inflammable air," respectively.

The early work of Lavoisier, Laplace, and Crawford was not followed up to any extent and it took several decades before thermochemistry, as distinct from calorimetry, was firmly established. One of the founders of this branch of chemistry was the Swiss-born Russian chemist Germain Henri Hess, a professor in St. Petersburg, according to whom the proper measure of chemical affinity was given by heats of dilution and combination. As he stated in an important paper of 1840, he realised that, "the atom which is held most strongly also develops the most heat, and that the quantity of developed heat could serve as a measure of affinity."[4] Based on a series of experiments involving mixtures of sulphuric acid with water, neutralisation of acids and bases, and the formation of ammonium sulphate in water, he demonstrated empirically what is known as the law of constant heat summation, or just Hess's law. According to this empirical law or generalisation, the amount of evolved heat is independent of the individual reaction processes. In the words of Hess: "A combination having taken place, the quantity of heat evolved is always constant whether the combination is performed directly or whether it takes place indirectly and in different steps."[5]

As an example, consider the consecutive neutralisation of phosphoric acid with sodium hydroxide, which proceeds in three steps:

---

3. The work of Crawford and other contributors to calorimetry and early thermochemistry is carefully described in Médard and Tachoire (1994).

4. Quoted from Schelar (1966), p. 103. Parts of Hess's 1840 memoir are translated in Leicester and Klickstein (1968), pp. 329-332. For background, see Leicester (1951).

5. Leicester and Klickstein (1968), p. 331.

(I) $H_3PO_4 + NaOH \rightarrow NaH_2PO_4 + H_2O,$ heat = X

(II) $NaH_2PO_4 + NaOH \rightarrow Na_2HPO_4 + H_2O,$ heat = Y

(III) $Na_2HPO_4 + NaOH \rightarrow Na_3PO_4 + H_2O,$ heat = Z

According to Hess's law the total amount of evolved heat will be X + Y + Z. As Hess realised, his law was of considerable practical significance as it allowed the calculation of heats which could not be directly measured. Thus the formation of carbon monoxide from carbon and oxygen will always be accompanied by the formation of carbon dioxide and for this reason the heat evolved in the first process alone cannot be directly established. But it can be found indirectly from the two processes

$$2\,C + 2\,O_2 \rightarrow 2\,CO_2 + 189 \text{ kcal}$$

$$2\,CO + O_2 \rightarrow 2\,CO + 135 \text{ kcal}$$

From Hess's law it follows by subtraction that

$$2\,C + O_2 \rightarrow 2\,CO + 54 \text{ kcal}$$

At about 1840 the concept of heat was under transformation, but Hess still referred to the caloric conception and heat as a conserved quantity; he did not relate his results to the new mechanical theory of heat. From the point of view of the slightly later principle of energy conservation Hess's law was merely a somewhat trivial consequence of this principle, such as Helmholtz pointed out in *Über die Erhaltung der Kraft*, his classical 1847 memoir on the conservation of energy.[6] In any case, with the work of Hess and the contemporary

---

6. Reprinted in *Ostwald's Klassiker der Exacten Wissenschaften*, no. 1 (Leipzig: Engelmann, 1889), where the reference to Hess is on p. 25. Remarkably, Helmholtz's masterpiece was not accepted for publication in *Annalen der Physik und Chemie* and consequently it appeared as a privately printed booklet. Still in the mid-1850s Thomsen seems to have been unaware of it.

contributions of Pierre Dulong in France, Thomas Graham in England, and Thomas Andrews in Ireland, thermochemical studies were slowly gaining an important position in chemistry. These early studies were mostly concerned with reactions which readily occur at moderate temperatures, such as the heats of neutralisation of acids and the heats of hydration of salts.

The first systematic and large-scale series of calorimetric determinations of heats involved in chemical reactions, after those of Hess, were carried out in Paris by Pierre Antoine Favre and Johann Theobald Silbermann. The two chemists initiated their collaborative thermochemical research programme in 1843 and six years later, after hundreds of accurate experiments and numerous publications, most of them in the journal *Comptes Rendus*, they terminated the collaboration. In 1852-1853 they recapitulated their very extensive work in a couple of voluminous memoirs in the *Annales de Chimie et de Physique*.[7] The work of Favre and Silbermann was well received and for more than a decade their measurements constituted the main body of data for thermochemistry. In 1849 the French Academy of Science announced a prize competition for the best study of heat given out in chemical reactions; the first prize was awarded to Favre and Silbermann and the second prize to Andrews.

In 1846 Favre and Silbermann constructed a new type of calorimeter which was better suited for measuring large amounts of heat. Their "mercury calorimeter" consisted of an iron bulb filled with mercury connected to a horizontal graduated tube. By measuring the expansion of the mercury they could find the heat evolved. The two Parisian chemists took pride in their instrument, which they argued was more precise and reliable than earlier calorimeters. Perhaps it was, but the mercury calorimeter and the method based on it was criticized by later chemists, including Thomsen and Berthelot (see Section 5.1). Favre and Silbermann made use of the "calorie" as a unit of heat, and it is sometimes stated that they introduced the unit. However, this is a mistake as the calorie unit goes farther

---

7. Favre and Silbermann (1853). The work of the two French chemists is covered in detail in Médard and Tachoire (1994), pp. 97-114.

back in time and was well-known in France by the mid-1840s.[8] During the second half of the nineteenth century two units of heat were employed in thermochemical and other studies. One was the "small calorie" (cal) and the other was the "large calorie" or kilo-calorie (Cal or kcal). The two units are related as simply

$$1 \text{ Cal} = 1000 \text{ cal} = 4184 \text{ J}$$

Favre and Silbermann performed experiments over a wide range, from various heats of combinations to heats evolved by dissolution. For example, they obtained for the first time an accurate value for the heat of combustion of hydrogen gas. Translated into the modern unit they reported 68.95 calories per mole, which is only slightly less than the correct value. Also for the heat of formation of zinc chloride

$$Zn + Cl_2 \rightarrow ZnCl_2$$

they determined a value in fair agreement with the modern one. In this case they made use of indirect processes and Hess's law. In a study of the heat of formation of nitrous oxide (N_2O) they found to their surprise that the compound *absorbed* rather than liberated heat energy. This was one of the very first examples of endothermic processes, a class of processes that soon would become important as well as problematic in thermochemical theory. Many years later, in 1880, Thomsen investigated the endothermic formation of nitric oxide (NO) from its constituents and measured the amount of heat involved in the process.[9]

The two French chemists were primarily occupied with measurements and data; they hesitated in drawing theoretical conclusions from their many experiments on chemical heat. By the early 1850s the mechanical theory of heat and the law of energy conservation were well established but Favre and Silbermann made no use of

---

8. See Hargrove (2006) for a history of the calorie. The unit was possibly first used by the French physicist Nicholas Clément in lectures of 1824.
9. Thomsen (1880e).

these important ideas in their work. They were not foreign to the concept of affinity, of course, and at one point they even motivated their research by referring to "the laws of chemical affinity ... [which are] still enveloped in deep enough obscurity."[10] But Favre and Silbermann were unwilling to define what they meant by affinity except that they tended to identify it with molecular stability. They thought that this property manifested itself in the heats of formation: the more exothermic the formation of a compound was, the more stable it was. In one of their last memoirs they wrote: "It seems difficult not to admit that there is quite a close connection between the energy of the affinities of different bases for the same acid, or the degree of stability of the compound formed, and the quantities of heat evolved in the act of combination."[11]

Like most other workers in thermochemistry, Favre was convinced that calorimetry was the key that would eventually unlock the secret of the nature of affinity. As he wrote: "To study chemical reactions, taking into account the quantities of heat involved, is, in our opinion, the best, perhaps the *only* way by which one may arrive at a correct conception of the force designated by the name of affinity."[12] In Copenhagen, Thomsen agreed.

4.2. A thermochemical system

In 1852 Julius Thomsen published an extensive memoir in the proceedings of the Royal Danish Academy in which he ambitiously aimed at establishing a rational basis for thermochemistry. Dissatisfied with the state of art of this branch of experimental chemistry the 26-year-old chemist wanted to formulate it as a logical system of definite propositions. The immediate reason for Thomsen's publication was a perceived competition from the British chemist and medical doctor Thomas Woods who the same year had proposed a thermal theory of chemical affinity. Thomsen had originally wished

---

10. See Levere (1971), p. 204.
11. Favre and Silbermann (1853), p. 491. The terms "exothermic" and "endothermic" were coined by M. Berthelot in lectures he gave at Collège de France in 1865.
12. Quoted in levere (1971), referring to a paper of 1860.

to base his theory on a large number of experiments and only then, armed with a wealth of experimental data, to formulate a theory in agreement with the data. But faced with Wood's paper he felt forced to adopt a more deductive-theoretical approach instead of the inductive-empirical approach he had first had in mind.[13]

At the end of his Danish memoir Thomsen pointed out that chemical processes do not depend solely on "the forces inherent in the substances" but also on external forces such as electrical actions. He illustrated the point by referring to the reactions involving zinc and copper in a Daniell cell, a case he would return to in his later work. He also called attention to the chemical actions of light which he conceived as analogous to those of electricity. Thomsen suggested that his thermochemical theory, when fully developed, would be applicable to the vegetable world: "The organic substances contained in the plants are formed by the reduction of water vapour, carbon dioxide and ammonia under the action of light and absorption of heat; this may in part explain why we have been unable to synthesise in our laboratories vegetable substances by means of the ordinary chemical operations."[14]

Thomsen quickly followed up on his 1852 memoir with a series of articles published in Poggendorff's *Annalen der Physik und Chemie*. These were in part a German translation of his 1852 memoir but in some respects went beyond it. The series of papers consisted of five parts and appeared 1853-1854 in the form of four consecutive papers with the common title *Grundzüge eines Thermochemischen Systems* (Basic Features of a Thermochemical System).[15] Altogether it covered 88 pages. The series of publications between 1852 and 1854, a veritable *tour de force*, formed the basis of Thomsen's theory of thermochemistry such as he would develop it in his later works. In a Danish paper of 1861 he sharpened some of his original formulations but without changing them substantially. Throughout his life

---

13. Thomsen (1852), p. 155; Woods (1852).

14. Thomsen (1852), p. 165. In the popular book Thomsen (1853b) he dealt in detail with the composition and chemical characteristics of plants.

15. Thomsen (1853c, 1854a, 1854b, 1854c). The last paper promised a continuation of the series, but no such paper appeared in print (1854c, p. 57).

he basically kept to the thermochemical system as he had presented it in his youth. Thus, most of the ideas introduced in *Grundzüge* can be found also in Thomsen's English summary volume *Thermochemistry* published 1908, more than half a century later.

For the energy involved in a chemical reaction Thomsen used various and more or less synonymous names such as "chemical force," "affinity" and "thermodynamic equivalent." He did not use the term "energy" (or "force," corresponding to the German *Kraft*). The chosen terminology may have made his articles difficult to read and appear more obscure than they actually were. For the heat evolved or absorbed in a process he spoke of the reaction's *varmetoning* or the German equivalent *Wärmetönung*, which in English can be translated to "thermal effect."[16] The *varmetoning* could thus be either positive or negative, corresponding to exo- and endothermic processes. Thomsen was convinced that "a study of the thermal effect in chemical actions will lead to the determination of the absolute values of the chemical forces in various substances."[17]

Thomsen based his theory on the postulate that there is inherently in nature a certain direction of evolution which matter follows spontaneously; it only moves in a different direction if an external force is applied. For example, he said that the natural "motion" of phosphorus was to unite with other elements such as oxygen. This motion corresponded to the strongly exothermic nature of the process

$$P_4 + 5\,O_2 \rightarrow P_4O_{10}$$

On the other hand, Thomsen noted that the process

$$N_2 + O_2 \rightarrow 2\,NO$$

---

16. Thomsen's neologism *Wärmetönung* (or the Danish *varmetonung*) was widely used in Scandinavian and German chemical literature but not elsewhere. I shall generally use the term "thermal effect." In a few cases, as in Merz (1904-1912), the literature in English made use of the hybrid term "heat-toning."

17. Thomsen (1852), p. 118.

was endothermic. Adopting a teleological terminology, he interpreted it as the "purpose" of nitrogen, in so far it was part of the
Earth's atmosphere, was the element itself. He realised that this purpose depended on the circumstances, for if nitrogen was mixed with
hydrogen it would in some cases follow the exothermic process

$$N_2 + 3 H_2 \rightarrow 2 NH_3$$

As he said, "If the Earth had an atmosphere of hydrogen, the purpose of nitrogen would have been ammonia."[18]

It should be noted that in his early work from 1852 to 1854
Thomsen used a nomenclature that differs substantially from the
present one. In accordance with a notation proposed by Berzelius
in the late 1820s he represented oxygen atoms in a compound by
dots above the chemical symbol and used barred symbols to represent two atoms. A couple of examples will illustrate the kind of symbols used by young Thomsen and many of his contemporaries:

$$\dot{H} = H_2O \; ; \quad \ddot{\overline{SH}} = H_2SO_4 \; ; \quad \overline{NH}^3 = 2 NH_3$$

Thomsen did not adopt Berzelius's notation in his textbook of
1850, where he used formulae such as N^2O, Fe^2O^3 and (BaO, N^2O^5),
with the latter representing barium nitrate or $Ba(NO_3)_2$. For the
sake of intelligibility I shall transcribe most formulae into modern
notation.

To express the heat effects in a general and systematic manner
Thomsen made use of a new nomenclature, which he had first proposed at the meeting of Scandinavian scientists held in Stockholm
in 1851. If a compound X_aY_b is formed by its constituents X_a and Y_b
he denoted the thermal effect (X_a, Y_b) or sometimes $[X_a, Y_b]$. For
example, lead sulphate may be formed by its elementary constituents in which case the thermal effect will be (Pb, S, O_4); but it may
also be formed by an oxidation of lead sulphide,

$$PbS + 2 O_2 \rightarrow PbSO_4$$

18. Thomsen (1852), p. 156.

In the latter case the thermal effect was written as (PbS, O_4). For the total amount of energy associated with one equivalent of X_aY_b he used the formula (X_aY_b), and for the heat he adopted a unit corresponding to the calorie, although without using the name. Thomsen further wrote the heat effect of the compound X_aY_b with c equivalents of water as

$$\left(X^a, Y^b, \dot{H}^c\right) \quad \text{or} \quad (X_a, Y_b, aq_c)$$

If the compound was fully dissolved in water he used the notation

$$(X_a, Y_b, Aq)$$

The symbol Aq denoted an amount of water so great that a further addition of water would have no thermal effect.

Among the results Thomsen derived from his formal system of thermochemistry was Hess's law. Suppose that the substances P, Q, R and S unite directly to form the compound $P_aQ_bR_cS_d$ and that this compound is then split up into P_aQ_b and R_cS_d; suppose further that these two compounds are again resolved in their constituents. Then Thomsen showed that

$$(P_a, Q_b) + (R_c, S_d) + (P_aQ_b, R_cS_d) = (P_a, Q_b, R_c, S_d)$$

That is, "the total thermal effect for the formation of a compound is always the same, whether it is formed directly or successively from its constituents."[19] As Thomsen noted, this was just what Hess had found experimentally. He was apparently unaware of Helmholtz's earlier demonstration that Hess's law follows straightforwardly from energy conservation. Although Thomsen knew about the principle of energy or "force" conservation, in his publications from 1852-1854 he did not refer directly to it. On the other hand, he used it implicitly and described it in words; it was indeed at the base of his thermochemical system. Thomsen's notation became widely used in German literature in particular. As late as 1897 the famous

---

19. Thomsen (1852), p.123; Thomsen (1853c), p. 356.

physicist Max Planck wrote: "J. Thomsen ... denoted by the formulae for the atomic or molecular weight of the substances enclosed in brackets, the internal energy of a corresponding weight referred to an arbitrary zero of energy. Thus [Pb], [S], [PbS] denote the energies of an atom of lead, an atom of sulphur, and a molecule of lead sulphide respectively."[20]

Having laid out the formal scheme of his thermochemical system Thomsen went on to consider the experimental data obtained by previous researchers. He was not impressed. In the case of acid-base neutralisation processes he pointed out that the values reported by Hess, Andrews, and Favre and Silbermann varied considerably. "There is no agreement at all!" he lamented. "Since I could not use such deviating results I was forced to repeat all the experiments to get results that at least approximately express the truth."[21] In his extended series of neutralisation experiments, undoubtedly made at the chemical laboratory of the Polytechnic College, Thomsen used a relatively simple calorimeter. He took great care in analysing the apparatus and the unavoidable experimental errors associated with it. Thomsen found thermal values that roughly agreed with those reported earlier but, he stressed, his own values were more precise and reliable. The novelty of Thomsen's series of papers 1852-1854 did not lie in his experiments but in the proposed theoretical system of thermochemistry and its associated definition of affinity.

In *Grundzüge* Thomsen added an interesting section on speculative atomic theory which is not to be found in his Danish memoir. "The atomic theory has proved exceedingly useful for chemistry," he wrote, "[and] I shall here try to apply it to the inherent forces of substances." For liquids Thomsen assumed that the inherent force manifested itself in the atoms or molecules performing circular motion with a characteristic radius and angular velocity. He did not address the question of the centripetal force necessary for the rotating motion. When two liquids were mixed, their molecules were supposed to attain the same angular velocity but keep their original

20. Planck (1897), p. 63.
21. Thomsen (1852), p. 131.

radii. "By this equalisation of angular velocity a loss in kinetic energy will occur, proportional to the amount of evolved heat."[22] For a body of mass m moving with velocity v, or in the case of uniform circular motion $v = \omega r$, the kinetic energy was known to be

$$E_{kin} = \frac{1}{2}mv^2 = \frac{1}{2}m\omega^2 r^2$$

However, Thomsen used the older definition of "living force" or *vis viva* and consequently ignored the factor ½. Let the aqueous solvent be characterised by atoms of mass M revolving in circles with radius r and angular velocity ω; for the substance mixed with the solvent the symbols are M_1, r_1 and ω_1. Thomsen showed from the mechanical theory that the heat Q evolved in the mixing process would be

$$Q = \frac{Mr^2 M_1 r_1^2}{Mr^2 + M_1 r_1^2}(\omega - \omega_1)^2 W \,,$$

where W denotes the conversion factor between heat and kinetic energy. To connect his atomic hypothesis to measurable quantities Thomsen considered a single "atom" of, for example, sulphuric acid dissolved in n atoms of water. He derived the following expression for the evolved thermal energy:

$$Q = \frac{n}{n + x} C$$

The two quantities x and C were given only in terms of the atomic hypothesis but could be estimated by comparing this expression with experimental data. For the heat evolved in aqueous solutions of sulphuric acid he found with good approximation that

$$Q = \frac{1085n}{n + 1.745}$$

Although Thomsen found a satisfactory agreement between theory and data, he realised that this alone did not justify his hypothesis of

---

22. Thomsen (1854a), p. 275. Thomsen used the term "lebendige Kraft" (living force) for the kinetic energy and did not distinguish sharply between atoms and molecules.

circularly moving atoms or molecules. "Some other hypothesis might possibly lead to the same result," as he noted. As we shall see in Section 6.2, he would later return to the heat of dilution in the system made up of varying amounts of H_2SO_4 and H_2O.

Thomsen's attempt to explain thermochemical processes in terms of atomic motions was meant to be an analogy and not a realistic model. And yet, when he applied it to the volume contraction of sulphuric acid in water he was able to derive contraction effects in agreement with experiments. The agreement made him express himself optimistically: "The proposed hypothesis may to some extent correspond to the truth; it is interesting that the general laws of motion are applicable to molecular motion and can contribute to find the laws [of molecules], if not their true causes."[23] Thomsen did not further develop his atomic-molecular theory of liquids and solutions. It is of interest mainly because it illustrates Thomsen's belief in a mechanical foundation of chemistry and also his considerable knowledge of mechanical physics. Such knowledge was not common among chemists in the mid-nineteenth century when chemistry was predominantly an experimental art.

As Thomsen was aware, he was not the first one to investigate the contraction and heat effects in the system of sulphuric acid and water. An investigation of this kind had earlier been reported by the Norwegian physicist Lorentz Christian Langberg, a newly appointed professor at the University of Christiania. Langberg found theoretical expressions of the effects similar to those published by Thomsen, and he verified the heat expression by means of Hess's data for the heat evolved in mixtures of sulphuric acid and water. He even suggested a definite relation between the changes in volume and the evolved heat. Not unlike Thomsen, Langberg assumed that "changes in volume as well as in heat are both effects of a higher cause, namely the striving of the chemical or molecular forces towards a new equilibrium position."[24]

---

23. Thomsen (1854a), p. 284.
24. Langberg (1845), p. 328. Langberg (1810-1857) had studied physics under Ørsted in Copenhagen and in 1844-1845 been on an extensive study tour in Europe. See https://snl.no/Lorentz_Christian_Langberg. He was editor of *Nyt Magazin for*

Thomsen considered his new system a step on the road that would change chemistry from a predominantly empirical science to an exact science governed by definite laws. This was a theme he would often return to, such as he did in a lecture to the Royal Danish Academy in 1861. On this occasion he asserted:

> The proposed theory is in perfect agreement with the basic laws of mechanics and it allows us, at least to some extent, to apply the mathematical method to chemical processes. By considering the evolution of heat caused by the formation of a chemical compound as a measure of the affinity ... it becomes possible to establish general laws for the chemical processes and to replace the older and uncertain doctrine of affinity with a new one built on the sure basis of numbers.[25]

With the new notion of affinity, he continued, "chemistry approaches ever more closely the exact sciences, and many apparent contradictions will appear as necessities."

Thomsen's ambition was to determine the absolute values of chemical forces by means of thermochemical measurements and thus supply the vague concept of affinity with a precise and operational meaning. In his Danish memoir of 1852 he stated his basic thesis of chemical actions being accompanied "by an evolution of force which in general manifests itself as an evolution of heat." Two years later he offered a more elaborate definition of affinity based on heat effects:

> The force which unites the components parts of a chemical compound is called affinity. ... In order to split up a compound, to overcome the affinities, a force is necessary the quantity of which can be measured as the thermal effect in the formation of the compound from its constituents in question. The affinity of two bodies manifests

---

Naturvidenskaberne, a journal of the same genre as the Thomsen brothers' *Tidsskrift for Physik og Chemie*. At the 1847 meeting of the British Association in Oxford he lectured on his work on the density of sulphuric acid at different degrees of dilution. Thomsen referred to his Norwegian colleague in Thomsen (1854a), p. 284.

25. Thomsen (1861), pp. 103-104. An abbreviated version of the paper was published in *Archiv for Pharmacie* **18** (1861): 433-440, 481-495.

itself in the ability of direct combination between them; if they do combine, an amount of heat will evolve corresponding to the affinity of the bodies. ... By considering the amount of heat evolved by the formation of a chemical compound as a measure of its affinity, or of the work required to again resolve the compound into its component parts, it must be possible to deduce general laws for the chemical processes, and to exchange the old theory of affinity, resting on an uncertain foundation, for a new one, resting on the sure foundation of numerical value.[26]

Conceiving a chemical process to be an exchange of affinities in which the weaker affinities were replaced by stronger ones, Thomsen concluded that "Every simple or complex action of a purely chemical nature is accompanied by evolution of heat." This general statement, sometimes known as "Thomsen's principle," became the much dis-cussed backbone of thermochemistry in the decades to follow.[27]

The phrase "of a purely chemical nature" was crucial in Thom-sen's formulation, where it was introduced as a kind of protection against the objection that although most chemical processes evolve heat, some do not. They are endothermic and thus may seem to contradict the principle. Thomsen was well aware of endothermic processes and included them in his concept of thermal effect (*varme-toning*) which referred to endothermic as well as exothermic process-es. He argued however that heat-absorbing processes were not of a "purely chemical nature" and hence outside the realm of his princi-ple. To qualify as a chemical action, a process would have to be governed exclusively by the forces inherent in the interacting sub-stances. "I see only chemical actions in such processes where the substances combine in definite proportions, according to their number of equivalence, and I will only consider these in order to test the theory by experience."[28] Most salts are dissolved in water under the absorption of heat, but not in definite proportions, and accordingly Thomsen conceived solution processes to be of a phys-ical rather than chemical nature. While Thomsen found this crite-

---

26. Thomsen (1854c), p. 34. See also Thomsen (1861), p. 104.
27. Thomsen (1854c), p. 36. See also sections 4.4 and 4.5.
28. Thomsen (1854c), p. 36, and similarly in Thomsen (1861), p. 103.

rion to be satisfactory, several other chemists considered it to be ad hoc and nothing but a means to protect the general principle against refutation.

Moreover, Thomsen's definition of a pure chemical process changed over time. In the mature version stated in *Thermochemische Untersuchungen* a process of purely chemical nature was said to be one which "proceeds without the expenditure of external energy and is accomplished only through the striving of the atoms towards a more stable equilibrium."[29] Thomsen referred to the "right of the stronger" as one of the important principles in the dynamics of chemical processes, meaning that processes would be governed by the saturation of the stronger affinities. He later added to this principle another one which he called the "maintenance of the status quo." With this he referred to a natural resistance of any change of molecular configuration.[30]

It was important for Thomsen that his theory not only agreed with known facts but also had explanatory as well as predictive power. For this reason he supplied his papers with numerous examples of chemical reactions and their thermal effects. Some of them related to the question of which metals would interact with acids under the evolution of hydrogen gas. In the case of hydrochloric acid Thomsen found that the limit was between lead and iron. Based upon his thermal theory he argued that whereas diluted hydrochloric acid would not dissolve lead, a concentrated acid would. "I made the experiment and the result agreed perfectly with the theory," he triumphantly commented. Thomsen similarly derived from his theory what he considered a novel prediction, namely that lead would dissolve in a diluted solution of sulphuric acid:

$$Pb + H_2SO_4 \rightarrow PbSO_4 + H_2$$

This he derived from the inequality

29. Thomsen (1882-1886), vol. 1, p. 16.

30. Thomsen (1854c), p. 35; Thomsen (1908), p. 203. The "right of the stronger" principle would later be associated with the "right of the fittest" motto in Darwinian evolution theory (see Section 7.2).

$$(Pb, O, SO_4, aq) > (H, O),$$

where the first quantity was 4690 heat units and the second 4300 heat units. Such a reaction had not been observed, but when Thomsen boiled pulverized lead with diluted sulphuric acid he noted the evolution of hydrogen gas. Again, "the result confirmed my theory completely."[31] Yet another phenomenon that he derived from the thermal theory and only subsequently verified was the decomposition of concentrated sulphuric acid into hydrogen sulphide. As Thomsen saw it, these and many other experiments strongly suggested that his thermochemical theory was more than just the summary result of experiments. He considered the principle of heat evolution as a measure of chemical affinity to have confirmed consequences and for this reason to be the proper foundation of a deductive thermochemistry.

Thomsen's *Grundzüge* did not attract much immediate attention. Among the few who did find it interesting was the young British chemist Henry Roscoe who at the time studied under Bunsen in Heidelberg and who would later become one of Great Britain's most distinguished chemists. In a letter of February 1854 to his cousin, William Stanley Jevons, he reported on current problems in chemistry. One of these problems was the nature of affinity about which he had spoken with Bunsen. The German chemist, he wrote, "said one day to me that all these phenomena of affinity are most complicated – depending on so many conditions & actions which we can in no way estimate or control – & I believe this is very true." Roscoe continued:

> I have been lately studying the subject of the Heat of Chem: Combination – & the connection of the evolved heat in combination with that absorbed in Decomposition – It is a most difficult subject – but there is a great deal to be done – I have read Thomson's papers – & Joule – & lots of others – also one by Thomsen a Dane – in Pogg: on the Formation of a Thermo Chemical system – vy [very] interesting.[32]

---

31. Thomsen (1852), pp. 161-163.
32. Roscoe to Jevons, 21 February 1854, as quoted in Jevons (1973), p. 66. W. S. Jevons became famous for his work in economy, logic and philosophy of science.

4.3. An experimental tour de force

After having become professor in Copenhagen with a well-equipped chemical laboratory at his disposal, Thomsen engaged vigorously in a comprehensive series of accurate experiments in thermochemistry. The main result of his efforts was a series of papers with the common title "Thermochemische Untersuchungen" (Thermochemical Investigations) published between 1869 and 1880. Altogether the series comprised no less than 32 communications published in German in either Poggendorff's *Annalen* (nos. 1-13) or *Journal für Praktische Chemie* (nos. 14-32). At the same time he published most of the work in Danish, typically in the proceedings of the Royal Academy of Sciences. Thomsen went systematically through a large number of inorganic compounds, determining their heats of formation and other thermal effects. He only investigated few organic compounds, but on this area of chemistry he reported in separate publications from the early 1880s. In the third communication Thomsen reported that hydrogen sulphide dissolved in water behaved as a monobasic acid similar to hydrochloric acid. This he concluded from measuring the heats of neutralisation of the two acids with respect to sodium hydroxide. Thomsen consequently suggested that the rational formula for hydrogen sulphide should be HSH rather than H_2S. Based on the chemical similarity between oxygen and sulphur he similarly proposed to rewrite water as HOH and consider it a monobasic acid.[33]

In his experiments on the heat evolution associated with the formation of nitrogen oxides, hydrates of sulphuric acid, and other compounds, Thomsen noted that the heats apparently appeared as multiples of a common constant, an observation which fascinated him. Experimentally he found for the smallest amount of heat

---

The refererence to Thomson is to William Thomsen, the later lord Kelvin. Roscoe is best known for his contributions to photochemistry and chemical spectrography; in 1869 he succeeded in isolating the metallic element vanadium.

33. Thomsen (1869d). His results were noticed in *Nature* **2** (1870): 407. The pK_A value for H_2S is 7.2 and 12.9 for HS^-. The corresponding values for H_2O and HO^- are 15.7 and 24.

$$(N_2O_4 \; aq, \; O) = 18.30 \text{ kcal}$$

$$(H_2SO_4, \; aq) = 17.85 \text{ kcal}$$

$$(N_2, \; O) = -18.32 \text{ kcal}$$

The regularity led him to suggest that the affinity was in general quantised as multiples of a common "caloric constant" given by approximately 18 kcal. Although Thomsen realised that the inference was hypothetical only, he found it interesting because it indicated a phenomenon that might throw light on the nature of the molecular forces. "I have no doubt," he wrote in 1872, "that on this basis there can later be developed a dynamics of the chemical processes ... similar to the one of the electrochemical theory."[34] He apparently had in mind that as there is a smallest amount of electrical charge, so there may be a smallest amount of thermal affinity. However, Thomsen's suggestion did not lead to the insight he had hoped for.

Thomsen measured the heat of formation of gaseous HCl directly by burning dry chlorine in a hydrogen atmosphere; for the other halide compounds HBr and HI he used indirect methods. His results showed that the thermal effect was greatest for HCl, much less for HBr, and negative for HI. This was as expected since it corresponded to the known stability and other properties of the compounds. He reported the values, all in kcal,

$$(H, Cl) = 22.00; \quad (H, Br) = 8.44; \quad (H, I) = -6.04$$

However, as Thomsen pointed out at several occasions, considered as measures of affinities these values were somewhat ambiguous since they related to molecules rather than atoms.[35] Strictly speaking, they did not prove that the halides' affinity to hydrogen decreased with their atomic weights. In the case of chlorine,

---

34. Thomsen (1872c), p. 174. For a couple of years Thomsen published several papers on his suggested affinity law.
35. Thomsen (1873d) and Thomsen (1884), p. 15.

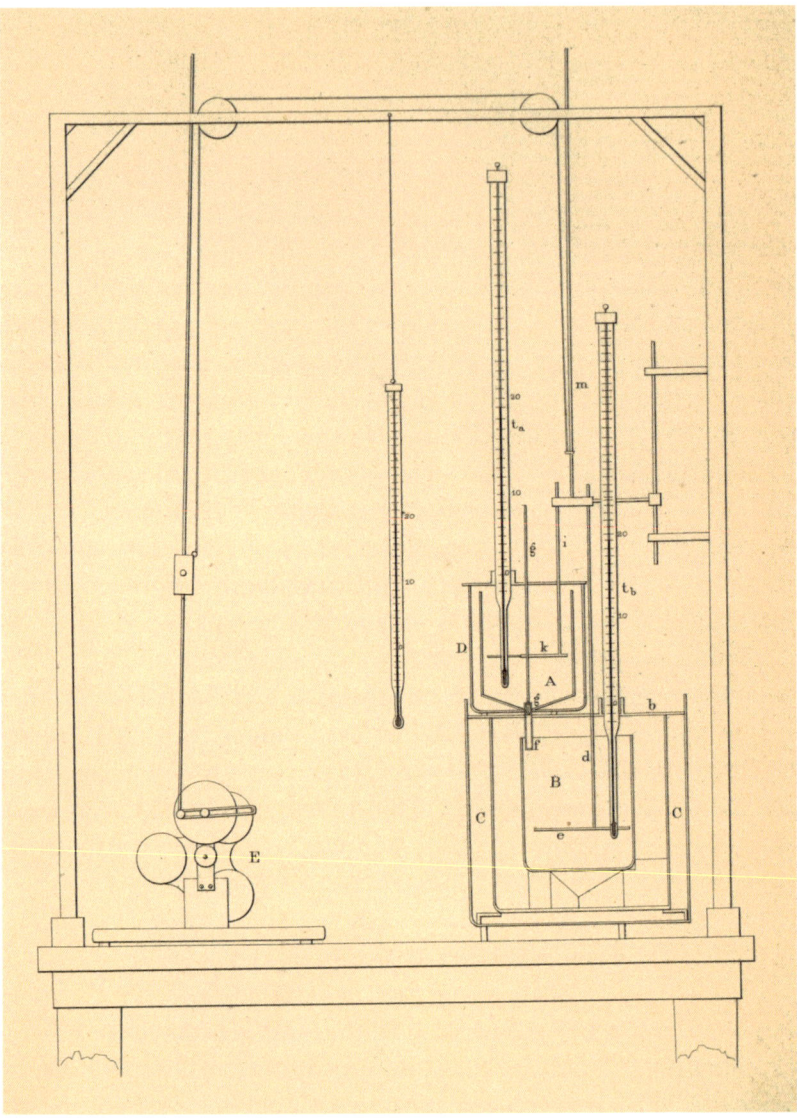

Figure 4.2. Thomsen's "mixing calorimeter" as he used it in measurements of the heat evolved in neutralisation processes. From paper of 1869 presented to the Royal Danish Academy of Sciences.

$$(H_2, Cl_2) = 2 (H, Cl) - (Cl, Cl) - (H, H)$$

The measured quantity is the one on the left side from which (H, Cl) is derived, and similarly for bromine and iodine. However, at the time the elementary heats of dissociation such as (Cl, Cl) and (H, H) were unknown, which made it possible to assume that the heats of dissociation increased with the atomic weight. In that case one might conclude that the affinity of the three halides to hydrogen did not after all vary in the expected manner. When Thomsen mentioned the possibility it was not because he believed in it. He merely wanted to illustrate that experimental results could in general be interpreted differently, depending on the assumptions.

The many papers in the series on thermochemical investigations appeared separately and in two different journals, and for this reason Thomsen felt a need to present his entire corpus of research collectively and systematically. As a result, between 1882 and 1886 the Leipzig publishing house J. A. Barth published what was unquestionably Thomsen's *opus major*, a four-volume work comprising a totality of 2,015 pages. Apart from volume 4, which was devoted to organic chemistry, the book mostly consisted of previously published material although in some cases with updated results. Volume 1 dealt with the thermochemistry of aqueous solutions, volume 2 with the formation of compounds of non-metals, and volume 3 with thermal effects relating to compounds between metals and non-metals. Volume 4 contained a detailed exposition of calorimetric methods and the instruments used for heat determinations of organic compounds. When the substance was not a gas it was burned in a "universal burner" designed by Thomsen (Figure 4.3). "By means of the universal burner it is possible to bring about the combustion of almost all volatile organic compounds of which the boiling point in not too high," he wrote.[36]

In the beginning of volume 1 Thomsen described his specially designed thermometers and the "mixing calorimeter" he used for measuring the heat in liquid neutralisation processes (Figure 4.2). He provided details about his laboratory room of volume 8,000 cu-

---

36. Thomsen (1908), p. 19.

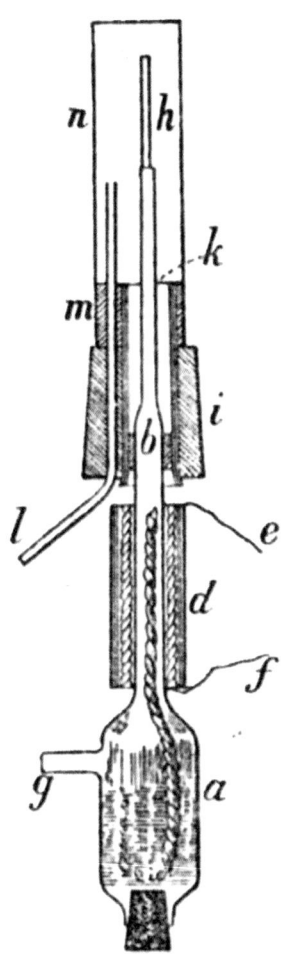

Figure 4.3. Thomsen's "universal burner" used for the combustion of organic compounds. Source: Thomsen (1908).

FIG. 2.

bic feet and with double doors and double windows. Moreover, he referred to his strict experimental routines, stressing that the room was kept at approximately the same temperature throughout the year. "The heat effects of chemical processes changes with the temperature," he wrote, "and comparable results are only achieved if the temperatures of the experiments are the same. In my work I have kept to a normal temperature between 18 °C and 20 °C."[37]

---

37. Thomsen (1882-1886), vol. 1, p. 23.

Thomsen further described how he kept his laboratory room at a fixed temperature by means of an evaporation device – essentially a thermostate of his own construction. "My working hours are from 12 to 4 o'clock and in this period the temperature only changed insignificantly," he added. He estimated the change to be less than 0.2 °C. In his science, as in his life generally, he wanted to be in control.

Obsessed by experimental accuracy Thomsen analysed in detail every possible sources of error connected with his calorimeter measurements. For example, was the reading of the thermometer a precise measure of the temperature of the substance in the calorimeter? Using Newton's law of cooling he investigated in 1868 the sensitivity of thermometers by means of mathematical methods. Thomsen concluded that the heat transport between thermometer and calorimeter was negligible: "Without making any appreciable errors one can assume that the readings of the thermometer in calorimetric experiments agree with the degree of heat of the liquid ... if it is kept in a state of constant motion."[38]

Thermochemische Untersuchungen was impressive by its sheer amount of data. It covered no less than 2,450 measurements on reactions of approximately fifty metallic and non-metallic elements with oxygen, halogens, sulphuric acid, and nitric acid, and more than 400 measurements on organic compounds. All the measurements were obtained by Thomsen working alone in his laboratory. The wealth of carefully systematised data presented in his four-volume work is today recognised as the earliest data base for thermochemistry. More than a century later a chemist at the U.S. National Bureau of Standards wrote about Thomsen's work that it "has continued to amaze me over the years."[39]

Although *Untersuchungen* (as I shall abbreviate the work) was throughout based on his thermochemical system it was predominantly experimental, with a large part of it consisting of data arranged in tables. There was but little discussion of theoretical issues, although Thomsen used the occasion to modify some of his earlier ideas. He now recognised that the relations between affinity

38. Thomsen (1868), p. 30.
39. Wagman (1992), p. 39.

and thermal data were more complicated than he had originally assumed. In volume 2 of *Untersuchungen* he explained that the way a chemical reaction proceeds depended on four factors: (i) the striving of the atoms towards positions of stable equilibrium corresponding to a maximum evolvement of energy; (ii) the resistance of molecules to change, which favours reactions attended by the minimum evolvement of energy; (iii) the temperature of the reaction; (iv) the stability of the possible products at the given temperature.[40]

In some contrast to his earlier formulations Thomsen was careful to point out that thermochemical data alone did not give direct information concerning the forces at work in chemical processes. The evolved heat was not, after all, an absolute measure of affinity. This he illustrated by the negative heat of formation of nitrogen oxide, which he expressed as

$$2 \, (N, O) = -43.15 \text{ kcal} + (N, N) + (O, O)$$

He similarly pointed out, as he had done earlier, that hydrogen did not react with nitrogen despite a heat of formation of

$$(N, H_3) = 11.89 \text{ kcal}$$

Thomsen formulated his general insight as follows:

> These data are merely expressions of the differences between the energies of the molecules which are decomposed and of those which are produced, and ... they are often affected by non-chemical reactions which accompany the chemical changes. Nonetheless they form the basis for further theoretical investigations. For the higher aim of thermochemistry is to establish the dynamical laws of chemical processes; to get insight in the enigmatic area of the constitution of the chemical compounds, meaning the molecule.[41]

As we have seen in Section 3.3, Thomsen had for long been aware of

---

40. Thomsen (1882-1888), vol. 2, pp. 460-474. See also Muir (1907), pp. 525-526.
41. Thomsen (1882-1886), vol. 1, p. 3.

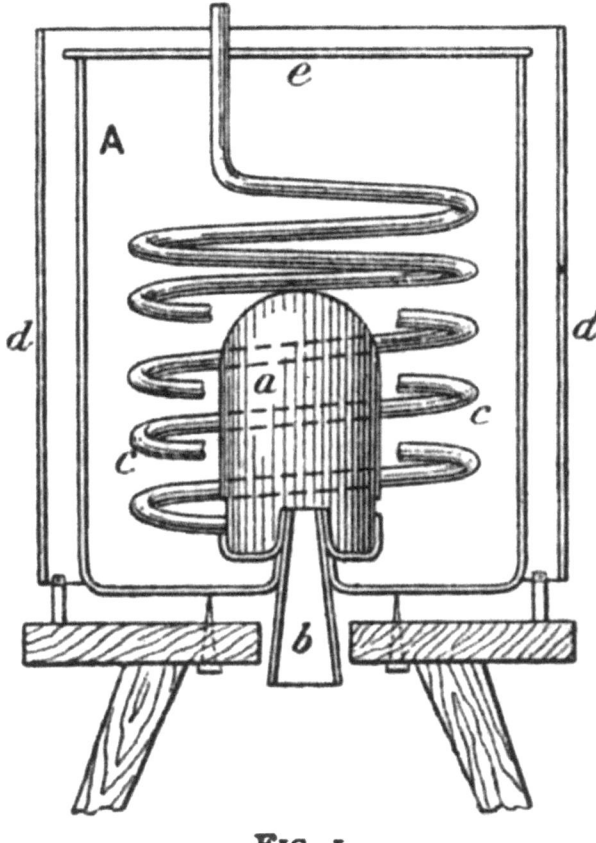

FIG. I.

Figure 4.4. Thomsen's calorimeter as he used it in the 1870s and 1880s.
Source: Thomsen (1908).

the second law of thermodynamics but without trying to incorpo-
rate it into his system of thermochemistry and without referring to
Rudolf Clausius's notion of entropy. In *Untersuchungen* he did refer
to Clausius's formulation, if only to rephrase his original principle
of evolved heat as the driving force of chemical processes. He wrote:

> Investigations in the domain of the mechanical theory of heat lead to
> the generalization that "the entropy of the world tends to a maxi-
> mum." In agreement with this statement is the experience that the

great multitude of chemical processes which are accomplished without the aid of foreign energy, and are free from by-reactions, are accompanied by production of heat.[42]

Contrary to some other scientists, Thomsen saw no real conflict between his thermochemical system and the second law of thermodynamics.

The first two volumes of *Untersuchungen* were positively reviewed by the Swedish chemist Otto Pettersson, at the Stockholm Technical University, who found them to be written in a "clear but condensed" style.[43] Pettersson stressed that whereas the thermochemical data would remain correct in the future, their interpretation would probably change. He highlighted Thomsen's experiments related to the Guldberg-Waage law of mass action as particularly valuable (Section 6.2). The bulky volumes of *Untersuchungen* were widely used by research chemists and in 1905 a much condensed single-volume version was published in Danish, to be followed by a German translation in 1906. An English translation appeared two years later.[44] The Danish summary volume was on Thomsen's request published by the Royal Danish Academy and the costs paid by the grant that the company J. P. Suhr and Sons had recently established in honour of Thomsen (see Section 2.5).[45] In the introduction Thomsen wrote: "It is my desire to provide in the Danish language a permanent record of the large amount of experimental work done at one of Denmark's principal scientific institutes, the value of which will surely continue to be appreciated as time goes on."

---

42. Thomsen (1882-1886), vol. 1, p. 12.

43. *Nordisk Revy* **1** (1883): 183-185.

44. Thomsen (1905b) and Thomsen (1908). The German version of 1906 was a literal translation made by Isidor Traube, a physical chemist: *Systematische Durchführung thermochemischer Untersuchungen* (Stuttgart: F. Encke). Traube visited Thomsen in 1904 to prepare the translation.

45. Lomholt (1942-1973), vol. 2, p. 73. The publication of volume 4 of *Thermochemische Untersuchungen* was supported by a grant from the Carlsberg Foundation established about a decade earlier.

The English version, translated by the young London chemist Katharine Burke, differed from the Danish one by including an introductory chapter from *Thermochemische Untersuchungen* in which Thomsen described his experimental methods and apparatuses used for his thermochemical investigations. It also slightly revised Thomsen's text in the light of the more recent knowledge relating to the ionic theory of dissociation. "Many of the statements which Professor Thomsen has made in these pages are now generally interpreted by the light of that theory," Burke pointed out in her preface.[46] On the other hand, *Thermochemistry* left out the postscript that Thomsen added to the Danish edition of 1905. Here Thomsen emphasised that "I have *personally* performed all the determinations of the heat effects of chemical processes; for only in this way could I be absolutely sure with regard to the reliability of the results." Moreover, addressing the younger generation of chemists the ageing Thomsen advised them not to "overflow scientific journals with immature communications." Unfortunately, he continued with unintended irony, "the spirit of time goes in the opposite direction ... Scientific communications appear in minimal doses and attention is more oriented towards the numerous details than towards the true progress of science."[47]

By the time that the three book publications appeared they were to some extent outdated and of value only because of their extensive numerical data. Burke's translation was reviewed by the young and brilliant American physical chemist Gilbert N. Lewis, who at

---

46. Thomsen (1908), p. vii. Katherine Burke (1875-1924) completed her B.Sc. in 1899 after which she worked at William Ramsay's laboratory at University College London. It was on Ramsay's request that she made the translation. I far as I can tell, she was not in contact with Thomsen. On the other hand, Thomsen was of course aware of the translation project of which he had been informed by Ramsay. At a meeting of 6 March 1908 in the Royal Danish Academy he presented the English book. For a brief account of Burke's career see Rayner-Canham and Rayner-Canham (2008), pp. 99-100.

47. Thomsen (1905b), p. 472. The irony is that Thomsen was a prodigious author of scientific articles, many of which were concerned with experimental details. He had himself overflown journals with fragmented doses of his research, some of these doses being small and unimportant while others were of a more subtantial nature.

the time was engaged in a comprehensive revision of chemical thermodynamics that had nothing in common with Thomsen's system. Although he praised the original *Thermochemische Untersuchungen* as "a work which has stood for twenty years as a classic," he saw no merit in the new publication with its "shrewd but antiquated theories of chemical affinity."[48] As Lewis pointed out, free energy and other concepts of chemical thermodynamics had made Thomsen's "esoteric principles" obsolete. He likened the octogenarian Danish professor to Rip Van Winkle, the fictional American who awakened after a twenty-year long sleep to discover to his surprise and dismay that everything had changed.

In the introduction to volume 1 of *Thermochemische Untersuchungen* Thomsen repeated what he considered to be the higher aim of thermochemistry, namely to elucidate the nature of affinity and ultimately the dynamics of chemical transformations on the molecular scale. His chosen branch of chemistry rested on two pillars, he said, the atomic-molecular theory of matter and the law of energy conservation. The latter pillar "forms the foundation of all quantitative thermochemical investigations." Thomsen elaborated:

> Up to the present time, an almost impenetrable veil has enveloped the internal structure of the molecules and the true nature of the atoms. ... We know almost nothing about the nature of the forces which dominate in the molecule and which cause the formation and decomposition of compounds. ... Chemical processes do not as yet admit of a mathematical discussion in their entire extent, as in the case, for example, with the phenomena of physics and astronomy; for the general mathematical discussion of chemical phenomena we lack that which is most important as a basis, namely, knowledge of the fundamental laws which govern the actions of the atoms. With each decade, however, chemistry approaches nearer and nearer the exact sciences, and already many laws of wider or narrower application are being established on the basis of experiment.[49]

48. *Journal of the American Chemical Society* **30** (1908): 1193-1194.

49. Thomsen (1882-1886), vol. 1, p. 3. Here quoted from *Popular Science Monthly* **23** (1883): 770-773, which brought an English translation of Thomsen's introduction under the title "The Aim of Thermo-Chemical Investigations."

The investigations of organic substances discussed at length in volume 4 deserve to be highlighted. Thomsen's general idea was to establish a connection between, on the one hand, the assumed structure of a compound and, on the other hand, its heat of formation. He argued that when such a connection was established, comparison between the calculated value and the experimental value of the heat of formation of the compound in question would provide a test for theories of molecular structure. This line of reasoning he had first suggested in a paper of 1869, his first contribution to *Chemische Berichte*, but at the time without developing it to any extent. As Thomsen acknowledged, he was inspired by an earlier paper by Ludimar Hermann, a German physiologist and neuroscientist working at the University of Zurich. Although Thomsen found Hermann's approach valuable, he criticized the conclusions that Hermann drew from it. At a first glance these conclusions might appear acceptable, but "a closer and more critical consideration shows that they are based on an illusion." [50] Thomsen was confident that he could do better and that his own precise thermochemical measurements would elucidate the structure of organic molecules and the forces binding atoms of carbon together.

In order to work out the idea Thomsen made use of the general principles of thermochemistry and assigned definite thermal values, or strengths of affinity, to the various types of chemical bonds, irrespective of their position in the molecule. These thermal values were determined from the heats of formation of molecules of known structure which again were based on the experimentally known heats of combustion. For example, the heats of formation at constant volume of ethane and methane were found to be (C_2, H_6) = 104.2 kcal and (C, H_4) = 59.6 kcal. Denoting the thermal value of a single carbon-carbon bond with v_1 and that of a carbon-hydrogen bond with r, Thomsen assumed that

$$(C, H_4) = 4r \quad \text{and} \quad (C_2, H_6) = 6r + v_1$$

---

50. Thomsen (1869f), p. 483. See also Thomsen (1882-1886), vol. 4, pp. 239-241. On Hermann and his little-known contribution to organic thermochemistry, see Kubbinga (2001), pp. 1090-1092.

From this it follows that $v_1 = 14.8$ kcal. His most controversial result concerned the thermal values of the double bond and triple bond, which we shall look at in connection with his proposal for a model of the benzene molecule (Section 5.2).

As another example, Thomsen considered series of homologous compounds such as the alkanes C_nH_{2n+2} or the alkenes C_nH_{2n}. He called the first member of the series M_1 and member number a he called M_a. For such a series,

$$M_a = M_1 + (a - 1)CH_2$$

Representing the heat of combustion of M_a by FM_a, this means

$$FM_a = FM_1 + (a - 1)D ,$$

where D is the constant difference found between the heats of combustion of two successive members. From measurements of heats of combustion of more than 100 substances, Thomsen concluded: "The difference between the heats of combustion of two neighbouring members in a series of homologous compounds is a constant which shows very small variations for the different series."[51] For alkanes he found $D = 158.23$ cal and for alcohols $D = 158.86$ cal. One of the organic compounds that Thomsen examined was cyclopropane or what was then called trimethylene (C_3H_6), a substance which showed a decidedly "peculiar behaviour." From its heat of combustion he found that the total thermal value of the bonds was 22.11 kcal/mole or, assuming three bonds, 7.37 kcal for each bond. But how could this agree with the value 14.8 kcal for a single bond? Thomsen was certain that the anomaly was real but consoled himself that "it remains as a solitary exception amongst a large number of observations."[52]

---

51. Thomsen (1908), p. 382.

52. Thomsen (1908), p. 450. The anomaly was explained by Adolf von Baeyer's strain theory of 1885 according to which the stability of a ring depends on the deviation of the chemical bonds from the value ca. 109° found in the tetrahedral structure. The deviation of the bonds in cyclopropane from the normal tetrahedral angle explains in part its unsaturated character. Modern structural chemistry confirms the basic

During the period from about 1870 to 1890, when Thomsen was intensely occupied with thermochemical measurements, he also published on subjects not directly related to thermochemistry, such as inorganic analytical chemistry and organic chemistry. The chemistry of the complex platinum salts, which earlier in the century had been investigated by William Zeise, Denmark's first professor of chemistry, was one of these subjects. Thomsen's interest was caused by experiments of 1867 in which he examined the reactions between copper and platinum chlorides in aqueous ammonia.[53] It was known at the time that reactions of this kind might result in double salts as described by Heinrich Magnus in Germany and George Buckton in England. Thomsen accidentally discovered a new double salt which, contrary to those known, was insoluble in water but soluble in hydrochloric acid. Based on careful experiments he concluded that the structure of the new salt, in his notation, was

$$PtCl, NH_3 \cdot HCl + NH_2Cu$$

He similarly wrote the salts of Buckton and Magnus as, respectively,

$$CuCl, NH_3 \cdot HCl + NH_2Pt \quad \text{and} \quad PtCl, NH_3 \cdot HCl + NH_2Pt$$

The mentioned salts are coordination compounds whose proper structures only became known several decades later. They are today written as

$$[Cu(NH_3)_4][PtCl_4], \quad [Pt(NH_3)_4][CuCl_4], \quad \text{and} \quad [Pt(NH_3)_4][PtCl_4]$$

The first one is Thomsen's, the second Buckton's, and the third is "Magnus' green salt" (as it is called).

Another of Thomsen's research interests concerned various com-

---

features of Baeyer's classical strain theory.

53. Thomsen (1867), with a German version in *Chemische Berichte* **2** (1869): 668-671. Ten years later Thomsen wrote a paper on the preparation of platinum compounds in *Zeitschrift für praktische Chemie* **15** (1877): 294-300. On the history and structure of these compounds, see Kauffman (1976).

pounds of gold, which he investigated in 1876 and returned to on later occasions.[54] The five papers he wrote on the subject in the period 1876-1888 were to some extent spin-offs of his thermochemical research programme. In his first paper Thomsen reported a new method of preparing auric chloride (AuCl), namely by thermal decomposition of aurous chloride ($AuCl_3$) in air heated to 185 °C:

$$AuCl_3 \rightarrow AuCl + Cl_2$$

He also prepared chloroauric acid ($HAuCl_4$) according to the process

$$2\ Au + 3\ Cl_2 + 2\ HCl \rightarrow 2\ HAuCl_4$$

In addition he reported preparation procedures of various gold bromides, including $AuBr$, $AuBr_3$ and $HAuBr_4$.

More interestingly, Thomsen produced a finely divided gold powder by means of sulphurous acid acting on $AuCl_3$, and when he exposed the powder to a stream of chlorine gas he obtained about 600 g of a product in which the chlorine-to-gold atomic ratio was close to two. As he observed, the process with chlorine suddenly stopped at 12 litre of Cl_2 per 100 g Au. This he interpreted as due to a new compound, which he described as either ($AuCl_3$, $AuCl$) or Au_2Cl_4. The new gold chloride was found to be highly hygroscopic and to decompose in water as

$$2\ Au_2Cl_4 \rightarrow Au_2Cl_2 + 2\ AuCl_3$$

The discovery claim attracted little attention until 1887, when two German chemists reported that they had failed in obtaining Au_2Cl_4 by Thomsen's method and consequently doubted its existence (and implicitly Thomsen's expertise). However, in response Thomsen proved experimentally that the compound was real; the experiments performed by the German chemists were in no way a refuta-

---

54. Thomsen (1876a). A Danish version of the paper appeared in *Tidsskrift for Physik og Chemie* **15** (1876): 289-297.

tion of his original results. He now reported a fixed atomic ratio of chlorine to gold of approximately 2.1. The new results, he wrote, "show beyond any doubt that a reaction between gaseous chlorine and gold leads to a compound with the composition Au_2Cl_4; the slight surplus of chlorine must be ascribed to the formation of a small amount of $AuCl_3$."[55] This time the discovery claim of a new compound consisting of gold in two different oxidation states was accepted, at least for a time. Thomsen's discovery attracted considerable attention. As the journal *Nature* commented, "Now that the work [of 1876] has been repeated and completely verified there is no longer any reason why Au_2Cl_4 should remain in the background."[56]

In 1874 Thomsen reported a new method of preparing larger amounts of hydrogen peroxide or what he called hydrogen hyperoxide (H_2O_2), a compound discovered by the French chemist Louis Jacques Thenard in 1818.[57] Thomsen's method consisted essentially in the preparation of moist barium hydrate peroxide BaO_2 from which the insoluble sulphate was precipitated by means of a very dilute sulphuric acid according to

$$BaO_2 + H_2SO_4 \rightarrow BaSO_4 + H_2O_2$$

In various modifications Thomsen's method was commonly used in chemical laboratories but it is now obsolete.

Thomsen was not much concerned with organic chemistry and yet what is likely the first example of an applied organic synthesis in Denmark must be ascribed to him.[58] In a brief paper of 1869 he suggested a new and more efficient synthesis of chloral hydrate, a compound traditionally produced by means of a method Liebig had

---

55. Thomsen (1888a), p. 106.

56. *Nature* **37** (1888), p. 398. Thomsen's compound corresponds to what in modern chemistry is called gold (I,III) chloride and ascribed the formula Au_4Cl_8. The Danish chemist K. A. Jensen (1983, p. 493) suggested that an investigation with modern structural methods of Thomsen's compound "would not be without interest." To my knowledge no such investigation has ever been made.

57. Thomsen (1874) and also in *Chemical News* **29** (1874): 156.

58. Thomsen (1869e). A German version was published in *Chemische Berichte* **2** (1869): 597. See also Kragh and Petersen (1995), pp. 263-264.

Figure 4.5. Julius Thomsen in the mid-1880s. Woodcut by H. N. Hansen. Source: Hansen (1886), p. 759.

demonstrated as early as 1831. Liebig's method was to add chlorine to ethanol,

$$C_2H_5OH + Cl_2 \rightarrow CC_3CHO + HCl$$

and then to let the aldehyde unite with water to form chloral hydrate:

$$CCl_3CHO + H_2O \rightarrow CCl_3CH(OH)_2$$

Thomsen's modification kept the first stage in Liebig's recipe but in the second step he boiled the chlorine-ethanol mixture, neutralised it with dried chalk and distilled it with melted calcium chloride. He reported a yield of approximately 140 % chloral hydrate relative to ethanol and with a consumption of chlorine about 4-5 times the weight of ethanol. When Thomsen published his paper he was still unaware that the German pharmacologist Oskar Liebreich in the same year had demonstrated the sedative and hypnotic properties of chloral hydrate and introduced it in medical therapy.

Alfred Benzon, a Danish pharmacist and founder of a successful chemical-medical business, adapted Thomsen's method for large-scale production but only to discover, undoubtedly to his dismay, that what came out of it was a different substance.[59] In other words, what Thomsen had announced as a method of producing chloral hydrate was a mistake. He most likely produced a chlorine alcoholate, the net process being

$$2\ C_2H_5OH + 4\ Cl_2 \rightarrow CCl_3CH(OC_2H_5)OH + 5\ HCl$$

As shown by examinations at the Copenhagen Municipal Hospital, the alcoholate had the same sedative effects as chloral hydrate but to a much smaller degree. Hence it was neither of medical nor commercial interest.

4.4. Contributions of Berthelot and others

At around 1870 classical thermochemistry was recognised as a most important part of chemistry, a status that would remain with it for about two decades. One indication of the status and popularity of thermochemistry was that in 1878 it was indexed under a separate heading in the *Chemisches Zentralblatt*, the period's leading abstract journal for chemistry. At the time the journal reported about fifty

59. Alfred Benzon (1823-1884) was a prominent and wealthy apothecary and businessman whose commercial laboratory was one of the first in Denmark. He published his work on Thomsen's method in *Tidsskrift for Anvendt Chemi* **2** (1870): 29-31.

papers on thermochemistry per year. Another indication was the appearance of several textbooks in thermochemistry in the 1870s and 1880s.

Although Thomsen was a pioneer of the field, he was not the only chemist who promoted it and contributed to thermochemical research. In the 1860s the theory was developed by Friedrich Mohr in Germany, Schröder van der Kolk in the Netherlands, and Berthelot, Henri Saint-Claire Deville and Jean-Baptiste Dumas in France. The first textbook in thermochemistry, *Grundriss der Thermochemie* from 1869, was written by Alexander Naumann, a professor at the University of Giessen. In accordance with the views of Thomsen in Copenhagen and Berthelot in Paris, Naumann based his account of thermochemistry closely on the mechanical theory of heat and the associated theory of affinity. He stated that "The mechanical theory of heat seems ... to be the most appropriate path in order to make chemistry approach its final goal, to formulate it as a mechanics of atoms."[60]

Most workers in classical thermochemistry agreed, but the theory did not necessarily presuppose atoms as real building blocks of matter. The "atoms" of the mechanical theory of heat could be conceived to be methodological rather than ontological entities, which was the view that Berthelot preferred. It was also the view of Hans Jahn, a chemist at the University of Vienna, whose 1882 exposition of thermochemistry was strongly indebted to the work of Thomsen. But contrary to Thomsen, he adopted a phenomenalist attitude. "Atomism is not a dogma," Jahn pointed out. Just as little as thermodynamics generally, thermochemistry "by no means states anything about the real constituents of matter but is a view which allows a description of the phenomena in complete correspondence with observations."[61]

Together with Thomsen, Marcellin Berthelot in Paris was undoubtedly the leading thermochemist in the second half of the nineteenth century. Berthelot began his research in thermochemistry in 1865, at first with theoretical work in which he used as experimental

---

60. Naumann (1869b), p. 2.
61. Jahn (1882), p. 205.

evidence the investigations of Favre and Silbermann; since 1873 he supplied them with extensive experiments of his own and of his pupils. In 1879 he published in two volumes *Essai de Mécanique Chimique* and in 1897, again in two volumes, *Thermochimie*. The latter work was primarily a compilation of experimental data determined by Berthelot and his students and can to some extent be regarded a French counterpart to Thomsen's *Untersuchungen*. The basic principles of thermochemistry as Berthelot exposed them in these and other publications comprised in their essence two statements which he derived as consequences of the mechanical theory of heat.

Berthelot assumed equivalence between the quantities of heat and what he called the *travail moléculaire* of the chemical reactions. According to Berthelot's "principle of molecular work," as he first stated it in 1865, the measure of chemical affinity was given by the quantity of heat evolved. In processes where external energy did not intervene, this quantity was a constant, depending only on the terminal states of the system. Berthelot considered the second principle – what he called the "principle of maximum work" – to be the proper foundation of rational thermochemistry. In one of its several formulations it stated that "Every chemical change accomplished without the intervention of external energy tends towards the production of the body or system of bodies which disengages the most heat."[62]

The validity of Berthelot's principle of maximum work was, in its earlier formulations, restricted to certain classes of chemical reactions but from 1875 onwards he stated it as a completely general principle of an almost a priori nature. As he wrote in his paper of that year, "every chemical change which can be accomplished without the aid of a preliminary work, and without the intervention of foreign energy, *necessarily* happens if it disengages heat."[63] In *Mécanique Chimique* he elaborated:

We can conceive of the necessity of this principle by observing that the system which gives out the largest possible amount of heat does

---

62. Berthelot (1875b), p. 52.
63. Berthelot (1875b), p. 212. Emphasis added.

not possess the energy necessary to accomplish a new transformation. Every new change requires the expenditure of work, and this cannot be done without the intervention of external energy. A system on the contrary which can give out more heat by further reacting has in itself the energy necessary to effect this change, without the intervention of external energy.[64]

Since Berthelot claimed that only those reactions occur spontaneously which are accompanied by evolution of heat it followed that endothermic processes could not take place spontaneously. Berthelot was as aware of endothermic processes as Thomsen was, but he thought that they were somehow influenced by external forces.

While the thermochemical systems of Berthelot and Thomsen were both based on the mechanical theory of heat and generally had much in common, the two chemists differed in their views on atomism.[65] As we have seen, Thomsen whole-heartedly supported the atomic theory of matter. Berthelot was willing to speculate that chemical reactions could be best explained in terms of interactions of minute particles, but he stressed that so far these so-called atoms and molecules were nothing but hypothetical constructs. His positivistic and strongly anti-metaphysical view of science prevented him from accepting atomism as a physical hypothesis. At least seen retrospectively his view on atomism was reactionary, such as illustrated by a controversy of 1877 in which he clashed with the French chemist Adolphe Wurtz, a staunch advocate of atomism. Not only did Berthelot reject Avogadro's hypothesis, he also denied that hydrogen and oxygen molecules consisted of two atoms in combination. As he objected, "Who has ever seen ... a gaseous molecule or an atom?"[66] In this regard, his views were quite different from those

---

64. Berthelot (1879), vol. 2, p. 421, as quoted in Schelar (1966), p. 114.

65. It is probably not a coincidence that the German-American positivist and anti-atomist John Stallo, in his classic *Concepts and Theories of Modern Physics* from 1881, gave much attention to Berthelot's thermochemical system whereas he ignored Thomsen's contributions. See Stallo (1960), pp. 305-310.

66. Quoted in Bensaude-Vincent (1999), p. 90. For Berthelot's anti-atomism, see also Nye (1981). Although he was indeed opposed to the atomic theory, on occasions he speculated on sub-atomic particles and the complexity of chemical elements.

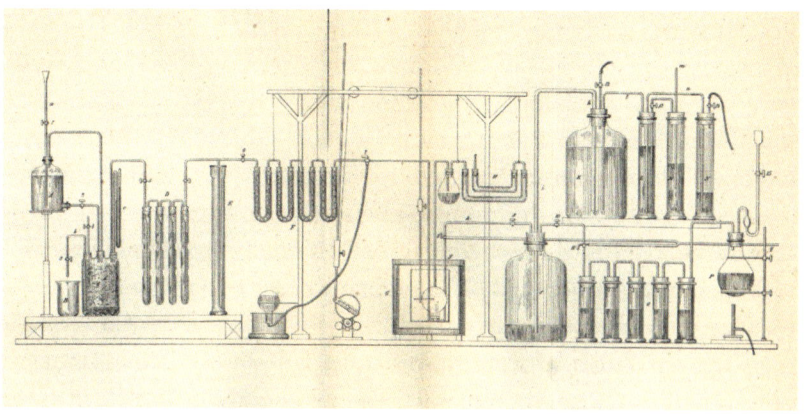

Figure 4.6. Thomsen's arrangement for measuring the heat of formation of hydrochloric acid. From paper of 1873 presented to the Royal Danish Academy of Sciences.

of Thomsen and the majority of chemists. Through most of his career Berthelot remained committed to a chemical notation based on equivalent weights, with the consequence that still in the mid-1890s he wrote the formula of water as HO rather than H_2O.

Berthelot maintained the general validity of his principle of maximum work for nearly three decades. Only in 1894 did he grudgingly admit that it was only approximately true by claiming that it was a primitive form of the second law of thermodynamics; for very low temperatures it would agree perfectly with the law of entropy increase. However, Berthelot was not really convinced that entropy was a theoretical tool superior to heat. It worked primarily for reversible reactions, he said, and these he considered as special and unrepresentative kinds of chemical processes. Although Berthelot half-heartedly admitted the value of the innovations of Gibbs and Helmholtz, he did not conclude that they superseded his own principles of thermochemistry. "Until the day arrives when one can apply purely thermodynamic definitions to the real mechanisms of physical phenomena, entropy will remain an obscure notion and an unknown quantity,"[67] he wrote. Berthelot did not realise that the day had already arrived.

---

67. Berthelot (1894), p. 1385. See also Dolby (1984), pp. 389-390.

4.5. Thermochemistry versus chemical thermodynamics

In a historical review of progress in nineteenth-century chemistry the British chemist William Tilden described in 1899 the thermochemical investigations of Thomsen and Berthelot. Although considering this line of work important he also pointed out that it had failed in revealing the dynamical laws that governed chemical processes. Tilden lamented: "Notwithstanding the labours of half a century, thermo-chemistry remains for the most part a mass of experimental results, which still await interpretation." It is plain, he continued, "that successful attempts at generalisation have been unsuccessful when considered in a strictly scientific sense."[68] Writing four years later Tilden's American colleague Frank W. Clarke remarked in a similar vein: "To many chemists ... the problems of thermochemistry have seemed to be hopelessly complex. Few general conclusions of unimpeachable validity have been developed by thermochemical research, and so, of late years, the entire subject has fallen somewhat in disfavor."[69]

Indeed, considered as front research classical thermochemistry had ceased to exist at the time of Thomsen's death in 1909. The once so exciting and progressive Thomsen-Berthelot approach to thermochemistry had turned into a degenerating research tradition already in the late 1880s, when it was met with insurmountable difficulties and increasing criticism. For a decade or so classical thermochemistry sought to maintain its former authority, coexisting with its young progressive alternative, physical chemistry, but keeping to its own standards. Eventually the inadequacy of thermochemistry was realised even by the few chemists who still worked within the tradition founded by Thomsen and Berthelot.

In the first volume of *Zeitschrift für physikalische Chemie* the eminent chemist Lothar Meyer, at the University of Tübingen, read the epi-

68. Tilden (1899), p. 34.

69. Clarke (1903), p. 2. The British veteran chemist Henry Armstrong (1927, p. 662) wrote: "Thermochemistry seems almost to be a subject of the past. Few to-day will realise how great was the interest we took in Berthelot's and Thomsen's results as they were published." This section is in part based on Kragh (1984) and Kragh and Weininger (1996).

taph over the approach of Thomsen and Berthelot. Meyer summarised: "One has to recognise that the fundamental hypotheses of the thermal theory of affinity have not been supported by observations. ... Many of the admirers of the thermal theory of affinity, which until a few years ago throned in undisputed majesty far above all facts, will perhaps find it hard to see it fade away."[70] Although Thomsen denied that classical thermochemistry was fading away, within a decade Meyer's evaluation proved right. The Thomsen-Berthelot approach was based solely on the first law of thermodynamics whereas the new chemical thermodynamics, an integral part of physical chemistry, made full use of the second law as well. The difference was crucial.

As early as 1864 the Dutch chemist Schröder van der Kolk used Clausius' formulation of the second law to interpret thermochemical data. He argued qualitatively that many chemical processes were not in accordance with the principle held by Thomsen and Berthelot and that the energy is but one component among others in the measure of chemical affinity. "Affinity and heats of combination are impossible to deduce from each other," he wrote.[71] In their investigations of what came to be known as the law of mass action the Norwegians Peter Waage and Cato Maximilian Guldberg came to a similar conclusion, namely that affinity is not the only factor which determines the behaviour of chemical processes. Without mentioning Thomsen by name, in 1867 they criticised the thermal theory of affinity: "The heat evolved during the reactions depends not only on the circumstances under which the reaction proceeds. ... If the circumstances can change the result of the reaction, then they necessarily cause, at the same time, changes in the evolved heat."[72] Guld-

---

70. Meyer (1887), p. 140 and p. 143. For a constructivist-sociological perspective on the dispute between classical thermochemistry and chemical thermodynamics in the late nineteenth century, see Dolby (1984) according to whom "the issue was not settled by the reconciliation of rival rationalities, but by the workings of a wider socio-historical process." He argues that "the outcome of the dispute could not have been determined purely by rational factors." In my view, although non-rational factors did play a role this is hardly a justified conclusion.

71. Van der Kolk (1864), p. 452.

72. Guldberg and Waage (1867), p. 15. They first presented their work in papers of

berg and Waage wondered how the thermochemical concept of affinity could possible account for the fact that in most reversible decompositions there is no net evolution of heat.

At the time when van der Kolk wrote his paper, the concept of entropy had not yet been introduced. This only happened the following year when Clausius coined the word for what he initially had called *Verwandlungseinhalt* (content of transformation). He defined the entropy difference ΔS between two states A and B of a physical system as

$$\Delta S = S_B - S_A = \int_B^B \frac{dQ}{T}$$

Q denotes the heat exchanged at the absolute temperature T. The first chemist to apply the law of entropy increase to chemical processes may have been August Horstmann, a lecturer in theoretical chemistry at the University of Heidelberg. In studies of dissociation processes of 1873, such as

$$PCl_5 \rightarrow PCl_3 + Cl_2 ,$$

he found that the condition for the equilibrium state corresponded to a maximum value of the entropy.[73] With α denoting the fraction of the un-dissociated substance he stated the equilibrium condition as

$$\frac{dS}{d\alpha} = 0$$

Horstmann considered the cause of dissociation to be given by the increase in entropy which implicitly was a denial of the thermal principles stated by Thomsen and Berthelot. The contributions of Horstmann were neither well known nor fully recognised by his

---

1864 written in Norwegian. See also Lund (1965). For Thomsen's contribution to the confirmation of the Guldberg-Waage law, see Section 6.2.

73. Horstmann's 1873 paper has been translated into English in *Bulletin for the History of Chemistry* **34** (2009): 76-82. On Horstmann and his early contributions to chemical thermodynamics and physical chemistry, see Kipnis (1997) and Jensen (2009).

contemporaries, and the same was the case with the works of van der Kolk and Guldberg and Waage. In a lecture to the Royal Institution in 1875, Lord Rayleigh, a highly respected theoretical physicist, criticized the thermal theory of chemical affinity. He deplored that "the chemical bearings of the theory of dissipation ... have not hitherto received much attention." Without mentioning any names, he alluded to Thomsen and Berthelot: "It is often stated that the development of heat is the criterion of the possibility of a proposed transformation, though exceptions to this rule are extremely well known."[74] Rayleigh's objection was elaborated by the Russian chemist Alexéi Potilitzin, an assistant to Mendeleev, who argued that the Thomsen-Berthelot principle could not possibly account for the existence of endothermic and reversible reactions.[75] The German chemist Bernhard Rathke agreed, criticising the principles of Thomsen and Berthelot for lacking in precision and facing too many exceptions. How, he asked, could these principles account for the fact that sulphur burned into SO_2 and not SO_3 when the heat of formation of the latter oxide was much greater than that of the first?[76]

The experimental objections to the Thomsen-Berthelot principle, such as incomplete dissociation, reversibility, and spontaneous endothermic processes, had been known for many years but were for a long time disregarded by workers in orthodox thermochemistry. Experiments which did not agree with the Thomsen-Berthelot principle were explained away, either by classifying them as exceptions, lying outside the range of the principle, or by forcing them to agree with it by means of more or less artificial assumptions. The British chemist Matthew Pattison Muir, at Cambridge University, complained that "Statements such as those quoted by Thomsen or Berthelot are true, only when an arbitrary separation is made of

---

74. Rayleigh (1875), p. 455. The "theory of dissipation" was William Thomson's name for the second law of thermodynamics. It was widely used by British scientists.
75. See report of meeting of the Russian Physical-Chemical Sociey in *Chemische Berichte* **12** (1879): 2369-2374. Potilitzin found that on heating a metallic chloride with bromine in a sealed tube an endothermic reaction occurs. See also Mendeleev (1892), p. 539.
76. Rathke (1882), p. 224.

chemical changes into two parts, and one of these parts is alone called chemical."[77] The attempts to rescue the universality of the Thomsen-Berthelot principle could not help to appear more and more unsatisfactory as counter-evidence and other objections accumulated. In 1873 Thomsen reluctantly admitted various anomalies indicating that his theory might not have general or absolute validity; or rather, that it was not sufficiently supported by experiments.[78] Only in 1882 did he suggest to modify it, although the modifications were then more cosmetic than substantial.

Looking back on the development of physical chemistry the Dutch chemist Jacobus van't Hoff, a Nobel laureate and one of the pioneers of that theory, wrote in 1905 about the thermal theory of affinity: "It seemed a decisive step when Thomsen and Berthelot declared that the heat developed in chemical change corresponds to the work that affinity can produce. ... Yet, this principle, however weighty, is not absolutely reliable." Van't Hoff considered the principle a stepping-stone towards the modern conception. Although it belonged to the past, it retained a connection to the present: "The Thomsen-Berthelot principle assumes a modified form in the rule that a fall of temperature induces the formation of the system which develops heat."[79] At about the same time Walther Nernst suggested that the principle – which he associated with Berthelot rather than Thomsen – might be formulated as a probabilistic rather than absolute law of nature. Sure, there were exceptions to the principle but it still was highly probable that a chemical reaction would lead to the evolution of heat. In this sense the principle contained "a genuine kernel of truth," Nernst suggested. His expressed his balanced view as follows:

> The claim of this rule to be an absolute law of nature must be rejected; and yet the rule does hold too often for us to ignore it entirely. It would be as absurd to give it complete neglect, as to give it absolute

---

77. Muir (1884), p. 299. Since 1879 Muir and Thomsen entered a correspondence which lasted for more than two decades (Royal Library, TSC).

78. Thomsen (1873a), p. 428. See also Section 4.3 and Section 5.1.

79. Van't Hoff (1905), p. 86 and p. 88. The correct spelling is van 't Hoff (with a space after the n), but the name is most often given as van't Hoff.

recognition. ... And so, in this case it is quite possible that, in a clarified form, Berthelot's principle may some time come to have some value.[80]

Of course, what Nernst said about Berthelot's principle was valid also for Thomsen's. Young Albert Einstein was familiar with the Thomsen-Berthelot rule and its limited validity in chemical thermodynamics. In 1905 – the same year in which he revolutionised the foundation of theoretical physics – he published a review of van't Hoff's revision of the rule of Thomsen and Berthelot.[81]

To make a long story short, chemical thermodynamics in its modern sense was largely established by Josiah Willard Gibbs and Helmholtz in the period from 1873 to 1883.[82] In monumental papers of 1873 and 1876 Gibbs argued that Clausius' entropy was a property equal in importance to the more intuitively clear quantities such as energy, pressure and temperature. He found it useful to operate not with entropy and energy as isolated variables but with functions or "potentials" that included both of them. The most important of these combinations is the so-called Gibbs energy, which in modern notation is written as

$$G = H - TS = (U + PV) - TS$$

U is the internal energy or the system's total energy, and $H = U + PV$ is today known as the enthalpy, a name coined in 1909 by the Dutch physicist Heike Kamerlingh Onnes. P and V denote pressure and volume, respectively, and the factor PV refers to the external work. The heat of reaction ΔH is thus a measure of the system's change in energy ΔU only when no external work is done.

Although Gibbs took pains to demonstrate that his abstract theory was also empirically useful, the theory initially had almost no

80. Nernst (1904), p. 689. Many years before quantum mechanics, Nernst believed that the laws of nature were generally of a probabilistic rather than absolute nature.
81. See Stachel (1989), pp. 128-130. Some of Einstein's earliest papers were in the area of physical chemistry.
82. For a summary review of the complex history of chemical thermodynamics, see Laidler (1993), pp. 83-130.

impact at all on either physicists or chemists. In this respect Helmholtz's later and independent theory was more successful. In 1882 Helmholtz introduced what he called the "free energy" (but is now known as the Helmholtz energy) as a quantity defined as

$$F = U - TS$$

The last term TS he called the "bound energy." Whereas F can be freely converted into other kinds of energy, TS will always appear as heat. Helmholtz commented:

> It appears unquestionable that, even in the case of chemical processes, a distinction must be made between the parts of their forces of affinity capable of free transformation into other forms of work, and the parts producible only as heat. ... I shall distinguish these two parts of the energy as the "free" and the "bound" energy. ... [It is] the value of the free energy, and not that of the total energy made known by the development of heat, which especially determines the direction in which chemical affinity can become active.[83]

According to Helmholtz a chemical reaction will proceed in that direction which involves a change of free energy into some other form of energy, which may or may not be heat. The quantity of heat produced when a system changes from its initial to its final state is given by the difference between the total internal energies of the two states. It is this difference which is measured in thermochemical experiments. However, the direction of the chemical change is determined by the work done by the free energy, and this work cannot be found simply by measuring the quantity of heat before and after a reaction. Helmholtz said that the older view of Thomsen and Berthelot – "which I have myself adopted in my earlier papers" – was justified in so far that it was approximately valid in a large number of chemical processes. He derived an important equation for

---

83. Helmholtz' 1882 paper "Die Thermodynamik chemischer Vorgänge" was published in the *Sitzungsberichte* of the Prussian Academy of Science. The quotation is from Ostwald (1980), p. 984.

the maximum free energy in a reversible and isothermal process taking place at constant volume:

$$T\frac{\partial F}{\partial T} = F - U \quad \text{or} \quad \frac{\partial(F/T)}{\partial T} = -\frac{U}{T^2}$$

It followed immediately from the equation that endothermic reactions might spontaneously occur, the only condition being

$$T\frac{\partial F}{\partial T} > F$$

In that case the internal energy would be negative, corresponding to absorption of heat. What is generally known as the Gibbs-Helmholtz equation quickly became a cornerstone of chemical thermodynamics.[84]

Helmholtz's chemical thermodynamics and the related work of Gibbs, Pierre Duhem, van't Hoff, and others signified the death of classical thermochemistry. Although the limited validity of the Thomsen-Berthelot principle had now been clearly demonstrated, classical thermochemists hesitated in adopting the new thermodynamics and abandoning the thermal theory of affinity. Thomsen was not a foreigner to mathematics but neither he nor most chemists of his generation had been trained in the kind of advanced mathematics which was the language of the new chemical thermodynamics. The theory's mathematical complexity alone would assumedly have been enough to keep him away from thermodynamics in the style of Helmholtz, Gibbs, and Duhem. In a letter of 1891 Helmholtz wrote: "Thermodynamic laws in their abstract form can only be grasped by rigorously schooled mathematicians, and are accordingly scarcely accessible to the people who want to do experiments on solutions and their vapour tensions, freezing points, heats of solution, &c."[85] Whatever the validity of Helmholtz's view, Thomsen belonged to the latter group.

---

84. The equation was due to Helmholtz and not to Gibbs. An earlier and more restricted version of it was obtained by Horstmann in 1872. For Helmholtz's route to chemical thermodynamics, see Kragh (1993).
85. Quoted in Kragh (1993), p. 429.

The thermal theory of chemical affinity was abandoned, but not because it was replaced by another and better theory of affinity. The entire concept of affinity, for a century and a half regarded as a crucial problem in chemistry, seemed to have lost its magic; at the turn of the century it was no longer considered interesting to search for mechanical models of chemical affinity. This is not to say that the concept disappeared or faded away from the chemical literature, but its meaning changed substantially. Ostwald wrote extensively about affinity, yet he thought of the concept in a way which was very different to Thomsen's ideas. Physical chemistry redefined the concept of affinity to be concerned with processes rather than the relationship of one substance and another. The question of the nature of affinity was of crucial importance to Thomsen and his generation, but with the redefinition it was sidestepped. "Most curious of all," wrote van't Hoff, "we can treat problems of affinity in an absolutely trustworthy way, so that calculations furnish a check upon experiment, without admitting anything about the nature of affinity or of the matter wherein the affinity is supposed to reside."[86]

The transformation of classical thermochemistry into chemical thermodynamics involved a generational conflict in which the younger generation developed its own paradigm which the older generation would neither understand nor accept. They did not consider van't Hoff's solution to be a climax in their endeavours to understand affinity, but an anti-climax. Leading thermochemists like Thomsen and Berthelot had invested too much of their scientific prestige in orthodox thermochemistry to change to a new paradigm. Julius Thomsen preferred to ignore the theory of Helmholtz and his successors.

---

86. Van't Hoff (1905), p. 89. See also Kragh and Weininger (1996).

CHAPTER 5

Scientific controversies

While some scientists do what they can to avoid controversies, Thomsen thrived with them. It was not part of his nature to admit errors of his own or pass over errors committed by other scientists within his fields of expertise. When he thought he was right, he *was* right. During his long and active life Thomsen was constantly involved in disputes of various kinds, some of them minor and some major; some of them related to his scientific work, while others were concerned with administration, business or local politics. He generally took these disputes very seriously and in some cases he used an inordinate amount of efforts on them. During the years 1869 to 1875 he was particularly busy with scientific controversies related to thermochemistry and other branches of the chemical sciences.

In this chapter I cover four controversies of a largely scientific nature, although the chosen cases differ widely and could have been extended with a few other cases. One of them, his disagreement with Hagemann concerning the theory of solutions and the strengths of acids, was of a minor nature; it hardly qualifies as a controversy in the proper sense of the term as it did not include a public exchange of views. The first case, on the other hand, was a heated controversy that involved both priority issues and the proper understanding of the laws of thermochemistry. Controversies come and go, but in this case it remained with Thomsen throughout his life.

Controversies in science are more the rule than the exception and they have been dealt with extensively by historians and sociologists.[1] It is generally agreed that in order for a scientific disagreement to qualify as a controversy it needs to be of some duration, be expressed in public, and take place by means of arguments and counterarguments. Although a *scientific* controversy must by defini-

---

1. See Engelhardt and Caplan (1987) for various cases of scientific-technical controversies and proposals of how to classify them.

tion relate to science, it almost always involves factors of a non-scientific nature. Moreover, a controversy is more than just a debate or a discussion. Whether individuals or groups, the parties must be committed to one of the opposing views and must attack the rival view. Only if the relevant scientific community considers the disagreement worth taking seriously will it evolve into a genuine controversy.

5.1. Thomsen versus Berthelot

During the latter half of the nineteenth century, Marcellin Berthelot exerted a huge influence on French chemistry and on French science generally (Figure 5.1).[2] A little prior to Berthelot's first works in thermochemistry a special chair was created for him at the Collége de France and in the years to follow he worked hard to build up the image he wanted of himself, as the unrivalled pioneer within key branches of modern chemistry. These branches included not only thermochemistry but also organic synthesis, chemical kinetics, physiological chemistry, the chemistry of explosives, and much more. Indeed, today Berthelot may be best known for his early syntheses of organic substances which included benzene, formic acid, and methane. Even history of chemistry, a subject that Berthelot seriously cultivated and wrote on at length, was considered a domain belonging to the versatile French chemist. As far as thermochemistry was concerned, he tended to represent it as a French invention with himself as the chief inventor assisted only by other French chemists. Berthelot's principle of maximum work and his general conception of thermochemistry came to be associated with the prestige of French science (see also Section 4.4). This view was bound to cause a collision with the ambitious, competitive and self-respecting Thomsen in Copenhagen, according to whom thermochemistry was neither a French science nor a Danish science. It was his science.

Thomsen had observed with increasing dissatisfaction that his own work on thermochemistry did not always receive the recogni-

---

2. See Crosland (1970-1980).

Figure 5.1. Marcellin Berthelot. Source: *Bulletin de la Société Chimique de France* **13** (1913).

tion it deserved and that it was sometimes seen as secondary relative to that of Berthelot. The almost complete lack of recognition in France annoyed him in particular. His dissatisfaction gave rise to a major controversy between two of Europe's most respected chemists.[3] For Thomsen it evolved into something like an obsession

3. The controversy between Thomsen and Berthelot is treated in Kragh (1984) and

which followed him throughout his life. As late as 1905 the 79-year-old chemist proclaimed the superiority of his measuring methods over those used by his French rival by means of his bomb calorimeter. Thomsen concluded that for volatile and gaseous substances the explosion with compressed oxygen in a bomb gave quite unreliable results.[4] He insisted that his own devise, known as the "universal burner," was superior when it came to most combustion studies. Of course, Berthelot flatly disagreed.

In March 1872, in a paper in *Chemische Berichte*, Thomsen opened his attack on Berthelot with a devastating critique of the latter's experimental work and, as he saw it, dubious use of data. The critique was methodological and he did not, at this stage, refer to questions of priority. Thomsen's attack was unusually blunt, not to say offending. The sad result of Berthelot's many mistakes and sloppy methods was, regrettably, that the French chemist had "loaded the scientific journals with a countless number of false and totally unusable numerical values."[5] In the case of the heat of formation of nitric acids (HNO_3 and HNO_2) Thomsen claimed that Berthelot's data were completely unreliable as he had failed to take into account the accumulation of experimental errors. Thomsen further charged that when Berthelot's experimental data failed to agree with his expectations the French chemist forced them to do so by means of unfounded hypotheses constructed for the purpose. At the end of his paper Thomsen even used the term "fraud" in connection with Berthelot's thermochemical works. It is, Thomsen said, "quite reprehensible to fill science with false data based on uncritical armchair work."

Berthelot did not reply directly to Thomsen's charges, but in the summer of 1873 he pointed out an error of observation that Thomsen had made in one of his measurements. He added: "When I reveal this error committed by M. Thomsen (and I could cite many more of the same order of magnitude) it is to indicate that the data of this author do not possess the absolute precision that he attrib-

---

Dolby (1984). See also Médard and Tachoire (1994), pp. 160-161.
4. Thomsen (1905c).
5. Thomsen (1872a), p. 181 and p. 185.

utes to them and in the name of which he confidently condemns the work of other scientists."[6] Berthelot relegated the critical remark to a footnote, as if to suggest that the dispute with the Danish chemist was of no great importance. But he was forced to change his view when he read another paper in *Chemische Berichte* in which Thomsen raised the priority issue concerning the foundation of thermochemistry. Though not directly accusing Berthelot of plagiarism, he concluded that Berthelot had done little more than restate the results that he had found himself twenty years earlier.

"For the last 6-7 years Mr. Berthelot has at any conceivable opportunity declared himself the originator of various laws of thermochemistry," Thomsen started his attack.[7] But this was a totally false claim as Thomsen's own work was published in the *Annalen* fourteen years before Berthelot began working on the subject. To prove his point – that Berthelot had merely repeated what he, Thomsen, had stated much earlier – he included long passages of his *Annalen* paper from 1853-1854 and compared them, line by line, with Berthelot's statements. Sure, the distinguished French savant had formulated the laws somewhat differently, but the substance was the very same. Thomsen summarized:

> As will be evident to any unprejudiced reader, twenty years ago I developed the well-known laws of thermochemistry; what is more, I tested them by means of numerous examples and applications, and in this way confirmed them. The careful reader of the mentioned memoirs will also observe that about fourteen years later Mr. Berthelot wrote on the same subject but without adding anything of substance. On the contrary, due to lack of critical attitude in his use of older observations he has let a large number of errors pass into the scientific literature.

Over the years Thomsen had noted various anomalies that necessitated a reconsideration of the thermochemical laws in the light of precise experiments. He contrasted his own attitude – the true and

---

6. Berthelot (1873a), p. 27.

7. See Thomsen (1873a). The quotations are from pp. 423, 427 and 428.

honorable scientific attitude – to that of Berthelot. "Instead of losing myself in futile speculations concerning the cause of these anomalies," he wrote, "I chose the more difficult but also safer way, namely to engage in a comprehensive series of experiments and thereby to collect a wholly new and reliable base of data." Berthelot, on the other hand, had chosen "speculations based on imprecise experiments" and a detailed study of his work was consequently nothing but "a waste of time." These were strong words that Berthelot could not ignore. His response came in a paper of 1873 in the *Bulletin* of the French Chemical Society.

Berthelot phrased his response in more restrained language than what Thomsen had used. As he pointed out, not unreasonably, "the polemics of the Danish professor is kept in a style which the scientists he respects do not ordinarily employ."[8] He avoided to attack Thomsen personally and pretended to speak on behalf of the French scientists and as a guardian of good manners in science. Berthelot had chosen to intervene only because it was "in the interest of science." His strategy was historical in so far that he suggested that most of Thomsen's original results could be found in earlier authors such as Hess, Andrews, and Favre and Silbermann. The central question concerned Thomsen's principle of heat evolved in chemical reactions which according to Berthelot was "neither original, nor conforming to all phenomena, nor identical to the principle I have proposed." No, it was a "veritable banality" which had been known to science for nearly a century. Berthelot stressed that his own two main principles of thermochemistry were quite different from those announced by Thomsen both with regard to their substance and with regard to their empirical consequences.

Berthelot's dismissal of Thomsen's principle as being of no scientific originality was also addressed to his compatriot Henri Saint-Claire Deville who in 1860, years before Berthelot, had proposed a

---

8. Berthelot (1873b), p. 485. See also Berthelot (1875a) in which the French chemist criticized Thomsen's measurements of the heats of formation of iodic acid, hypochlorous acid and some other substances. He argued that Thomsen's values were based on objectionable methods and that they were hence inferior to his own values.

thermochemical concept of affinity similar to that of Thomsen.[9] The exchange of views of 1872-1873 between Thomsen and Berthelot was only the beginning of an extended and embittered controversy that would last for more than two decades. The subject of the controversy was not only scientific data and their consequences for chemical theory. It also concerned, and perhaps even more importantly, which of the two scientists should be credited as the founder and doyen of the science of thermochemistry. Both Thomsen and Berthelot felt that this honour could not be divided between them. In what was perhaps an attempt to advertise his priority to the French scientific community, in 1873 Thomsen entered the contest for the Prix Lacaze of the French Academy of Science.[10] However, the prize was awarded for physiology applied to medicine, an area in which Thomsen's thermochemical work was largely irrelevant. Moreover, his controversy with Berthelot made him unwanted for a prize. The elaborate application was a mistake.

Thomsen's attack of 1872-1873 on Berthelot's methods and scientific credibility was reiterated in 1878 in yet another paper in *Chemische Berichte*.[11] Thomsen claimed, as he had done earlier, that Berthelot was not only a poor experimenter but also a biased scientist who judged experimental values according to whether or not they agreed with his own doubtful hypotheses. Berthelot, on his side, expectedly criticized some of Thomsen's measurements for being inaccurate and uninteresting. In the case of the heat of dissolution of anhydrous sodium sulphate and other substances Thomsen had found values that Berthelot thought were wrong. His reasons for criticizing *le savant professeur de Copenhague*, Berthelot assured, was merely to fix the real values of thermochemical data "and not to diminish the merit of the work of Mr. Thomsen, whom I set as high as

---

9. According to a review article on "Chaleur" (Heat) in Wurtz (1869-1870), pp. 813-833. The extensive article referred to numerous French authors, including Berthelot, Favre and Silbermann, but not to Thomsen. As mentioned in Section 2.4, Deville was the first to develop a commercial method for the manufacture of pure aluminium.

10. The application, a detailed account of Thomsen's work in thermochemistry, is kept in the Thomsen archival material, the Royal Library.

11. Thomsen (1878). See also the "historical note" in Thomsen (1882-1886), vol. 1, pp. 78-79.

anyone."[12] Thomsen probably doubted the sincerity of his colleague and rival in Paris.

The persistent rivalry between Thomsen and Berthelot can be followed through many of their publications in the period 1872-1886 and in a few cases even later. Neither of the two chemists wasted an occasion to point out the poor methods, illegitimate conclusions, and sloppy measurements of the other; or, conversely, to claim their own priority and competence. Instead of applying the results and methods of each other they jealously stuck to their own work and referred to the other only sparingly or for the purpose of criticism. As mentioned, as a result of Thomsen's anti-French feelings Berthelot's bomb calorimeter for combustion in a high-pressure oxygen atmosphere was never introduced in Thomsen's laboratory in Copenhagen.[13] The apparatus was invented in the mid-1880s and caused many foreign chemists to come to Paris to acquire first-hand knowledge of it. Thomsen was not among them. Despite its undeniable advantages he was unable to accept the use of an apparatus developed by his Parisian rival. The only Danish chemist who worked on thermochemical subjects at Berthelot's laboratory was Johan Fogh, and he had, perhaps characteristically, no connections to Thomsen at all.[14]

While Thomsen was aggressive and published several anti-Berthelot papers, the powerful Berthelot could afford the more discrete tactics of ignoring his colleague in Copenhagen. In Berthelot's

12. Berthelot (1878), p. 452.

13. A primitive bomb calorimeter had been used by the Irish chemist Thomas Andrews as early as 1848, but it was only with the invention of Berthelot and his collaborator, the physicist Paul Vieille, that the apparatus was turned into a powerful and versatile instrument for chemical analysis. Although it was usually credited Berthelot, and him alone, it was actually Vieille who invented the apparatus which he described in an article of 1884, one year before his collaboration with Berthelot. See Barkan (1999), p. 145.

14. Fogh (1865-1925) graduated from Dresden Technical University and worked 1890-1891 with Berthelot in Paris, co-authoring a paper with the French chemist. He worked in particular with the thermal properties of thiosulphate salts. Since 1897 Fogh was employed at the chemical laboratory at the Agricultural College in Copenhagen. See Veibel (1943), pp. 143-144 and Médard and Tachoire (1994), p. 292.

voluminous writings on thermochemistry Thomsen did appear, but only as a very secondary figure. Thomsen retorted by omitting Berthelot's name from the condensed Danish version of *Thermochemische Untersuchungen* he published in 1905 (and therefore also from the German and English translations of it). The Thomsen-Berthelot controversy was well-known in the chemical community. As late as 1905 the British physical chemist James Walker referred to the ongoing "controversy regarding the accuracy of the methods employed by Thomsen and Berthelot respectively." The subject of the controversy, at the time relating to organic compounds, was this: "Thomsen contends that his method is more satisfactory than the bomb method as used by Berthelot, and points out that regularities are displayed in his heats of combustion for homologous series which are absent from those of Berthelot."[15]

Thomsen's public disagreements with French chemists did not begin with his attack on Berthelot in 1872. Three years earlier he had published a brief paper in which he criticized the measurements that Favre and Silbermann had conducted about 1850 on the basis of their mercury calorimeter.[16] As Thomsen explained, in the case of acid-base neutralisation heats and heats of dissolution of some salts, their results differed as much as 30 per cent from his own measurements. The measurements made in Copenhagen, Thomsen assured, were correct within a limit of only 1 per cent. Silbermann had passed away in 1865 and it was thus left to Favre, the Parisian veteran in thermochemical measurements, to defend his honour as an accurate experimenter. His reply did not satisfy Thomsen, who maintained that the mercury calorimeter was unsuited for precise thermochemical experiments.[17] During the period from 1869 to 1873 the two chemists exchanged several papers on the subject without agreeing on the substance of the dispute.

In one case Favre assisted Berthelot's cause by hinting that Thomsen's confirmation of the variation of the heat of reaction with temperature was not an original work as it illegitimately relied on

---

15. Walker (1905), p. 8.
16. Thomsen (1869c).
17. Thomsen (1891b) and Thomsen (1872b).

results first obtained by Favre.[18] However, Favre was not really Berthelot's ally, for he got engaged in another dispute, apart from that with Thomsen, where Berthelot questioned the reliability of his mercury calorimeter.[19] At least with regard this question, Thomsen and Berthelot agreed. They both considered Favre a potential rival and none of them accepted more than one founder of the science of experimental thermochemistry.

Berthelot's claim to be the founder of rational thermochemistry was generally accepted by the French scientific community. Outside France, Thomsen's merits were more fully accepted if not, and especially not in Britain, at the expense of the merits of Berthelot.[20] At the 1895 meeting of the British Association in Ipswich, the chemist Raphael Meldola briefly surveyed in his presidential address the history of chemistry during the last half-century. "Thermo-chemistry as a distinct branch of our science," he said, was founded by Hess, Andrews, Graham, and Favre and Silbermann in the 1840s. "But the elaboration of thermo-chemical facts and views in the light of the dynamical theory of heat was first commenced in 1853 by Julius Thomsen, and has since been carried on concurrently with the work of Berthelot in the same field which the latter investigator entered in 1865."[21] This was precisely the view of history that Thomsen fought to establish.

In 1883 the Royal Society decided to award one of its most prestigious medals, the Davy Medal, for pioneering contributions in thermochemistry. The honour was shared between Thomsen and Berthelot, no doubt to the dissatisfaction of both chemists.[22] As president of the Royal Society, the biologist and evolutionist Thomas Huxley motivated the award as follows:

---

18. Favre (1874). For Thomsen's work, see Chapter 6.

19. The intertwined controversies between Thomsen, Favre and Berthelot are described in Médard and Tachoire (1994), pp. 118-124.

20. See for example Muir (1884), pp. 297-299.

21. *Nature* **52** (1895): 477. See also Tilden (1899), pp. 30-34, for a similar assessment.

22. The first Davy Medal of 1877 was awarded to Kirchhoff and Bunsen for their work in spectral analysis, and in 1882 the medal was shared by Mendeleev and Meyer for their contributions to the periodic system.

The thermo-chemical researches of Berthelot and Thomsen have extended over many years... Chemists had identified a vast variety of
substances, and had determined the exact composition of nearly all
of them, but of the forces which held together the elements of each
compound they knew but little. ...The materials for forming any general theory of the forces of chemical combination were but scanty and
imperfect. The labours of Messrs. Berthelot and Thomsen have done
much towards supplying that want, and they will be of the utmost
value for the advancement of chemical science.[23]

One might expect that as Berthelot was supported by the French
chemical community, so would Danish chemists support Thomsen.
However, this was not the case. They chose to ignore the controversy, which was simply absent from the Danish chemical literature
in the period.

Although the French chemical community sided with Berthelot
in his dispute with Thomsen, there was one noteworthy exception,
namely the physicist, chemist and polymath Pierre Duhem, since
1894 professor at the University of Bordeaux. Duhem nourished a
strong antipathy against Berthelot and his dominance over French
science, not only for scientific reasons but also for political, religious and philosophical reasons.[24] An expert in and advocate of
generalised thermodynamics he considered Berthelot's thermochemistry and principle of maximum work to be outdated, the principle being nothing but a "ridiculous tautology."[25] In a critical essay
review of Berthelot's *Thermochimie* he launched a full-scale attack on
the celebrated scientist and his complete misrepresentation of the
nature of thermochemical processes. Berthelot, he said, had been
led to commit himself to defend antiquated doctrines against attack

23. Presidential address, in *Nature* **29** (1883): 136-140, on p. 140. See also *Proceedings of
the Royal Society* **36** (1883): 76.

24. Berthelot was an atheist and a representative of the Third Republic, while Duhem
was an orthodox Catholic and strongly anti-Republican; both were positivists, but of
very different schools. For a full account of Duhem and his relations to Berthelot, see
Jaki (1984).

25. Duhem (1897), p. 370. See also Dolby (1984) and Berry (1968), pp. 33-35.
Duhem's review appears in an English translation in Duhem (2002), pp. 215-234.

from new ideas. "To this sterile and thankless task he employed all his ingenuity, all his time, all his labour, ... and today he is too perspicacious not to recognise that thermodynamics has created, without him and in spite of him, the chemical statics to which he had dreamed of attaching his name."[26]

Although Thomsen was no more a thermodynamicist than Berthelot, Duhem nonetheless found it opportune to enter the priority controversy on the side of Thomsen. Not only did he argue that Thomsen's principle was clearer and in better accord with experience than Berthelot's, he also concluded that the latter's conception added nothing that was not already contained in Thomsen's earlier work. As he sarcastically wrote about Berthelot's system, "These diverse propositions furnish the clear and complete statement of the thermochemical system; the only thing lacking is the name of Julius Thomsen."[27]

The controversy between Thomsen and Berthelot was to some extent due to differences in temper, scientific style and perspective. Thomsen, a citizen of a small country without a great scientific and patriotic tradition, was uninterested in forming a school around him and his work in thermochemistry. In this respect, Berthelot was quite different. Thomsen's scientific ambition was to gain recognition as the one who had provided thermochemistry with a solid theoretical foundation supported by precise and reliable experiments. By personality he was a fighting character, often arrogant and fiery and highly critical of views which differed from his own. He was unwilling to accept that other researchers could substantially improve the knowledge of thermochemistry which, he was inclined to think, was virtually completed with his own investigations. Berthelot was a fighter too, and hardly less arrogant, but he was far from a loner. Much of his work relied on his assistants and extended network of colleagues in France, and he did not identify himself as

---

26. Duhem (1897), p. 392.

27. Duhem (1893), p. 51. In Duhem (1897), p. 370, he supported Thomsen's "incontestable priority" over Berthelot. Thomsen was presumably aware of Duhem's support, but he never referred to the French scientist. As far as I know, there was no contact between Thomsen and Duhem.

the leader of a single branch of chemistry such as Thomsen did. The school of chemistry that Berthelot created included Paul Sabatier (a Nobel laureate of 1912) as well as many other French chemists of distinction.

Despite their differences, Thomsen and Berthelot had much in common. Both scientists advocated an empiricist method and subscribed to positivist norms of science; they emphasised that scientific laws should be the result of observation and that hypotheses should be used cautiously and only if they were closely linked to experiments. Perhaps more than Berthelot, Thomsen tended to conceive the exact determination of thermochemical quantities as an end in itself and consequently judged experimental accuracy as the prime virtue of his science. Berthelot's attitude was less empiricist in so far that he conceived the significance of thermochemical measurements to lie primarily with their theoretical consequences. According to Harry C. Jones, a physical chemist at Johns Hopkins University, Thomsen was "the type of mind that delights in accurate experimental work." He admitted that Berthelot's measurements were not "as accurate as those of Thomsen," but then "Berthelot made thermochemical measurements for a definite purpose, and that was to see to what far-reaching conclusions they would lead."

While Jones considered Thomsen a relatively minor figure in the development of thermochemistry, he praised the work of Berthelot. The French chemist, he said, "was the first to recognise clearly the importance of the study of energy changes, for the foundation of development of a real science of chemistry. ... He connected chemical activity in general with thermal changes, and thus gave, in my opinion, an epoch-making contribution to chemistry."[28] Thomsen may not have read Jones' book, but if he did he would not have been pleased. He would have been much more pleased had he lived to read the section on thermochemistry that Walker wrote for the 1911 edition of *Encyclopædia Britannica*. According to this source, "Julius Thomsen was the first investigator who deliberately adopted

---

28. Jones (1903), pp. 35-36 and p. 42.

the principle of the conversation of energy as the basis of a thermo-chemical system."[29]

The rivalry between Thomsen and Berthelot may have been re-lated to the political situation in Europe after the Franco-Prussian war, which led to the Third Republic and a seat in the Senate for Berthelot. In France there was a wide-spread hostility against Ger-man science and what was felt to be Germany's attempt to obtain a monopoly in science. The prestige of French science was a constant preoccupation of leading French scientists, among them many chemists. A part of this prestige was Berthelot's principle of maxi-mum work and French thermochemistry in general. Although Thomsen was not a German, and although Denmark in 1864 had had its own traumatic experience with the German military forces, he and his work in thermochemistry was closely linked to German science. Most of his major publications, including the attacks on Berthelot, appeared in the *Berichte* of the German Chemical Society or in other German journals.

5.2. The structure of benzene

Thirty years after Michael Faraday first isolated and identified ben-zene in 1825, the structure of the increasingly more important sub-stance was still completely unknown. About the only thing known was that its stoichiometric formula, assuming $C = 12$ in units of $H = 1$, was C_6H_6. Before Kekulé's breakthrough in 1865 a few sugges-tions regarding the structure of benzene (or "benzol" as it was then generally called) were put forward. However, they lacked empirical evidence and were scarcely taken seriously. For example, Josef Loschmidt's proposal of 1861 that the structure was $H_2C=C=CHHC=C= CH_2$, clearly disagreed with the chemical prop-erties of benzene. In 1865 Friedrich August Kekulé presented his famous closed-chain formula with alternating single and double bonds which a decade later had won broad acceptance among or-ganic chemists.[30] However, it was realised – and Kekulé himself real-

---

29. *Encyclopædia Britannica* (1911), vol. 26, p. 805.
30. On Kekulé's theory and the problem of benzene's structure, see Rocke (1985)

ised – that the hexagonal formula was not entirely satisfactory. For one thing, the three double bonds would expectedly result in benzene being strongly unsaturated, which in fact it was not. For another thing, if the compound were hexagonal there should be two mono-halogen substitutes (C_6H_5X) and four di-halogen isomers ($C_6H_4X_2$). Experiments indicated that there were only one of the former and three of the latter.

Several chemists found Kekulé's formula to be unconvincing and consequently suggested alternatives. For example, in 1867 Adolph Claus, at the time at the University of Freiburg, put forward a "centric" formula without double bonds; and two years later another of Kekulé's former students, Albert Ladenburg at the University of Heidelberg, suggested a prismatic model which also had no double bonds. Ladenburg's formula was innovative by picturing the benzene molecule as extending in three dimensions, the carbon atoms being connected by means of nine single bonds. Although it was generally in better agreement with empirical facts than Kekulé's hexagonal structure, few chemists considered it an attractive alternative. By 1880, when Thomsen turned towards organic structures based on thermochemical reasoning, Kekulé's theory was widely accepted and had entered most textbooks in chemistry.

As mentioned in Section 4.3, Thomsen believed that his general method of assigning thermal values to chemical bonds was applicable also to organic structural chemistry. By means of measurements of heats of formation he would be able to decide which types of carbon bonds entered any organic compound which could be brought into a gaseous state. Not only had he in this way founded organic structural chemistry on a solid experimental basis, he also believed that it paved the way for what he considered the ultimate goal of chemistry, namely an understanding of molecules from a mathematical point of view. This is what he said in Stockholm in July 1880, when he presented his ideas to the twelfth meeting of Scandinavian scientists: "This work has opened up a wide field for investigations which, more quickly than one should perhaps ex-

---

and Brush (1999). See also Russell (1971), pp. 242-257 for various alternatives to the Kekulé formula.

pect, can bring the chemistry of carbon compounds a big step clos-er to the mathematical sciences."[31]

First and foremost, the method allowed Thomsen to base the question of benzene's structure on the firm ground of experimental thermochemistry; or so he thought. He devoted the entire fourth volume of *Thermochemische Untersuchungen* to this ambitious research programme, which he first described in papers of 1880.[32] As a start-ing point for his investigations he determined experimentally the heat of combustion at constant volume of simple hydrocarbons such as methane (CH_4), ethane (C_2H_6), propane (C_3H_8), and ethyl-ene (C_2H_4). From this he concluded that the amounts of heat which corresponded to a single and a double bond were approximately the same; even more remarkable, when two carbon atoms united by a triple bond ($-C \equiv C-$), the additional thermal value was almost zero. In effect, he denied the existence of multiple bonds, as usually understood, between adjacent carbon atoms! Although he contin-ued speaking of double and triple bonds, this was merely a con-venient manner of speaking. Denoting the affinities or strengths of the bonds by v_1 (single), v_2 (double) and v_3 (triple), Thomsen found

$$v_1 = 14.57 \text{ kcal}, \quad v_2 = v_1, \quad v_3 = 0$$

Thomsen thus concluded that compounds which contained double or triple bonds according to the usual view were in fact saturated. He realised that the conclusion seemed strange as it ran counter to the intuition that the number of bonds represents the force required to separate two atoms in combination. However, he argued that the result, strange as it might seem, was actually supported by the known chemical relations of the series of hydrocarbons. Thomsen based his entire reasoning on the assumption that the molecule of gaseous carbon was diatomic (C_2). As critics were quick to point out, the assumption lacked experimental justification and for this reason alone his conclusions were dubious.[33]

---

31. Thomsen (1880f), p. 533.
32. Thomsen (1880a), (1880b) and (1880c). See also Kragh (1984).
33. Muir (1885), p. 71.

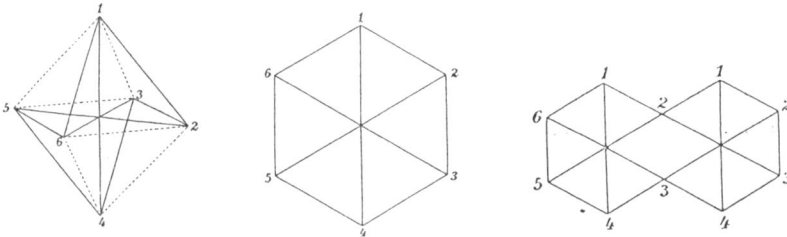

Figure 5.2. Thomsen's model of benzene in three different projections. Source: Thomsen (1886a).

It followed from Thomsen's view that the structure of benzene had to be reconsidered. The molecule could not contain double bonds, such as was also suggested by its resistance to addition reactions of the type

$$C_6H_6 + Br_2 \rightarrow C_6H_6Br_2$$

"The question concerning the constitution of benzol [benzene] can now be given a decisive answer by means of experiments," Thomsen confidently wrote. "I can be decided with certainty whether a hydrocarbon contains only single bonds or partly single and multiple bonds."[34] Apart from the six C-H bonds, benzene contained 18 bonds or valencies which could be arranged in different ways; only two of these agreed with the chemical properties of the compound. One of them was Kekulé's formula with three single and three double bonds, and another was a structure with nine single bonds. To determine the question Thomsen made use of an equation giving the heat of formation for a hydrocarbon of structure C_nH_{2m}, namely

$$(C_n, H_{2m}) = -nd + (2m + x + y)r$$

In this expression d is the heat of dissociation of carbon, r is the heat of formation corresponding to a C-H bond, and x and y denote the number of single and double bonds, respectively. Inserting experi-

---

34. Thomsen (1880a), p. 136.

mentally determined values for d and r, Thomsen found the following heats of combustion for the two possibilities:

$$(x, y, 2m) = (3, 3, 6) = 846.04 \text{ kcal} \quad \text{and}$$
$$(x, y, 2m) = (9, 0, 6) = 802.23 \text{ kcal}$$

On the basis of a series of carefully conducted experiments he arrived at 805.58 kcal for the heat of combustion of gaseous benzene. The conclusion followed directly: "The six carbon atoms of benzol are united to each other by nine single bonds, and the previous assumption of a structure of benzol with three single and three double bonds is *not* supported by experiment."[35] Thomsen's conclusion was evidently in agreement with Ladenburg's prismatic structure, but in 1880 he did not explicitly endorse this structure and he also did not suggest an alternative of his own. This he did only six years later. For heterocyclic compounds he similarly suggested single-bond structures. For example, he argued that pyridine, C_5H_5N, consisted of a pentagon of five CH radicals with the nitrogen atom outside the ring. The structure of thiophene, C_4H_4S, was a tetragon of CH radicals bound to a sulphur atom.

Thomsen's entry in the land of structural organic chemistry was not well received by specialists in the field. His theory attracted some attention, but mostly of the critical kind. In 1884 the young British chemist William Ramsay thought that he had found an error in Thomsen's reasoning; however, he was soon forced to admit that his objection was invalid. Encouraged by the answer from Thomsen, three years later 34-year-old Ramsay requested the Danish chemist to review his scientific work and recommend him for the position as professor at University College, London.[36] Thomsen complied with the request and Ramsay submitted his testimonial together with testimonials from other leading chemists, among them W. Ostwald in Riga and P. Waage in Christiania. In 1887 Ramsay succeeded Alexander Williamson in the chemistry chair in Lon-

---

35. Thomsen (1880a), p. 138.
36. Ramsay to Thomsen, 30 October 1884, 10 November 1884, and 5 March 1887 (Royal Library, TSC). On Ramsay's nomination and later career, see Davies (2012).

don where he would make his celebrated discoveries of the inert gases. Ramsay and Thomsen continued to be on friendly terms.

Despite a general disbelief in Thomsen's new thermo-structural theory, a few chemists of repute found it to be valuable, if not necessarily correct. Hans Jahn at the University of Vienna adopted parts of Thomsen's theory, and Lothar Meyer in Tübingen wrote approvingly to his colleague in Copenhagen:

> If your experiments in mass action, in neutralisation etc. have already shown that thermochemistry is suitable for something else than just lengthy conversions á la Berthelot of negative heats of reaction into positives ones, then your structural researches now open up a very wide perspective which even the most dense aromatic fog of colour will not be able to obscure.[37]

The British chemist Walther Noel Hartley, a pioneer of chemical spectroscopy, investigated in a series of papers the absorption spectra of organic substances. He found that whereas absorption bands in the ultraviolet region were present in aromatic compounds, they were absent in open chains of carbon atoms. According to Hartley, each carbon atom in benzene and its derivatives must be in direct union with three other carbon atoms. This result, which contradicted the Kekulé model, he considered as support of Thomsen's thermochemically based conclusion that "the six carbon-atoms in benzene are nine times singly linked with each other."[38]

In an extensive and generally sympathetic review the American chemist Josiah P. Cooke, at Harvard University, judged Thomsen's theory to be important; but his praise was mixed with more than one dose of reservation. "The interest of this investigation depends not so much on its results as on its method," he concluded. "It is a bold push beyond the beaten tracks of science, and although the first results of the venture must be accepted with caution, the skill

---

37. Jahn (1882), p. 147. Meyer to Thomsen ca. 1887, as quoted in Bjerrum (1909), p. 4983, who gives no date. The letter may no longer be extant.
38. Hartley (1881), p. 161.

displayed calls forth our admiration."[39] Other chemists, including Mendeleev and Muir, criticized Thomsen's confidence in his method and objected to the logic of his argumentation. Mendeleev, who had performed his own measurements of the heat of combustion of hydrocarbons, suggested that the single-bond benzene formula did not follow from Thomsen's data.[40] He doubted if the structure of benzene could be inferred from thermochemical measurements.

A perplexed Muir went further. He found Thomsen's theory, if taken at face value, to be nothing but "absurd" and "meaningless."[41] In England, Thomsen's thermo-structural theory was also critically received by Henry Armstrong and Spencer Pickering. As Armstrong pointed out, "if we accept Thomsen's conclusions in their entirety ... this would be to acknowledge that our entire system of constitutional formulae is based upon a false conception, to which there is no possible key."[42] Yet Armstrong agreed with Thomsen that Kekulé's benzene model was probably wrong and that a better alternative was a model without double bonds. In contrast to the majority of German chemists, Cooke, Muir, Pickering and Armstrong did not dismiss Thomsen's reasoning completely; they found his theory suggestive and sought to modify it in order to bring it into accordance with accepted views.

Thomsen's theory was sharply criticized by some of the German chemists, in particular Julius Wilhelm Brühl from Freiburg and Friedrich Stohmann from Leipzig. Brühl favoured an alternative method of determining the structure of organic compounds, namely based on optical measurements of molecular dispersion and refraction. Optical methods of this kind attracted much interest at the time. Let the refraction index of a substance be n and its density d. The British chemist John Gladstone showed in the early 1860s that a large number of liquids satisfied the relation

39. Cooke (1881), p. 98.

40. Mendeleev (1882), which was an extensive German abstract of an article Mendeleev had written in Russian.

41. Muir (1884), p. 174 and Muir (1885), p. 72.

42. Pickering (1888). Armstrong (1887), p. 102. On Armstrong's centric benzene model, which was inspired by Thomsen's theory and had features in common with Claus' model of 1867, see Russell (1971), pp. 252-253.

$$\frac{n-1}{d} = \text{constant}$$

In 1869 the Danish physicist Ludvig V. Lorenz derived from his phenomenological theory of light another relation between the two quantities. This relation was independently and on a different basis derived by the Dutch physicist H. A. Lorentz in 1878 and is therefore often known as the Lorenz-Lorentz formula.[43] According to this formula,

$$\frac{n^2-1}{n^2+2}\frac{1}{d} = \text{constant}$$

Moreover, Lorenz showed that the refractivity is additive and the formula thus valid also for mixtures of gases and liquids. For a mixture of r components it reads

$$\sum_{i=1}^{r} \frac{m_i^2-1}{(m_i^2+2)d_i} = \text{constant}$$

Realising that the refractivity depends on the molecular structure Lorenz developed his theory into a method for estimating the size of molecules. In a later paper he found in this way for the radius R of molecules in the air that $R \geq 1.4 \times 10^{-10}$ m. Lorenz's method did not allow him to estimate an upper bound of molecular size.

Brühl used the Lorenz-Lorentz formula and the Gladstone formula to show that the molecular refraction of a compound could be calculated by summation of the atomic refractions and in this way give information of the structure of the compound. By means of this and related spectrometric methods he believed to have confirmed Kekulé's formula for benzene. Not only did Brühl argue forcefully and at length against Thomsen's conclusions concerning benzene, he also dismissed his entire thermo-structural theory as based on "speculations" and totally out of contact with chemical reality. "Thomsen's so-called theory of heat of formation is unsuited as a method of investigating the atomic constitution of chemical compounds; the arguments based on it regarding Kekulé's benzene for-

43. See Kragh (1991).

mula ... have no weight."[44] Thomsen was aware of Brühl's approach to the benzene problem but claimed that it did not contradict his view of nine single bonds.[45] He also knew of his compatriot Lorenz's theory, which he had presented to the Royal Danish Academy, but it is unknown if Thomsen discussed the theory and its chemical consequences with Lorenz.

Stohmann had worked in Berthelot's laboratory in Paris, where he specialised in experimental thermochemistry. Based on measurements with the new bomb calorimeter he questioned Thomsen's measurements of the heat of combustion of benzene. While Thomsen's value for liquid benzene, as published in volume 4 of *Untersuchungen*, was 791.9 kcal per mole, Stohmann reported a value in much better agreement with Berthelot's earlier determination of 776.0 kcal. Stohmann and his collaborators in Leipzig found 779.53 kcal (the modern value is 781.15 kcal). The redetermination was an attack on what Thomsen was most proud of, his superb competence as an experimenter, and he consequently spent no time before responding to Stohmann's challenge.[46] The disagreement between the two chemists resulted in a brief but vigorous dispute in which Thomsen maintained the superiority of his measurements.[47] Stohmann was clearly irritated over what he considered to be Thomsen's stubbornness. He claimed that the quarrelsome Danish chemist was unwilling to enter a scientific dialogue and discuss matters impartially. In 1887 he declared that it was pointless to continue the discussion. Several years later and with the benefit of hindsight, Nernst commented:

> It happened by chance that Thomsen, in using his universal burner, obtained heats of combustion which were a little too high, and so his old values for the heat of combustion of gaseous benzene, when corrected, coincide quite well with the value calculated on the assump-

44. Brühl (1887), p. 236. On Brühl's and others' method of molecular refractivity, see Muir (1884), 307-318 and Nernst (1904), pp. 306-313.

45. Thomsen (1880d).

46. Thomsen (1886b), a reply to Stohmann, Rodatz and Herzberg (1886).

47. Stohmann (1886); Stohmann (1887). See also the exchange of views between Thomsen and Stohmann in *Journal für Praktische Chemie* **34** (1886): 55-56.

tion of nine single bonds. This circumstance, viz. that the values first determined by him seemed for the time to speak against the Kekulé formula, gave opportunity for much controversy.[48]

In their later publications on the benzene problem Stohmann and Brühl chose to ignore Thomsen. Although the two German chemists agreed in this respect, they disagreed with regard to the formula of benzene. While Stohmann maintained that thermochemical data ruled out three equivalent double bonds in the molecule, Brühl supported Kekulé's oscillation hypothesis according to which a carbon-carbon bond in benzene changed rapidly between a single and a double bond.

Still at the time when Thomsen completed volume 4 of his *Thermochemische Untersuchungen*, he did not propose a structural formula of benzene but merely repeated his earlier results, which he summarised as follows:

> In none of the [organic] substances which I have investigated – and their number amounts to about 120 – do we find an example of a substance in which two carbon atoms are bound together more strongly than the force of a single bond. ... A carbon atom cannot bind to another one with more than one valency; what we usually call double bonds are really single, and the compounds must consequently be regarded as unsaturated.[49]

About a year later, at a meeting of the Royal Danish Academy of 22 October 1886, he proposed a new and, to his mind, more satisfactory benzene formula. This formula or model was based on an octahedral structure and it satisfied the requirement of nine single bonds that he thought to have proved experimentally.

The new structure was somewhat similar to Ladenburg's earlier model, but Thomsen objected that the prismatic structure, contrary to his own, did not have the necessary spherical symmetry that would place the six carbon atoms on equal footing. He placed the

---

48. Nernst (1904), p. 319.

49. Thomsen (1882-1886), vol. 4, pp. 270-271. Preface dated October 1885.

carbon atoms at the corners of a regular octahedron in such a way that each of the atoms was connected to three others by one axial and two peripheral bonds (Figure 5.2). As Thomsen demonstrated, the model explained the number of di-substituted isomers and it could easily be extended to a similar model of the naphthalene molecule $C_{10}H_8$. "There is hardly any doubt," he said, "that this hypothesis is to be preferred over the one commonly adopted [Kekulé's], which does not satisfy either the requirements of a spherical distribution of the carbon atoms or their connection by means of nine single bonds."[50]

Although Thomsen's octahedral alternative was known and discussed by a few chemical authors, it did not arouse a great deal of interest. At the time the eminent German chemist Adolf von Baeyer, at the University of Strasbourg (which was then Strassburg), had suggested a revision of Armstrong's planar-centric formula. Baeyer's formula, which was supported by Stohmann, attracted much more attention than Thomsen's. For example, in an extensive review of 1894 Brühl dealt in detail with the alternatives of Baeyer and others while he chose to disregard Thomsen.[51] Among the few chemists who commented on the octahedral formula was Alexander Miller, a British chemist, who pointed out that it faced various chemical difficulties that did not occur in Kekulé's conception of benzene. Miller concluded that these difficulties of the Thomsen formula "outweigh the advantages claimed for it by its author, and on this account it may be doubted whether it will commend itself to chemists."[52]

Yet, as late as 1911 Thomsen's octahedral model was alive and known if not widely accepted. The British physicist and chemist Charles Everitt included it as one of several stereo-chemical alternatives in his extensive article on chemistry published in *Encyclopædia*

50. Thomsen (1886a), p. 185.
51. Brühl (1894). Baeyer received the 1905 Nobel Prize in chemistry for his work on organic synthesis of dyes and other substances.
52. Miller (1887), p. 215. See also Vaubel (1903, pp. 463-464) which included Thomsen's formula together with a dozen other alternatives to the Kekulé formula. Jones (1903, p. 21) considered Thomsen's model to be interesting and as a possible alternative to the one of Kekulé.

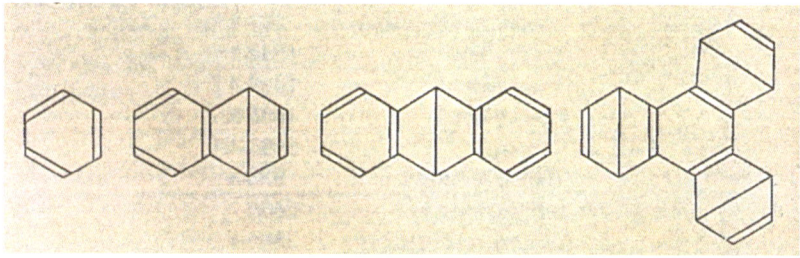

Figure 5.3. Thomsen's 1891 structural formulae for benzene, naphthalene, anthracene, and chrysene. Source: Thomsen (1891).

Britannica. As he pointed out, other chemists had proposed spatial structures inspired by and closely resembling that of Thomsen. One of them was James E. Marsh, who in 1888 proposed a modification of Baeyer's model by adopting features of Thomsen's octahedral formula.[53] However, Marsh's model was peculiar to benzene and contrary to Thomsen's it excluded naphthalene and phenanthrene from the class of benzene derivatives.

In 1891 Thomsen reconsidered the structure of aromatic compounds on the basis of his thermal theory and Stohmann's data for the heat of formation of five compounds in their crystalline state (Figure 5.3). The compounds were benzene (C_6H_6), naphthalene ($C_{10}H_8$), anthracene and phenanthrene (both $C_{14}H_{10}$), and chrysene ($C_{18}H_{12}$). Thomsen's calculations showed "beyond any doubt that not all the bonds in benzene can have the same value, and the same holds for other typical hydrocarbons."[54] This implied in effect a reintroduction of the double bond in benzene, but Thomsen stressed that the so-called double bonds in the aromatic compounds differed widely from those in the alkene group. For the strengths of the first group he obtained

$$v_1 = 18.58 \text{ kcal} \quad \text{and} \quad v_2 = 29.87 \text{ kcal}$$

and for the second group

53. *Encyclopædia Britannica* (1911), vol. 6, p. 57; Marsh (1888).
54. Thomsen (1891), p. 70.

$$v_1 = 14.59 \text{ kcal} \quad \text{and} \quad v_2 = 13.71 \text{ kcal}$$

The aromatic double bond was thus stronger and might be conceived as the sum of two single bonds of unequal strength. Although his 1891 model for benzene looks very much like the Kekulé model, the two were different. Thomsen offered still another picture which retained the nine single bonds of his earlier model. But in this picture six of the bonds were peripheral and the other three were centric, more or less like in the earlier centric models.

Thomsen may reluctantly have admitted to himself that his octahedral benzene model was a failure. He did not return to it and in his Danish 1905 summary volume of his thermochemical investigations he did not refer to the model. On the other hand, he stuck to his method of thermo-structural organic chemistry and maintained that "benzene does not contain double, but nine single bonds between the six carbon atoms."[55] Thomsen never converted to the Kekulé structure.

In 1903 the American chemist Frank W. Clarke proposed a new thermochemical theory of organic substances which to a large extent was based on Thomsen's data but with conclusions that in some respects differed from those of the Danish chemist. Although Clarke was careful to praise Thomsen's work for its accuracy and consistency, he also pointed out that "much of Thomsen's reasoning depends upon hypotheses which have been more or less questioned, so that although his conclusions are highly interesting, they have not won universal acceptance."[56] With regard to the carbon-carbon bonds his results differed completely from those held by Thomsen. Clarke argued that the thermal values of the three types of bond were the same, which "is directly opposed to the conclusions reached by Thomsen, whose opinions are always entitled to the highest respect." Despite Clarke's expressed respect for Thomsen, the ageing Danish chemist was not pleased. He totally rejected

55. Thomsen (1905b), p. 444.
56. Clarke (1903), p. 3. Frank Wigglesworth Clarke (1847-1931), Chief Chemist of the U.S. Geological Survey, is best known as a pioneer of geochemistry, but he also contributed to organic chemistry and atomic weight determinations.

the theory of the American chemist as speculative and useless. According to Thomsen, Clarke's attempt to reform thermochemistry was "completely hypothetical and, since it has not been confirmed experimentally, also without any value whatsoever."[57]

Two years later, now 79 years old, Thomsen was again on the warpath to defend his thermochemical system. This time he responded vehemently to an alternative theory of thermochemistry proposed by Daniel Lagerlöf, an undistinguished Swedish chemist. In a couple of articles Lagerlöf had argued that Thomsen's theory of carbon compounds was erroneous, which caused an angry Thomsen to respond that the theory of the Swedish chemist was wrong, confused and devoid of scientific value. The work of Lagerlöf was "based on a hypothetical foundation which is demonstrably wrong, whereas my theory of the heat phenomena is developed in an exact manner and in full agreement with experimental data."[58] Lagerlöf complained that Thomsen expressed himself categorically without specifying or even justifying his critique, and he repeated that Thomsen's theory rested on an inconsistent foundation. Wanting the last word in the debate, Thomsen replied that unfortunately Lagerlöf was "not open for instruction as he just maintains his hypotheses and postulates." He characterised his Swedish opponent as "a somewhat callous type" who demonstrated a "juvenile arrogance."[59]

Thomsen went in his grave firmly convinced that his own theory was basically correct, but contemporary chemists disagreed. In an obituary notice of 1909, Muir, referring to Thomsen's work on the thermochemistry of carbon compounds, suggested that one should distinguish between the data and the theoretical interpretations: "The data are sure. Personally, I think his theoretical conclusions are inadmissible."[60] In fact, not all of Thomsen's data were sure.

As early as 1887, in the first volume of *Zeitschrift für physikalische Chemie*, Wilhelm Ostwald opined that Thomsen's thermochemical

---

57. Thomsen (1903), p. 493.
58. See Thomsen (1905d), p. 181 and Lagerlöf (1905). Daniel Lagerlöf (1855-1919) mostly worked as a chemistry teacher in Helsinki. He published his "Thermochemische Studien" in several papers of 1904 in *Journal für praktische Chemie*.
59. Thomsen (1905e). The "juvenile" Lagerlöf was fifty years old.
60. Muir (1909), p. 47.

approach had met its Waterloo in its failed attack on organic molecules. Referring to Stohmann's superior measurements, Ostwald
wrote: "It appears almost unbelievable that Thomsen, to whom
thermochemistry owes so many exceptionally precise data, could
have made such serious errors. Would also this scientist, like once
Berzelius at the end of his career, have met in organic chemistry a
rock at which his art and science was doomed to fail?"[61] Thomsen of
course denied that this was the case. Referring to the omission of
some of his organic data from Landolt's and Börnstein's authoritative *Physikalisch-Chemischen Tabellen* Thomsen protested in a letter to
Ostwald of 1905 against "such an outrageous treatment from the
side of one of the main works in German literature."[62]

5.3. On Avogadro's law

Still in the mid-nineteenth century there was a great deal of uncertainty regarding chemical formulae and molecular weights, and a
corresponding uncertainty with regard to the reality of atoms and
molecules. What physicists and philosophers called atoms were not
necessarily the chemists' atoms. Some ten years later, in the 1860s,
much of the uncertainty had disappeared. A major reason for the
clarification was the general acceptance of the ideas that the Italian
physicist Amedeo Avogadro had first proposed in 1811. According
to what became known as Avogadro's rule or law, equal volumes of
all gases at the same temperature and pressure contain the same
number of particles.[63] Confusingly, in his paper of 1811 Avogadro

---

61. Ostwald, in a review of one of Stohmann's papers, in *Zeitschrift für physikalische Chemie* **1** (1887): 201.
62. Thomsen to Ostwald, 27 January 1905, quoted in Kragh (1984). The comprehensive work generally known as "Landolt-Börnstein" was first published in 1883, edited by Hans H. Landolt and Richard Börnstein. Thomsen referred to the third edition of 1905 which actually contained Thomsen's heats of combustion of organic compounds (on pp. 425-426). See also W. Meyerhoff, co-editor of the third edition, to Thomsen, 9 November 1903 (Royal Library, TSC).
63. See Fisher (1982) for the history of Avogadro's law and its diverse interpretations by chemists and historians of science. The number of gas molecules in a unit volume, or what is known as "Avogadro's number" (ca. 6×10^{23} per mole), was of course

did not use the term "atom" but instead coined the word "half-molecule" and spoke of several different kinds of molecules. Whatever the terminology, he suggested that elementary gases such as hydrogen and oxygen did not consist of atoms in the sense of Dalton but of molecules combining two or possibly more elementary atoms of the same kind (such combinations were ruled out in Dalton's theory). Instead of writing the synthesis of water as

$$H + O \rightarrow HO,$$

Avogadro effectively described it as

$$2\,H_2 + O_2 \rightarrow [H_4O_2] \rightarrow 2\,H_2O$$

The symbol $[H_4O_2]$ denotes a hypothetical intermediary molecule. It followed from Avogadro's reasoning that the weights of most gaseous elements had to be doubled. Whereas Dalton had found the atomic weight of oxygen to be 7.5, according to Avogadro it was 15, later to be revised to 16. Alas, his brilliant insight was either ignored or rejected. Half a century elapsed until it changed the scene of chemistry. When it happened it was largely due to another Italian scientist.

   In a paper of 1858 Stanislao Cannizzaro, at the time professor at the University of Genoa, argued forcefully that a number of chemical problems could be solved on the basis of Avogadro's hypothesis. He showed that the hypothesis provided a safe route to the calculation of atomic and molecular weights. Based on this kind of reasoning Cannizzaro changed the atomic weights of oxygen, carbon, sulphur and sixteen other chemical elements; the change was important to the slightly later attempts to formulate a periodic relationship between the elements. In early September 1860 a large meeting, the first international conference of chemistry worldwide, convened in Karlsruhe, Germany. It was attended by about 140 of the world's most eminent chemists. As a result of this famous meeting a drastic shift in attitude occurred with respect to Avogadro's hypothesis and

---

unknown to Avogadro and only became estimated at the turn of the century.

its significance for chemistry.[64] When Lothar Meyer, who attended the meeting, published his influential *Die Modernen Theorie der Chemie* in 1864, he based it to a large extent on Avogadro's hypothesis. And when Walther Nernst published his no less influential textbook *Theoretische Chemie* in 1893, he subtitled it *vom Standpunkte der Avogadro-schen Regel und der Thermodynamik*.

The Karlsruhe meeting was dominated by German chemists but also attracted many chemists from other countries, especially from France, Italy, Austria, Russia, and Belgium.[65] Several Britons crossed the Channel to join the congress. Whereas three Swedes participated, there were no attendees from Denmark and only one from Norway (namely A. Strecker, who was German). At the time chemistry in Denmark was at a low level as seen from an international perspective. The country's only professor Edward Scharling was no longer scientifically active and Forchhammer had for long focused on geology rather than chemistry. Julius Thomsen was still a relatively unknown physics teacher at the Military High School who may not have been considered for the Karlsruhe congress. On the other hand, the subjects discussed at the congress were of interest to him. In regard of the late recognition of Avogadro's hypothesis it is noteworthy that Thomsen included it in his elementary textbook of 1850. Without referring to Avogadro by name, he wrote: "Equal volumes of the gaseous elements contain the same number of atoms; it follows that the atom-numbers [atomic weights] of individual gases must be proportional to their densities."[66]

At about the same time as chemists revived Avogadro's hypothesis, mathematical physicists established the kinetic or mechanical theory of gases. This kind of theory was first presented by the German physicist August Krönig in 1856 and in a more elaborated and sophisticated form by Rudolf Clausius the following year. In his important 1857 paper in the *Annalen der Physik* Clausius assumed

---

64. For the Karlsruhe meeting and its impact, see Ihde (1961).

65. De Milt (1948) includes a list of the 127 known attendees. See also the original account of the French chemist Charles-Adolphe Wurtz which can be found online as https://web.lemoyne.edu/giunta/karlsruhe.html.

66. Thomsen (1850a), p. 87. A more extensive exposition of Avogadro's hypothesis and its consequences was given in Jørgensen (1860).

Avogadro's law in order to deduce the equipartition of kinetic energy among the molecules of a gas. A few years later James Clerk Maxwell improved on Clausius' theory by employing a statistical distribution of molecular energies and velocities. In memoirs of 1860 and 1867 the Scottish physicist emphasised the relation of the kinetic (or dynamical) gas theory to Avogadro's law, albeit without referring to it by name. In his 1867 memoir Maxwell stated that he could deduce from "purely dynamical considerations ... that for every kind of substance the number of atoms, or molecules, in the gaseous state ... must necessarily be the same."[67] He referred to gases of the same volume and at the same temperature and pressure. Although Maxwell's papers were not well known to the chemists, many were aware of those of Clausius and Krönig. As Alan Rocke has observed, in the decade before the 1860 Karlsruhe congress physicists had come up with a handful of derivations of Avogadro's hypothesis based on the mechanical theory of heat.[68]

But could Avogadro's law really be derived as a strict consequence of the mechanical theory of gases? In a paper of 1869 the physical chemist Alexander Naumann, one of the attendees of the Karlsruhe congress, claimed to have done just that. Thomsen, an avid reader of the chemical literature, disagreed and pointed out what he believed was an elementary error in Naumann's alleged proof.[69] The reply led to an extensive exchange of views concerning the subject that primarily involved Naumann and Thomsen but also attracted the attention of a few other chemists. In a rejoinder to Thomsen's critique Naumann denied any error of his own and instead criticized the reasoning of the Danish chemist. The reply clearly annoyed Thomsen, who in a paper in *Chemische Berichte* extended his criticism and detailed his own view concerning the relationship between Avogadro's law and the mechanical gas theory. Thomsen emphasised that his critical remarks were in no way an

---

67. Quoted in Brush (1986), p. 197. Brush's book offers a complete history of the kinetic theory of gases.
68. Rocke (1984), p. 291.
69. Naumann (1869); Thomsen (1870b). Naumann to Thomsen, 28 February 1871, Royal Library (TSC).

attempt to discredit Avogadro's law, the importance and chemical fertility of which he fully recognised. His aim was to clarify the law's epistemic status and relation to the new mechanical theory of heat. According to him the two theories were different rather than Avogadro's being simply a special case of the physicists' general theory of gases. What the dispute with Naumann was about, Thomsen wrote, was this:

> It only concerns two physical theories, the mechanical theory of heat and Avogadro's law, and the question is whether or not Avogadro's law is a necessary consequence of the mechanical theory of heat. So far the answer is this: The statement that equal volumes of different gases under the same external conditions [temperature and pressure] contain the same number of molecules does not presuppose the mechanical theory of heat; on the other hand, this assumption [Avogadro's] does not contradict the mentioned theory.[70]

To derive Avogadro's law one would have a means to determine the kinetic energy of molecules, Thomsen argued, and no such means existed. For this reason the law had the status of an independent hypothesis that could be wrong or at least of limited validity without consequences for the kinetic-molecular theory of heat.[71] Indeed, he suggested that the law might not be applicable to all chemical phenomena and in particular not to the diffusion of gas mixtures. Thomsen considered his disagreement with Naumann to be methodological in nature:

> Such anomalies are of great importance to the further development of the theories and for this reason should not be regarded as immaterial. Theories are the results of human thought whereas phenomena are the language of nature; they are truths which we do not yet fully comprehend. When two theories are in disharmony, one or both of them must be revised; should it be proved that Avogadro's law is uni-

---

70. Thomsen (1870c), p. 955.

71. In a Danish review article Thomsen (1871a, p. 69) repeated his view of Avogadro's law being a hypothesis rather than an experimentally proven fact.

versally valid the chemical theories must be changed accordingly.[72]

Thomsen believed that Avogadro's law was valid, but not for all kinds of chemical phenomena and processes; he consequently saw no need of changing what he called the chemical theories.

In 1871 Lothar Meyer entered the dispute, stating that he was in full agreement with Thomsen's view of the status of Avogadro's law. He called the law "theoretical chemistry's brightest lodestar" but added the rhetorical question, "who will claim that even it cannot also possibly shine with a false light?"[73] Despite his general support of Thomsen in his dispute with Naumann, Meyer pointed out some weaknesses in Thomsen's arguments. Both of the two discussants had assumed that all molecules in a uniform gas have the same kinetic energy and hence velocity. For example, Thomsen stated that the pressure of a gas "according to the mechanical gas theory of Krönig and Clausius" could be written

$$P = \frac{1}{3}nmv^2,$$

where n is the number of molecules per unit volume, and m and v denote the mass and velocity of the molecules. As Meyer pointed out, the mechanical theory of heat only referred to average energies and average values of the square of the velocity, meaning that the correct formula was

$$P = \frac{1}{3}nm\overline{v^2}$$

Thomsen replied that he was well aware of the difference and that he had only used the simplified and wrong formula in order to keep to Naumann's arguments.[74]

The Thomsen-Naumann dispute covered about ten exchanges in the scientific literature, including two or three written by other chemists. Although Thomsen referred to it as a "fight" (*Streit*) – and

---

72. Thomsen (1870c), p. 950.

73. Meyer (1871), p. 32.

74. Thomsen (1870c), p. 950 and Thomsen, *Chemische Berichte* 4 (1871), p. 185.

a fight he had won – it was limited to a scientific disagreement and did not involve attacks of a more personal nature. The episode demonstrates that Thomsen was at the time acquainted with the difficult work of Krönig and Clausius (neither he nor Naumann referred to Maxwell). Although deeply absorbed in his experimental programme of thermochemistry, Thomsen apparently found the kinetic gas theory and its relation to Avogadro's law to be of such interest that he wanted to spend much of his time on the question. He did not return to it in his later work.

5.4. Hagemann and the concept of avidity

In his important memoir of 1869 relating to Berthollet's affinity theory and the Guldberg-Waage law of mass action, Thomsen reported thermochemical data on the action of diluted acids and bases (see further in Section 6.2). One of his aims was to find a measure for an acid's relative strength. He realised that the heat of neutralisation would not do and consequently devised an indirect and more involved method based on the heat evolved or absorbed when an acid was added to an aqueous solution of a salt of another acid.[75] The first might be sulphuric acid or nitric acid, and the latter soda. From the known heats of neutralisation of the two acids with the same base Thomsen was able to calculate the proportion in which the base was divided between the acids. This was an old problem which had first been attacked by Claude Louis Berthollet in his *Essai de Statique Chimique* from 1803. According to the French chemist the base was divided in the ratio of the acids' chemical equivalents or what he called their "active masses."[76] However, Thomsen proved that this was incorrect and that the division instead followed the Guldberg-Waage law of mass action. He found that the final distribution of the base, after equilibrium had been established, was the

75. Thomsen (1869a); Thomsen (1869b); Thomsen (1882-1886), vol. 1, pp. 97-148. Thomsen had outlined the method as early as 1854, see Thomsen (1854b).

76. Berthollet's "law" had previously been criticized by other chemists. For a careful analysis of Berthollet's theory and its impact, see Holmes (1962). See also Lindauer (1962).

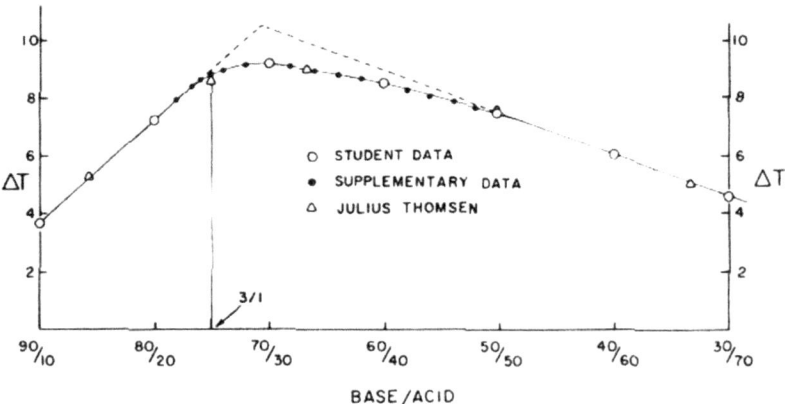

Figure 5.4. Temperature increase in the neutralisation process between NaOH and H_3PO_4 when mixed in various ratios. The modern data are in good agreement with Thomsen's. Notice that the maximum $\Delta T = 9.1$ °C does not occur at the ratio 3/1, reflecting that the dihydrogen and monohydrogen phosphoric acids are not strong. Source: Mahoney et al. (1981).

same as that which resulted when equivalent quantities of the two acids and the base reacted mutually.

For example, if equivalent amounts of sulphurous soda, nitric acid and sulphuric acid (H_2SO_4, 2 HNO_3, Na_2SO_4) reacted in dilute aqueous solutions the distribution was the same as in the system (Na_2SO_4, 2 HNO_3) and also in the system (2 HNO_3, H_2SO_4). Thomsen found that two-thirds of the soda combined with the nitric acid, and one-third with the sulphuric acid. From this he concluded that the "striving" (*Bestreben*) of nitric acid towards neutralisation was twice as great as for sulphuric acid. He found a similar result for hydrochloric acid. This Thomsen interpreted as nitric acid (or hydrochloric acid) being twice as strong as sulphuric acid. In regard of sulphuric acid having a much greater heat of neutralisation than the two other acids, it was a novel and bold interpretation.[77]

Thomsen introduced the concept of *avidity* as a measure of the

---

77. For a modern version of Thomsen's experiments made by college students, see Mahoney et al. (1981) and Figure 5.4. Referring to Thomsen, they comment: "There is something very pleasing in the fact that ... freshmen armed with expanded polystyrene cups can collectively harmonize with the master calorimetrist himself."

relative strength of an acid, thus ascribing nitric acid and hydrochloric acid an avidity which was twice that of sulphuric acid. "This
peculiar striving of acids towards neutralisation ... I describe by the
name 'avidity'," he wrote.[78] Consider an equivalent of the acid A',
which reacts with one equivalent of the salt AB of another acid A and
where B refers to a base. The salt will decompose into x equivalents
of A and $(1 - x)$ of AB. Thomsen argued that the total thermal change
written as $[AB, A']$ could be expressed as

$$[AB, A'] \cong x[(A', B) - (A, B)] + [(1 - x)AB, xA] + [xA'B, (1 - x)A']$$

The formula was sometimes referred to as "Thomsen's avidity formula."

In 1869 and later works Thomsen determined the relative avidities of a large number of acids, taking nitric and hydrochloric acids
to have avidities equal to 1.00 (Table 4). He further demonstrated
that the avidity was independent of the acid's heat of neutralisation
and also independent of the concentration; it only varied slightly
with the temperature.

We find in Thomsen's avidity the first measure of an acid's
strength as a constant belonging to the acid itself and not to its degree of acidity or what later would be called its pH value. Somewhat
anachronistically it can be regarded as precursor of the later concept of acid strength (commonly denoted K_A) as given by its degree
of dissociation in a highly diluted solution. Its full meaning was
only recognised after the emergence of the ionic theory of dissociation and Ostwald's law of dilution dating from 1888. According to
this law, for an aqueous solution of an acid of molar concentration
C and with a degree of dissociation α, the ratio

$$C\frac{\alpha^2}{1 - \alpha} = C\frac{\Lambda^2}{\Lambda_0(\Lambda_0 - \Lambda)} = K_A$$

will be approximately constant. The quantity K_A is a dissociation
constant characteristic for the acid and the symbol

$$\Lambda = \alpha\Lambda_0$$

---

78. Thomsen (1869a), p. 90.

denotes the electric conductivity, where $\Lambda = \Lambda_0$ at infinite dilution. The order of the acids as determined by Ostwald, but not their relative strengths, agreed in part with those found by Thomsen.

Name	Formula	Avidity
Nitric acid	HNO_3	1.00
Hydrochloric acid	HCl	1.00
Hydrobromic acid	HBr	0.89
Hydroiodic acid	HI	0.79
Sulphuric acid	H_2SO_4	0.49
Selenic acid	H_2SeO_4	0.45
Trichloracetic acid	CCl_3COOH	0.36
Oxalic acid	$(COOH)_2$	0.24
Monochloracetic acid	$CH_2ClCOOH$	0.09
Citric acid	$(CH_2)_2COH(COOH)_2$	0.05
Acetic acid	CH_3COOH	0.03
Hydrogen cyanide	HCN	0.00

Table 4. Some of Thomsen's avidity values. Apart from the strong acids, the order of the strength of the acids agrees with modern data.

The concept of avidity was well known in late-nineteenth chemistry but many authors, including Ostwald, preferred the name "affinity" rather than avidity to characterise the strength of an acid. On the other hand, he recognised Thomsen's theory to be an anticipation of his own, more advanced acid-base theory based on the ionic dissociation hypothesis.[79] The notion of avidity as an acid constant was somewhat controversial. For example, it disagreed with Berthelot's principle of maximum work according to which there could be no equilibrium distribution of HCl and H_2SO_4 when mixed with a base. For this and other reasons the French chemist dismissed Thomsen's theory of avidity. As he saw it, acids could not be ascribed a definite strength independent of the amount of water in which they were dissolved. Pickering, an independent and original chemist who was often involved in controversies, found it "incom-

---

79. Ostwald (1902), pp. 777-784.

prehensible how these ideas of 'avidity' could have been accepted almost without question."[80] By the early years of the new century "avidity" was rarely used any longer, although Thomsen continued to do so until his death. It still exists in modern chemistry, but with a different and more specialised meaning.[81]

Thomsen's close friend and former business associate Gustav Hagemann was among those who disagreed with Thomsen's solution theory and the concept of avidity. As mentioned in Section 2.5, Hagemann started his shining business career in the production of cryolite soda; he subsequently moved into the Danish sugar beet industry (De Danske Sukkerfabrikker), which he and others developed into a large and profitable corporation. A successful engineer and entrepreneur, he was not a research chemist and yet he maintained an interest in problems of pure science and theoretical chemistry. These problems occupied much of his time in the period 1886-1889, when he unsuccessfully tried to enter the world of academic research chemistry.[82] In 1886 he submitted to the University of Copenhagen a dissertation on "molecular volumes" in the hope of qualifying for a doctorate. The university appointed Thomsen and S. M. Jørgensen as examiners. But Hagemann never became a doctor of philosophy as he chose to withdraw the dissertation, most likely after informal consultations with Thomsen who might have wished to spare him the humiliation of having it dismissed.

Hagemann's ill-fated memoir, which he published at his own cost in both Danish and German editions, was rather critical to Thomsen's thermochemically based work and no less critical to Ostwald's volumetric method.[83] Among other objections he criticized the two distinguished chemists of ignoring the role of water in solution processes, which according to Hagemann were of a chemical and not of a physical-mechanical nature. His memoir was critically reviewed by Ostwald in *Zeitschrift für physikalische Chemie*, where the ris-

---

80. Pickering (1889), p. 29. Like Thomsen, Pickering did all his experimental studies by himself, without any laboratory assistant or attendant. See Dolby (1976).

81. Avidity is used in biochemistry to characterise the strength of certain multiple molecular interactions, such as between a protein receptor and its ligand.

82. Hagemann's work is discussed in Vinding (1942), pp. 190-209.

83. Hagemann (1886).

ing star in German chemistry listed a series of Hagemann's wrong ideas and unjustified deductions. He concluded that the memoir was "completely dilettantish." Neither Ostwald's sharp criticism nor the ill fate of the dissertation discouraged Hagemann, who continued publishing tracts on physical chemistry based on the idea that volume change was the proper measure of chemical energy changes. He did not supply his arguments with new experiments.

In a memoir of 1887 Hagemann criticized Thomsen's work on avidity in general and his conclusion that the avidity of sulphuric acid was half that of nitric acid in particular. He suggested that some of Thomsen's calorimetric measurements were wrong. Moreover, "The avidity formula is substantially false and all the avidity values derived from it are wrong."[84] Thomsen reviewed the memoir in restrained language, concluding that it was confused and badly argued; Ostwald agreed, but expressed himself less restrained and much more polemically.[85]

In other publications from the period Hagemann complained about the unfair reception of his work and went on developing a speculative hypothesis of chemical forces that rested on what he called "molecular oscillations." In order to explain thermal and volumetric chemical changes he suggested that heat energy was stored in the atomic constituents of matter in the form of wave-like oscillations with different frequencies and amplitudes.[86] When two chemical substances combined, their oscillations would interfere constructively. Hagemann believed to find support for his speculations in a mathematically complex hydrodynamic theory that the Norwegian physicist Carl Anton Bjerknes had developed to explain electromagnetism and gravitation on the basis of ether oscillations.[87] From Bjerknes' calculations and his own experiments with mechan-

---

84. Hagemann (1887), p. 8.

85. Thomsen (1887b). Ostwald's critique appeared in *Zeitschrift für physikalische Chemie* **17** (1887), pp. 199-200.

86. Hagemann (1888).

87. Bjerknes' theory was not generally accepted, but it inspired a few physicists and chemists in the late nineteenth century to take up speculations similar to Hagemann's, that is, to conceive atoms and molecules as oscillating systems governed by the laws of continuum physics.

Figure 5.5. G. A. Hagemann at his desk with the statue of Thomsen in front of him. Photograph from ca. 1905, as reproduced in Vinding (1942), p. 161.

ical models he concluded that two oscillating atomic spheres would interact with an inverse-square force of the same type as in Newton's law of gravity. Depending on the relationship between their phases, the force would be either attractive or repulsive.

Hagemann was convinced that in this way a variety of physical and chemical phenomena could be fully explained. His theory of molecular oscillations was as ambitious as it was speculative. It was dismissed as amateurish nonsense by Ostwald and Horstmann. Thomsen politely ignored it, but one may assume that he most likely shared the view of his German colleagues. He was aware of Bjerknes's theory, to which he had been exposed at the 1880 Scandinavian meeting in Stockholm and which he found to be fascinating but also enigmatic.[88] A decade later Thomsen got involved in another attempt by a Danish scientific speculator, baron S. P. Wedell-Wedellsborg, to generalise theoretical chemistry and physics into a unified theory (see Section 7.3).

It took Hagemann most of five years until he realised that his attempt to contribute to chemical theory was a failure. He felt that he had been treated unfairly by the German chemists and presumably also by Thomsen, who at no stage supported him. Despite their disagreements concerning the theories of chemistry, and despite Thomsen's role in the evaluation of his dissertation (whatever that role was), Hagemann and Thomsen remained on friendly terms and would continue to do so until Thomsen's death. To say that Hagemann respected Thomsen would be an understatement. He almost worshipped him. "Throughout his life," Hagemann's biographer says, "his feelings toward Thomsen were like those of a younger pupil; he did everything possible to please Thomsen, who could at times appear tyrannical."[89] In his office Hagemann had placed next to his writing desk a large bronze statue of Thomsen, so he constantly could be inspired by the great man (Figure 5.5).

---

88. Thomsen (1880f), pp. 533-534.
89. Vinding (1942), p. 45.

CHAPTER 6

Contributions to physical chemistry

Experimental thermochemistry was the branch of science to which Thomsen devoted the major part of his scientific life. On the other hand, although he had a reputation of a brilliant experimenter and narrow specialist he also contributed to other fields of the chemical and physical sciences. These contributions were less significant than his great work in thermochemistry but still they were of some importance and deserve to be recalled. In the period from about 1858 to 1866, in particular, Thomsen was much occupied with questions of a physical and technological nature. These included the energy equivalent of light and the construction of batteries and other electric devices. The latter area of research was as much or more oriented towards practical applications as towards scientific insight. Although not an applied scientist Thomsen always had in mind possible applications in technology and industry. This was the background for his early work on the manufacture of cryolite soda, and he never forgot the lesson based on his success in this area.

What is today called physical chemistry emerged in the 1880s but with roots longer back in time. One of the roots was thermochemical research and Thomsen was also engaged in other of the roots, in particular equilibrium processes and the theory of solutions. Experiments from the early 1880s led him to doubt the then popular hydrate theory of solutions, and in an important paper of 1869 he examined in detail the new law of mass action which had been proposed by Guldberg and Waage in Norway. These topics would soon acquire a new meaning within the framework of physical chemistry as established by Ostwald, Arrhenius and van't Hoff. Thomsen stood at the side-line in this development, unable or rather unwilling to accept the new ideas based on osmotic pressure, thermodynamics, and the theory of ionic dissociation in solutions. And yet he was involved in the development, if only indirectly and somewhat peripherally. Ostwald, who was initially inspired by Thomsen's thermochemical work on affinity, took him seriously and met with him personally. When Thomsen died in 1909 general or physical chemistry had undergone a major transformation, and some would even say a revolution. But not according to Thomsen who preferred to close his eyes to the new approaches that threatened to make his life-work obsolete or at least less important.

6.1. The energy of light

Although Thomsen was very much a chemist, at a few occasions he contributed to physical research with work in optics and electricity in particular. After all, for several years he served as a teacher of physics at the Royal Military High School and so was part of Denmark's small community of physicists in the post-Ørsted era. One of his contributions, a pioneering paper of 1865 on physical optics, merits attention because it was the first determination ever of the mechanical equivalent of artificial light. The famous Austrian physicist-philosopher Ernst Mach praised Thomsen for having established photometry "on quite a new basis" and for having measured "in absolute physical units the quantity of light emitted from a light source in unit time."[1] Curiously, Thomsen's important paper is missing not only from Stig Veibel's bibliography of 1943 but also from Thomsen's own list of his publications dating from 1905.[2] Most likely Thomsen's investigation was inspired by his more practical interest in the problems of public gas lighting in Copenhagen which caused him to examine the luminosity and economic efficiency of various types of gas burners.[3] Another inspiration was his general interest in the unity and convertibility of the forces of nature. As heat and chemical change were closely connected, so were heat and light.

Thomsen's investigation was quite independent of the physical nature of light, a question which was under transformation at the

1. Mach (1926), p. 30.

2. Veibel (1943); Thomsen (1905a).

3. Thomsen (1863c) and also Thomsen (1863d). On this line of practical research see also Section 8.2 and Hyldtoft (1994), pp. 143-144. Danish historians of science have apparently been unaware of Thomsen's work of 1865, which today is almost unknown.

time. It was also independent of the tradition in physics to measure the pressure that light possibly exerts on a body and which goes back to the eighteenth century. Only in 1902 did the Russian physicist Peter Lebedev prove conclusively the feeble light or radiation pressure which was predicted theoretically by Maxwell's electromagnetic theory nearly three decades earlier. But all this was of no relevance to Thomsen who was solely concerned with the energy of light and not with its constitution or mechanical pressure.

Investigations of the thermal energy associated with light go back decades before the discovery of energy conservation and was initially restricted to solar radiation. In 1800 the famous British astronomer William Herschel examined the heating effect of sunlight and in this way discovered what he first called "calorific waves" or "invisible light" but eventually became known as infrared rays or just IR. With the wave theory of light and the Italian physicist Macedonia Melloni's development in 1831 of the thermopile – a detector based on the thermoelectric effect – the infrared spectrum was understood as an extension of the visible spectrum at long wavelengths. Shortly after Herschel's discovery, and inspired by it, Johann Wilhelm Ritter in Germany, a close friend of Ørsted, found the "chemical rays" that characterised the new ultraviolet or UV part of the solar spectrum.

While Ritter's discovery marked the birth of photochemistry, a branch of chemical research that attracted intense attention in the nineteenth century, chemists saw no reason to deal with the infrared rays. It was an area of science primarily investigated by astronomers and solar physicists, who were interested in measuring the total amount of solar heat received at the surface of the Earth. The first approximately accurate value of this quantity, known as the solar constant, was determined by the French physicist Claude Servais Pouillet in 1838, who obtained the value 1.73 cal^{-1} cm^{-2} min^{-1} (the modern value is 1.95 cal^{-1} cm^{-2} min^{-1}). As it turned out later in the century, most of the Sun's heat is in the infrared area. Based on Pouillet's value, in a paper of 1852 William Thomson calculated the mechanical equivalent of sunlight but without considering other sources of light.

Thomsen's approach was very different. He was not interested in the Sun but in the amount of heat represented by the visible light emitted by any kind of artificial light source. As Mayer, Joule and Colding had measured the mechanical equivalent of heat, so he wanted to measure the mechanical equivalent of light. He first communicated his experiments on the issue at the meeting of Scandinavian scientists held in Stockholm in July 1863, and the same year he published his result in an article written in Danish. A briefer version in German and English followed two years later. As he wrote: "That the luminous ray exerts a mechanical action cannot be doubted; but as to its magnitude, there are at present no determinations. I have proposed to determine this, at all events approximately."[4] This he did by first filling a glass bulb with hot water; he measured the heat emitted from the bulb by means of a thermopile and compared it to the reduced temperature of the water. As he noted, the loss in temperature arose partly from radiation and partly from cooling by contact with the air, and Thomsen was interested only in the first factor. To isolate this factor he compared the reading of the thermopile with the reading obtained when the water was replaced by a sperm oil candle, a light source commonly used at the time. Similar experiments with gas flames convinced him that the unit of light energy per minute corresponded to the burning of 8.2 g of sperm oil in an hour.

However, Thomsen wanted to measure the effect of the luminous rays only, and for this purpose he filtered out the infrared heat rays by letting the radiation pass a 20 cm thick layer of water. He tested the method by means of the gas burner that Robert Bunsen had invented ten years earlier, using it with and without access of air. Further investigations led him to conclude that a layer of water of the mentioned thickness completely absorbed the infrared rays and only transmitted visible light (he did not refer to the ultraviolet light). Thomsen could now measure the heating effect of the visible light and found that a light source with an intensity corresponding to a candle burning 8.2 g per hour would in one minute radiate vis-

---

4. Thomsen (1865b), p. 246. The paper was originally published in German as "Das mechanische Aequivalent des Lichts," *Annalen der Physik und Chemie* **125** (1865): 348-352. The Danish version was Thomsen (1863d).

ible light energy of the magnitude 3.6 cal. By taking into account the reflection of light in the glass walls the result was refined to 4.1 cal per minute. Expressed in mechanical terms he stated his conclusion as follows:

> The unit of the quantity of work in a second, that is, 1 kil. [kg] raised to a height of 1 metre in a second, is equal to that which the luminous rays contain which issue from a source of light whose luminous intensity is 34.9 times as great at that developed by a candle which burns 8.2 grms. [g] of spermaceti in an hour.[5]

This rather convoluted statement was later translated as follows: a surface of 1 cm^2 receives per second 23.5 ergs from the normally incident luminous rays from a sperm oil candle, which burns 8.2 g of sperm oil per hour.[6] Thomsen suggested that this was the maximum value and that its true value might be somewhat lower if of the same order of magnitude. At the end of his paper he promised to "continue the investigation with more intense light, as solar and the electric light." However, this did not happen, possibly because he was about to change his physics position at the Military High School to a chemistry professorship at the University.

Thomsen's paper, published in both German and English, attracted considerable attention in the physics community. Moses Farmer, an American inventor and electrical engineer, reported on experiments with strong electrical arc light made in Boston by means of a large battery of Bunsen cells, and from these experiments he derived a value for the mechanical equivalent of light which was in "remarkable close agreement with the results of Professor Thomsen."[7] Friedrich Mohr, one of Germany's leading chemists and an early advocate of the doctrine of energy conservation, dealt in critical detail with Thomsen's determination of the mechan-

---

5. Thomsen (1865b), p. 249.

6. Mach (1926), p. 30. 1 J = 10^7 erg = 0.239 cal.

7. Farmer (1866). Moses G. Farmer (1820-1893) made several inventions in electrical technology, including multiplex telegraphy, new designs of dynamos, and forms of incandescent electric light.

ical equivalent of light. Although he found it to be interesting, he concluded that light as a form of energy was far less important than heat; as he pointed out, whereas heat was an expression of molecular motion, this was not the case for light.[8] In 1889 the German physicist Ottokar Tumlirz, an assistant of Mach at the German University in Prague, reconsidered Thomsen's experiments and calculated a better value of the mechanical equivalent of light on the basis of improved experiments.[9] He stated the value as 0.0056 cal s^{-1} = 0.34 cal min^{-1} and related it to contemporary measurements of the solar constant made by the American astrophysicist Samuel Langley.

Still at the turn of the century Thomsen's method attracted interest among physicists, astronomers and engineers. For example, it was further developed by the Swedish physicist Knut Ångström, a specialist in the study of solar radiation (and the son of Anders J. Ångström after whom the length unit 1 Å = 10^{-10} m is named). As two American physicists noted in 1903, "Ever since the publication of the paper of Julius Thomsen on the Mechanical Equivalent of Light ... it has been the general practice of students of artificial illumination to estimate radiant efficiency by some modification of his method."[10] However, with later developments the method and the very notion of a definite mechanical equivalent of light became obsolete.

While Thomsen did not include economic and engineering aspects in his German paper of 1865, he did so in the earlier Danish version. As he pointed out, the energy efficiency of the commonly used light sources was very low, with the heat-to-light ratio varying between approximately 300 (sperm oil) and 1000 (gas). "The way we use light is extremely wasteful," he wrote, suggesting that in the future new low-temperature sources of light would replace the ones currently in use. Perhaps, he speculated, the illumination system of the future might be based on chemically phosphorescent substances. The nature of such substances was not well known at the time,

---

8. Mohr (1868), pp. 54-57.
9. Tumlirz (1889).
10. Nichols and Coblentz (1903), p. 267.

but Thomsen suggested that phosphorescence was due to changes in the molecular configuration of the substance. "Should it become possible to make phosphorescent light with a high luminosity the lighting devices will surely be radically transformed," he wrote.[11] Low-temperature and economy-efficient light technologies of a new type would become a reality much later, if in the form of electric LED and CFL light bulbs based on fluorescence and not on naturally occurring phosphorescent bodies.

6.2. Chemical phenomena: dilution and equilibrium

From time to time Thomsen contributed to areas of chemistry which became important in and were classified as belonging to physical chemistry. These contributions were not part of a definite line of work aimed at physical chemistry but results that he derived from his general research programme of thermochemistry. In this section we discuss three such contributions, one relating to the hydrate theory of solutions, another to the Guldberg-Waage law of mass action, and the third dealing with the influence of temperature on heats of reactions.

In his extensive work on acid-base neutralisation processes Thomsen found that when equivalent quantities of different acids were neutralised by the same strong base, such as NaOH, the amount of heat evolved by strong acids was not significantly greater than that evolved by weak acids.[12] For example, his measurements resulted in an evolved heat of 13,740 calories when one mole of NaOH was neutralised by HCl and 13,450 calories for formic acid (HCOOH). This and other data obviously conflicted with the generally held view that the amount of evolved heat was a measure of the strengths of the affinities. Arrhenius' ionisation hypothesis offered an explanation for the apparently anomalous behaviour, for according to this hypothesis the neutralisation process was essentially independent of the particular chemical compounds, namely of the form

---

11. Thomsen (1863d), p. 206.
12. Thomsen (1882-1886), vol. 1, pp. 149-308.

$$H^+ (aq) + OH^- (aq) \rightarrow H_2O$$

However, Thomsen refrained from interpreting his neutralisation data as support for ionic dissociation. From his studies of the heats of hydration of salts containing water Thomsen suggested in 1885 that the molecular weight of liquid water might be twice that of water in vaporous form.[13] He realised that the suggestion was incompatible with the ordinary valence ascribed to oxygen and probably for this reason he did not pursue the idea. However, other contemporary chemists suggested that liquid water should be represented as $(H_2O)_2$ rather than H_2O and others again that water molecules existed in complex isomeric structures.

The theory of solution processes played a crucial role in physical chemistry as it emerged in the late 1880s. Although Thomsen never accepted Arrhenius' ionic theory of dissociation he was involved in the earlier attempts to understand what happens when a salt or acid, or some other compound is dissolved in water. This question he investigated by means of thermochemical methods in the third volume of *Untersuchungen*, where he reported about 300 calorimetric experiments on the thermal effects due to the union of an anhydrous salt with varying amounts of water. He found that the first molecule of water was stronger bound than the subsequent molecules, as shown by the different thermal effects. Following up on this line of research Thomsen investigated systematically the change of the heat of solution with varying amounts of water in other strong electrolytes. One of his aims was to establish whether definite hydrates were formed in aqueous solutions. He reasoned that if this were the case, the heat of solution should vary discretely with the added water, indicating the formation of hydrates; on the other hand, if no hydrates were formed one would expect a nearly continuous variation.

The theory of solution was a hot topic of chemistry in the 1880s, where a "hydrate theory of solution" was accepted by many chem-

---

13. Thomsen (1885) and Thomsen (1882-1886), vol. 3, p. 181. The same conclusion was independently argued by F.-M. Raoult in France, who based it on his studies of the freezing points of solutions.

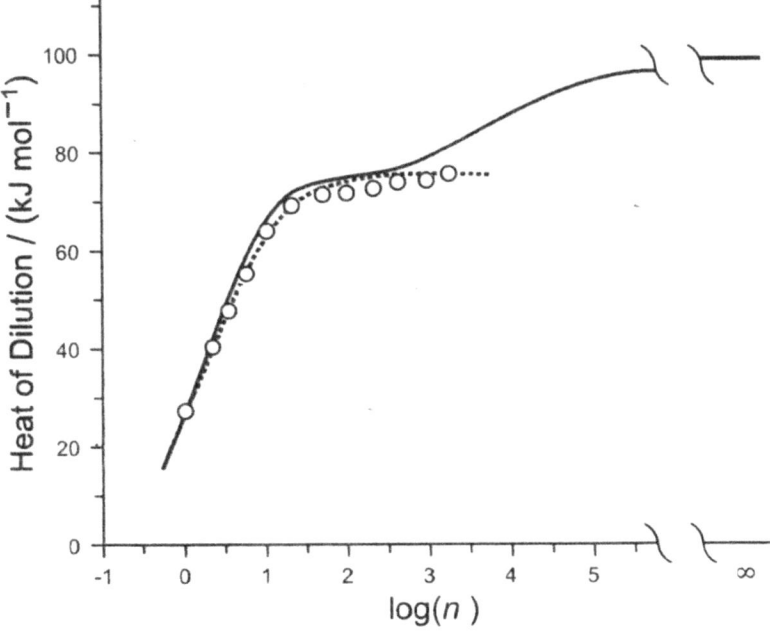

Figure 6.1. Heat evolved when mixing one mole of sulphuric acid with n moles of water. The circles are Thomsen's experimental data and the dotted line for $n > 1600$ represents the values calculated from his empirical formula; the solid line shows modern results. Source: Leenson (2004).

ists. According to this theory, when a salt or a strong acid was dissolved in water the solvent first formed hydrates and these would be dispersed throughout the liquid. Mendeleev, who had favoured a theory of this kind since 1865, suggested that all the water combined with the dissolved compound; at different concentrations the density and other physical properties of the solution would change as one hydrate replaced another hydrate. Another prominent chemist who endorsed the hydrate theory was Berthelot in France, and in one version or other it was also defended by several British chemists. However, after 1887 the various hydrate theories were criticized by Arrhenius and other supporters of the new physical theory who regarded them as contrary to the ionic theory of dissociation. According to Arrhenius and most other "ionists," water played only a passive role in the solution process. Although ionic dissociation was generally accepted by the turn of the century, in the case of strong electrolytes versions of the hydrate theory continued to attract attention. When Thomsen entered the subject in the mid-1880s, the hydrate theory was popular if somewhat controversial.

Thomsen performed more than 400 dilution experiments with weak and strong electrolytes, among which his work on sulphuric acid merits particular attention.[14] When concentrated sulphuric acid is mixed with water, a large amount of heat evolves. This had been known for long but Thomsen was the first to examine the phenomenon systematically and in detail. As mentioned in Section 4.2, he had begun investigations on the subject already in the early 1850s; what he published thirty years later was to some extent an elaboration of his earlier work. Thomsen showed that the maximum rise in temperature was no less than 159 °C and that it occurs when one part by weight of the acid is mixed with 0.330 part of water, or approximately one mole acid mixed with five mole water.

Starting with the anhydride SO_3 Thomsen formed the acid H_2SO_4 and from there added water in proportion to the acid. For the dilution process

$$H_2SO_4 \cdot 4\,H_2O + 5\,H_2O \rightarrow H_2SO_4 \cdot 9\,H_2O$$

he found an evolved heat of 2.559 kcal, which agreed almost perfectly with the measured heat capacities before and after the process. To measure the very small temperature changes at high dilution he used an optical tube which he claimed could measure the temperature with an accuracy of 0.001 °C. If n represents the number of gram-molecules (or moles) water he found that the evolved heat measured in kilo-calories for dilutions $n = 1$ to $n = 3$ were 6.379, 9.418, and 11.137. All his data could be approximately represented by an expression which Thomsen stated as

$$(H_2SO_4, nH_2O) = Q = \frac{17.860n}{n + 1.7983}$$

---

14. On this issue, see Leenson (2004).

This was in qualitative agreement with what he had found in 1854, but based on much more precise experiments. No signs of discrete jumps were visible, which made Thomsen conclude:

> One sees that the fixed points can be united by a regular curve, and that throughout, there are nowhere any signs of an irregularity, such as would indicate the formation of definite hydrates. ... The heat evolved on diluting liquid sulphuric acid with water is a continuous function of the water used, and excludes absolutely the acceptance of definite hydrates as existing in the solution.[15]

According to Thomsen the only hydrate was $H_2SO_4 = SO_3 \cdot H_2O$ whose heat of formation amounted to more than one half of the total heat. He thought that other hydrates such as $H_2S_2O_7$ might be formed, but on solution in water it would be transformed into the ordinary acid:

$$H_2S_2O_7 + H_2O \rightarrow 2\,H_2SO_4$$

In general Thomsen concluded that hydrates higher that those known in solid form did not exist in aqueous solutions. For acids he proposed the general formula

$$Q = \frac{n}{n + x} C$$

In this expression C and x are two empirically determined constants, C corresponding to Q at infinite dilution where $n \rightarrow \infty$. However, as Thomsen was aware, some of his data for highly diluted solutions did not agree with his formula (see Figure 6.1). He even reported measurements at $n = 400$ in which the heat decreased slightly with dilution, but suggested that these aberrations were unimportant and might be explained as a contraction effect.

$$Q = \frac{n}{n + x} C$$

---

15. Thomsen (1882-1886), vol. 3, p. 8.

"The generally accepted 'hydrate theory' of solution can scarcely be expected to survive the dissemination of Thomsen's researches," wrote Muir in a textbook on thermochemistry published in 1885.[16] Another British chemist, Spencer Pickering, called Thomsen's results "indeed remarkable," pointing out that the results were "in direct opposition to Berthelot's, and have gained ready acceptance among chemists."[17] One can imagine that Thomsen was pleased to have shaken the authority of his rival in Paris. However, Pickering did not accept Thomsen's conclusions. Favouring his own version of the hydrate theory as an alternative to ionic dissociation he argued that Thomsen's curve was in fact irregular and that it did not exclude the formation of higher hydrates.

Indeed, Thomsen's measurements could be interpreted both as arguments for and against the hydrate theory. This, as well as the enduring influence of his work, is illustrated by a paper by the American physicist William Francis Magie, at Princeton University, who as late as 1907 wrote:

> The evidence which proves an efficient interaction between the solute and solvent, and a modification of the properties of the solvent, was discovered by Julius Thomsen. On investigating the heat capacities of many aqueous solutions of electrolytes, Thomsen found that ... the heat capacity of a dilute solution is less than that of the water which enters into it. It is manifestly inadmissible to ascribe a negative heat capacity to the solution. ... We are therefore forced to believe that the solute, or some part of it, interacts with the surrounding water in such a way as to diminish the heat capacity of the water.[18]

Magie was in favour of an "association theory of solution" which incorporated aspects of the ionic theory with the hypothesis of formation of unstable hydrates.

---

16. Muir (1885), p. 167.

17. Pickering (1890), p. 94. On Pickering's critique of parts of Thomsen's thermochemical work, see Section 5.4.

18. Magie (1907), pp. 138-139. Jones (1903) defended another ionic version of the hydrate theory. According to Jones's "solvate theory of solution" dissolved substances combined with more or less of the solvent. See also Servos (1990), p. 77.

Even more important for the future development of physical chemistry was Thomsen's paper of 1869 on the law of mass action developed by Guldberg and Waage to which we have already referred in sections 4.5 and 5.4. Cato Guldberg, a mathematician at the Royal Norwegian Military Academy, and his brother-in-law Peter Waage, professor of chemistry at the University of Christiania, developed their theory of mass action in three stages, at first in splendid isolation from European mainstream chemistry.[19] In whatever of the stages the theory was highly abstract and mathematical and the law of mass action appeared in forms quite different from the modern textbook version. In 1864 they presented their theory to the local Academy of Science and Letters and three years later they developed it into a publication in French entitled *Études sur les Affinités Chimiques*. In 1879 they published a modified version in German which was in part based on the theory of molecular collisions and reaction velocities. Thomsen, who was undoubtedly aware of the Norwegian paper of 1864, referred only to *Études*, which for this reason is the most relevant source in the present context.

In 1867 Guldberg and Waage stated as their opinion that "nothing can so soon bring chemistry into the class of the truly exact sciences as just the researches with which this investigation deals ... a branch of chemistry which, since the beginning of the century, has unquestionably been far more neglected than it deserves."[20] Thomsen very much shared their wish of turning chemistry into an exact science. Guldberg and Waage suggested that the affinity between substances might be described not by the evolved heat but by the forces acting between molecules; these unknown forces might sometimes be balanced by forces governing the reverse reaction and thus lead to a state of equilibrium. For the affinity forces they argued that they were proportional to the product of the concentrations or what they called the "active masses" of the substances. For each substance, its active mass was the concentration raised to a certain power to be determined by experiment; in their

---

19. For the history and content of the Guldberg-Waage law of mass action, see Muir (1907), pp. 400-409, Partington (1964), pp. 588-595, and Lund (1965).
20. Quoted in Partington (1964), p. 591.

first publications they did not relate the exponents to stoichiometry.

For the resultant force responsible for a reaction such as A + B → A' + B' Guldberg and Waage coined the name "coefficient of affinity." Assigning the affinity coefficient of this process the symbol k and describing the active masses with the letters p, q, p' and q' (referring to A, B, A' and B', respectively), the force of affinity between A and B would be the product kpq. For the reverse process it would be given by $k'p'q'$, meaning that the condition for the state of equilibrium could be written as

$$kpq = k'p'q'$$

As Guldberg and Waage wrote: "After having determined the active masses of p, q, p' and q', one can find the ratio between the coefficients k and k'. On the other hand, having found the quotient k'/k, one can calculate in advance the result of the reactions for an arbitrary initial state of the four substances." With P, Q, P' and Q' representing the number of molecules before the reactions starts, and x the number of molecules of A and B transformed into A' and B' when equilibrium is established, they found

$$(P - x)(Q - x) = \frac{k'}{k}(P' + x)(Q' + x)$$

"With the aid of this equation, one easily determines the value of x." Guldberg and Waage satisfied themselves that their law was at least approximately correct by comparing it with own experiments and also with results obtained by Berthelot and his collaborator Pean de Saint-Gilles, who in 1862 had studied the equilibrium involved in the esterification of acetic acid by ethanol, that is

$$CH_3COOH + C_2H_5OH \rightarrow CH_3COOC_2H_5 + H_2O$$

The work of Guldberg and Waage offered for the first time some promise of a quantitative evaluation of affinity, but still by the late 1860s it was almost unknown to the international chemical community. In his paper of 1869 Thomsen did much to make it better known, to clarify its meaning, and to substantiate the law of mass action by means of thermochemical experiments.[21] Moreover, although his interest focused on the problem of affinity, his work was also an important contribution to the development of the concept of chemical equilibrium.[22] Thomsen first rewrote what he called "Guldberg's theory" in the form

$$k(\alpha - x)(\beta - x) = \frac{1}{k_1}(\gamma + x)(\delta + x)$$

With $n = kk_1$ he then derived a complicated equation giving x in terms of n, α, β, γ and δ.[23] To find the value of n he used as an example the case where one equivalent of sodium sulphate reacts with one equivalent of nitric acid in a dilute aqueous solution ($\alpha = \beta = 1$, $\gamma = \delta = 0$). As mentioned in Section 4.5, he had shown that in this case $x = 2/3$, from which $n = 2$ followed. The general formula for x could then be reduced to

$$6x = 4(\alpha + \beta) + \gamma + \delta - \sqrt{[4(\alpha + \beta) + \gamma + \delta]^2 - 12(4\alpha\beta - \gamma\delta)}$$

This equation could be used to find the amount of chemical change as given by x in all reactions involving sulphuric acid, nitric acid and sodium sulphate. As Thomsen showed, this could be done by measuring the thermal values of the reactions, and these experimental values could then be compared with the ones derived from the law of mass action. The result was an unequivocal verification of the law: "It turns out that in all cases where we have made use of Guldberg's formula we obtain a close agreement between the quantity of decomposition calculated according to this formula and the value derived from experiment."[24] Thomsen found that the discrepancy

---

21. Thomsen (1869a), pp. 94-101, and similarly in Thomsen (1882-1886), vol. 1, pp. 118-123, and Thomsen (1908), pp. 150-153. For some reason Thomsen always referred to Guldberg's theory or law, whereas he consistently left out the name of Waage, his chemical colleague in Christiania.

22. As noted in Lindauer (1962).

23. The equation in Thomsen (1869a), p. 94 contains an error, as the quantity n^2 appearing on the left side should be x.

24. Thomsen (1869a), p. 101.

between theory and observation in his series of experiments only varied between two and three per thousand. He did not comment on the critical remarks that Guldberg and Waage had made against his thermal theory of affinity (see Section 4.5).

Thomsen's confirmation of the law of mass action, published in a widely read German science journal, was important in making the theory of the two Norwegians better known. For example, it was Thomsen's 1869 paper that alerted young Ostwald to the theory and thus contributed to his career path in physical chemistry.[25] The paper was extensively reviewed in the German popular science journal *Gaea*, which found Thomsen's critique of Berthollet's principle, his notion of avidity, and his confirmation of the Guldberg-Waage law to be of great importance.[26] Many years later the famous British physicist Joseph J. Thomson, director of the Cavendish Laboratory, published a treatise in which he applied general dynamical methods to a number of physical and chemical phenomena.[27] Thomson was at the time much interested in the borderland between physics and chemistry, including topics such as dissociation, chemical equilibrium, and diluted solutions. He was well aware of Thomsen's *Untersuchungen*, to which he referred. When sulphuric acid reacted with sodium nitrate, the process was usually written as

$$H_2SO_4 + 2\,NaNO_3 \rightarrow Na_2SO_4 + 2\,HNO_3$$

But it was also possible that the "molecule of sodium nitrate" was represented by $Na_2N_2O_6$ and the one of nitric acid by $H_2N_2O_6$. To decide between the two possibilities Thomson made use of the thermal effects that the Danish chemist had first measured in 1869 for different proportions of the involved substances.[28] By interpreting

---

25. See Servos (1990), p. 22, who describes Ostwald's reading of Thomsen's paper as "a formative encounter."

26. Emsmann (1870).

27. See Thomson (1888). The book was based on a lecture course at the Cavendish Laboratory in the autumn of 1886.

28. Thomson (1888), pp. 226-227, referring to Thomsen (1882-1886), vol. 1, p. 121. Thomson did not believe that as much as 90 per cent of a dilute solution of HCl was dissociated (p. 213). When he became aware of Arrhenius' ionic theory of

Thomsen's data dynamically Thomson concluded that in very dilute solutions the molecules were probably of the form HNO_3 and $NaNO_3$, but in stronger solutions they were represented by $H_2N_2O_6$ and $Na_2N_2O_6$.

In a book of 1912 Arrhenius referred to Thomsen as an early advocate of the law of mass action:

> Julius Thomsen, the renowned Danish thermochemist, was very well acquainted with the work of Guldberg and Waage and probably he was influenced by it when he concluded the first volume of his *Thermochemische Untersuchungen* (1882) with the following words: "The aqueous solutions of substances contain them in a condition which, just as the gaseous state, reveals their physical qualities in the simplest manner, so that a direct comparison of the two states is permissible."[29]

The close analogy between the gaseous and the liquid states of matter, both obeying the laws of the mechanical theory of heat, became a cornerstone in the physical chemistry. The Norwegian originators of the law of mass action appreciated the value of Thomsen's contributions. As Guldberg wrote, "Your thermochemical investigations promise to solve several of the problems which we have examined experimentally but which could not be solved previously, when the [idea of] avidity was unknown."[30]

A few years after Thomsen had examined the Guldberg-Waage law it was also taken up by Horstmann, who in 1873 applied it to the dissociation of gases. However, it took more than another decade until the significance of the law was generally recognised. In its modern formulation the affinity constants k and k' no longer appear but are replaced by rate constants. For a reversible reaction at constant temperature between the systems $aA + bB$ and $cC + dD$ the equilibrium condition is

$$K = \frac{v_\rightarrow}{v_\leftarrow} = \frac{[C]^c[D]^d}{[A]^a[B]^b},$$

dissociation, he reacted strongly against it. See Dolby (1976), p. 325.

29. Arrhenius (1912), p. 79. Thomsen (1882-1886), vol. 1, p. 449.

30. Guldberg to Thomsen, 28 September 1877 (Royal Library, TSC).

where the squared brackets refer to molar concentrations. An equation of this type was first proposed by van't Hoff in 1877, who at the time was unaware of the earlier work of Guldberg and Waage, and then by Guldberg and Waage in their 1879 paper in which they defended their priority. It was only in this paper that the esterification equilibrium of Berthelot and Saint-Gilles was expressed in a form recognisable today:

$$\frac{[CH_3COOH][C_2H_5OH]}{[C_2H_5COOCH_3][H_2O]} = \frac{1}{4}$$

As van't Hoff pointed out, and as also Guldberg and Waage realised more vaguely, the law of mass action is valid only at constant temperature. By taking into account the second law of thermodynamics van't Hoff showed that the equilibrium constant K varies with the temperature T and the involved heat Q according to

$$\frac{d}{dT}\ln K = \frac{Q}{2T^2}$$

The equation is commonly referred to as the van't Hoff equation. With van't Hoff's exposition in his *Études de Dynamique Chimique* from 1884 the modern understanding of the law of mass action was essentially established.

In a paper of 1873 Thomsen pointed out that the heat effect of a chemical reaction depended on the temperature at which it was measured. In order to compare different results, "it is necessary to perform the various experiments at the same temperature, such as I have done in my own experiments, where I have kept to 18 °C."[31] To investigate the problem he made a large number of neutralisation and dilution experiments at different temperatures. For the case in which substances A and B at initial temperature t reacted to form substance C at final temperature T, he derived for the change of heat ΔQ the expression

$$\Delta Q = Q_T - Q_t = \int_t^T (q_a + q_b - q_c)dt$$

---

31. Thomsen (1873c).

The quantity q_a denotes the heat capacity of A or what Thomsen called its "calorimetric equivalent" (and similarly for q_b and q_c). Although the heat capacities change somewhat with the temperature, the change is very small and the equation can consequently be stated as

$$\Delta Q = Q_T - Q_t = (T - t)(q_a + q_b - q_c)$$

Thomsen thus found the heat variation per degree to be given by

$$\frac{Q_T - Q_t}{T - t} = q_a + q_b - q_c$$

By measuring the thermal effects at different temperatures and comparing them to measurements of the specific heats he showed that the approximation was valid. Berthelot did not believe that the temperature dependence of the heat effect was measurable and he denied that Thomsen's experiments verified it. The agreement was only apparent and due to Thomsen's inaccurate experiments, he claimed. As Thomsen saw it, Berthelot's response was yet another proof of the French chemist's dishonesty.[32]

Thomsen's 1873 paper was important but not quite as original as he presented it to be. From general arguments based on thermodynamics Kirchhoff had derived, as early as 1858, the same relationship between ΔQ and the heat capacities. What is sometimes known as Kirchhoff's law or equation can, in the case of constant pressure, be written

$$\frac{\partial(\Delta Q)}{\partial T} = \Delta C \, ,$$

where ΔC refers to the mean heat capacity of the reactants. From Kirchhoff's equation or the similar one of Thomsen one can calculate the heat of reaction at a particular temperature from the heat of reaction at some other temperature. Thomsen did not cite Kirchhoff's result and he also did not refer to earlier work on the subject done by Naumann in 1869 and Pfaundler in 1870. Whether the ne-

---

32. Thomsen (1882-1886), vol. 1, p. 79.

glect was deliberate or not, it resulted in a couple of minor priority controversies adding to the other and more serious controversies Thomsen was involved in at the time. According to Thomsen, what mattered was not the formua relating heat and temperature but that "the agreement between theory and experiment was first fully established by my investigation."[33]

6.3. Thomsen and the new physical chemistry

Although Kant categorically claimed that chemistry could never become a science on par with physics, chemists were not so sure.[34] Since the mid eighteenth century they successfully used a variety of physical theories and techniques to study chemical processes and the interrelationship of chemical and physical properties of matter. Physical or general chemistry, or sometimes "theoretical chemistry," existed well before it was institutionalised in the 1880s. During the period 1819-1824 what was originally published as *Annalen der Physik* appeared under the title *Annalen der Physik und der physikalischen Chemie*, the predecessor of Poggendorf's *Annalen der Physik und Chemie*. The eminent chemist Hermann Kopp served as professor of physical chemistry at Heidelberg since 1863 and there were a few other chairs of this kind. Electrochemistry, thermochemistry, crystallography, spectroscopy, and optical studies of substances were among the branches of classical general chemistry, but although they existed as research fields they were fragmented and without disciplinary unity. In his magisterial work on nineteenth-century science and philosophy, the German-British chemist John Merz named Thomsen as "one of the founders of physical chemistry," but this is to overrate the significance of his work on classical thermochemistry.[35]

A coherent *discipline* of physical chemistry only came into existence in the late 1880s, not only by merging existing branches of general chemistry but also by establishing it on a new theoretical

33. Thomsen (1876b), p. 307. See also Médard and Tachoire (1994), p. 304.

34. For Kant's pessimistic claim, see Section 7.2.

35. Merz (1904-1912), vol. 2, p. 153. As other founders Merz mentioned G. Hess, H. Kopp, H. Regnault, and M. Berthelot.

framework and, no less importantly, a new institutional infrastructure in the form of journals, research schools, and a community of practitioners.[36] The emergence of what today is called physical chemistry can conveniently be dated to 1887, the year in which Wilhelm Ostwald was appointed professor of physical chemistry in Leipzig. It was also the year in which and he and van't Hoff founded the journal *Zeitschrift für physikalische Chemie, Stöchiometrie und Verwandtschaftslehre* (Journal of Physical Chemistry, Stoichiometry and Affinity Studies). The highly successful and trend-setting journal was German but international in scope such as indicated by its board of associate editors which included a large number of high-ranking chemists. Among them figured Julius Thomsen together with colleagues such as Mendeleev from Russia, Ramsay from England, Lothar Meyer and Victor Meyer from Germany, and Berthelot and Henri Le Chatelier from France; also Guldberg and Waage from Norway were on the list. Arrhenius was missing, but his name was added in 1890.

As far as content is concerned, the new physical chemistry of the Ostwald school relied on two pillars, chemical thermodynamics and the theory of solutions. The first pillar, with its foundation in the theories of Helmholtz and Gibbs, was developed into a powerful chemical tool by van't Hoff in particular. The Dutch chemist also realised that the physical laws governing osmotic pressure could be understood in terms of thermodynamics and that osmosis provided a useful analogy to solutions. However, it was only with Arrhenius' radical hypothesis of ionic dissociation that a new and unifying framework was established. Arrhenius had proposed a limited version of the hypothesis in 1884 as part of his doctoral dissertation in Uppsala, but it was ignored until Ostwald recognised its value (Figure 6.2). A full version of Arrhenius' theory appeared in the first volume of *Zeitschrift für physikalische Chemie*. The three pioneers of physical chemistry based their work on the ionic hypothesis, which for a decade or so remained controversial in a large part of

36. For the early history of physical chemistry, see Laidler (1993) and Nye (1993). Sociological and historiographical perspectives on the origin of the discipline are offered in Dolby (1976b) and Barkan (1992).

Figure 6.2. Two of the pioneers of physical chemistry, W. Ostwald (left) and S. Arrhenius. Source: *Popular Science Monthly* **65** (1904). Wikimedia Commons.

the chemical community. While Ostwald, van't Hoff and Arrhenius were unquestionably the founders of the new physical chemistry, also the younger German chemist Walther Nernst belonged to the pioneers. All four received the Nobel Prize in chemistry. When Ostwald received the prize in 1909 – the year of Thomsen's death – the world of chemistry had fully recognised his and his allies' physical chemistry as an indispensable chemical discipline.

In a semi-historical account of 1903 the American chemist Harry Jones, a former assistant to Ostwald, enthusiastically sang the praise of the new physical chemistry. He separated sharply between chemistry before and after 1887. While in the earlier period chemistry had been essentially empirical and systematic, it was only with the advances of Ostwald and his allies that it became truly scientific. Without caring about the historical details and nuances, Jones described the development as follows:

> A system of chemistry is one thing and a science of chemistry is another; just as the making of brick is one thing and the building of the

brick into a piece of architecture is another. ... The period which ends about 1887 may be said to have been the "fact discovering" or "brick making" period. It was, of course, very important in itself ... [but it was] not a science of chemistry, as it is frequently thought to have been by those who have not followed closely the later development.[37]

Thomsen would not have recognised in this description the development of chemistry as he knew it. He believed that he was much more than a brick maker and that his thermochemical data had already contributed in making chemistry a science.

Despite figuring as an associate editor of the *Zeitschrift*, Thomsen never embraced the new trend in physical chemistry or just expressed serious interest in it. He used the journal as an outlet for several of his publications, contributing with nine papers in the period 1887-1905, but none of these were on subjects central to the Ostwald school of physical chemistry. As far as I know, Thomsen never used the term "ion" or referred to Arrhenius' theory of ionic dissociation. But of course he was aware of it and that from an early date. He may have met Svante Arrhenius, then a 21-year old science student, when the two participated in the twelfth meeting of Scandinavian scientists that convened in Stockholm in 1880. Thomsen lectured on his thermochemical theory of affinity in organic compounds, and Christian Blomstrand, a chemistry professor at the University of Lund, gave a lecture on chemistry as an atomic science.[38]

As pointed out in Section 4.3, in 1883 Arrhenius' mentor at Stockholm Technical University, Otto Pettersson, wrote a review of Thomsen's *Untersuchungen*, a work well known to Arrhenius. After Arrhenius had completed his ill-fated dissertation in 1884, a work published in French, he sent copies of it to a few distinguished scientists abroad. Among the recipients were Thomsen in Copenhagen to whose work Arrhenius had referred. Wanting his thesis to be better known Arrhenius suggested to Thomsen that he might write

---

37. Jones (1903), p. 17.
38. See Thomsen (1880f) which includes summaries of some of the talks given in Stockholm, including Blomstrand's and his own.

a critical review of it in a scientific journal, but this did not happen.[39] According to Arrhenius' biographer, Thomsen made his opinion known to his friend, the Uppsala chemist Per Theodor Cleve, telling him that there was nothing new in the dissertation.[40] Only from Ostwald in Riga did the young Swede receive an encouraging response. One may assume that Thomsen did not read Arrhenius' later so famous dissertation carefully and that although it dealt with topics in which Thomsen had an interest, such as equilibrium processes (including the Guldberg-Waage law) and a critical treatment of Berthelot's principle of maximum work.

Ostwald prepared his doctoral dissertation in 1877, at a time when he was employed as an assistant at the Riga Technical University in Latvia, then part of the Russian empire. Concerned with the problem of measuring chemical affinities he was much inspired by Thomsen's thermal approach; but he developed an alternative and more simple volumetric method which he thought might supplement or even replace the calorimetric method. The method, as he presented it in his doctoral dissertation of 1878, consisted in comparing the volume variations in chemical reactions between acids and bases in dilute solutions. Contrary to Thomsen, Ostwald concluded that the evolution of heat was not the proper measure of affinity. "To test my hypothesis," he wrote, "I repeated Thomsen's experiments according to the new [volumetric] method, and I had the pleasure of arriving at precisely the same results as the Danish scientist."[41] After having become a professor in Riga and studied Arrhenius' thesis, in the summer of 1884 Ostwald went for a tour to the Scandinavian countries. His primary target was Arrhenius in Uppsala. From Uppsala he went to Gothenburg and from there to Christiania, where he met with Guldberg and Waage. After his trip

---

39. Arrhenius to Thomsen, 4 June 1884 (Royal Library, TSC).

40. Crawford (1996), p. 50. The other recipients of Arrhenius' dissertation were Ostwald, Clausius, and Lothar Meyer. Arrhenius' dissertation included an electrolytic dissociation theory but not yet an ionic one. He only presented the crucial hypothesis of ions formed spontaneously by electrolytes in solutions, even in the absence of an electric current passing the solution, in 1887.

41. Ostwald's dissertation as reprinted as issue 250 of the series *Ostwald's Klassiker der Exakten Wissenschaften* (Leipzig: Akademische Verlagsgesellschaft, 1966).

to Norway he went by boat to Copenhagen to meet Thomsen and possibly other Danish chemists. In his autobiography Ostwald wrote about Thomsen, "whose work had served me as an ideal and whom I therefore wanted to pay my respect." He continued:

> My Swedish and Norwegian colleagues had prepared me for meeting an arrogant and unapproachable colleague. However, he had for some time expressed in public his appreciation of my work, so I was not particularly worried. In fact, I encountered a very dignified gentleman with a smoothly shaven face and a conspicuous excrescence on the left temple, every inch a privy councillor. As I left after an hour he warmed up from his objective attitude to such a personal level that he recommended me to spend the evening in Tivoli.[42]

After having visited the Tivoli Garden and the Thorvaldsen Museum in the centre of Copenhagen, Ostwald went on by steamer to Lübeck in Germany.

Although Ostwald considered Thomsen to be a scientist of the past and not of the present, he held him in esteem. In his voluminous textbook on general chemistry first published in 1887 Ostwald quoted extensively from Thomsen's early work on thermochemistry while at the same time leaving no doubt that the Danish chemist's thermal principle of affinity was essentially wrong. But Ostwald's critique of Thomsen was mild compared to his judgment of Berthelot's theory and the attempts of the French chemist to keep it alive by means of arbitrary and often artificial hypotheses. The authority that Berthelot's principle still held in France, he said, could only be explained by sociological factors. He even likened the distinguished French chemist to the fictional Baron von Münchhausen who pulled himself and his horse out of a swamp by his hair![43]

---

42. Ostwald (2013), p. 158, first published 1926-1927. Ostwald to Thomsen, 24 September 1885 (Royal Library, TSC). See also Crawford (1996), p. 53. Ostwald did not refer to other Danish chemists in his recollection.

43. Ostwald (1902), vol. II, p. 97 described Berthelot's attempt to save his principle of maximum work as "an operation of the same order as Münchhausen's famous escape from the swamp by a forceful pull in his own bootstraps." In fact, according to the original story of 1785 von Münchhausen did not perform his magic by a pull

Entering the still ongoing priority dispute between Thomsen and Berthelot, Ostwald argued in favour of the Danish chemist: "If one compares the content of Berthelot's statements and those published by Thomsen in 1854, one finds that they are in agreement and only differ in a formal sense. The true author is thus undoubtedly Thomsen to whom priority over Berthelot belongs with ten years." Moreover, "The technique used by Berthelot has not, in general, quite the same degree of accuracy that Thomsen has obtained."[44]

Ostwald argued from general and theoretical reasons that Thomsen's determinations from 1874 of the heats of formation of mercury halides were probably incorrect because they disagreed with Helmholtz's relationship between heats of formation and the electromotive force of a galvanic cell. He consequently asked Nernst, who at the time worked as Ostwald's assistant in Leipzig, to determine the heat evolved when mercury unites with bromine,

$$Hg + Br_2 \rightarrow HgBr_2$$

Confirming Ostwald's suspicion, Nernst found a heat of formation of 41.9 kcal or approximately 25 per cent less than what Thomsen had derived on the basis of his theory of thermal affinity. It turned out that the mercury Thomsen had prepared in 1874 and thought to be pure was not quite pure after all. Among the reasons that Nernst gave for doubting Thomsen's results, one stands out as curious as it related to seventeenth-century chemistry. He referred to "an older observation of Olaus Borrichius, according to whom mercury chloride was reduced by antimony under a distinct evolution of heat."[45] Nernst's point by mentioning a 200-year-old observation was that this reaction could not occur if Thomsen's data were correct.

Although Thomsen readily admitted his error made twenty-four years ago, he was much annoyed to see his thermochemical method

---

in his bootstrap but in his hair.

44. Ostwald (1902), vol. II, p. 65 and p. 67.

45. Nernst (1888), p. 23. Olaus Borrichius was the Latin name of Ole Borch (1626-1690), professor at the University of Copenhagen and an accomplished chemist and alchemist. One wonders if Nernst realised that Borch and Thomsen were compatriots.

defeated by the new physical chemistry – and by a 24-year old neo-phyte.[46] In a letter to Ostwald, Arrhenius reported: "I have recently received a small letter from Nernst; he will now travel even farther away from the seriously assaulted Thomsen. I had hoped that at least his inorganic works could rest in peace, but it seems not to be so."[47] Indeed it did not. As a result of Nernst's correction Thomsen changed to another method in which mercury was precipitated in a pure state by treating Hg_2I_2 with potassium iodide.

Thomsen was in most respects conservative and had no sympathy at all for the new wave in chemistry which threatened to make his life-work obsolete. As mentioned, he never used the term "ion" in connection with solutions. One may speculate that a further reason that discouraged him from taking an interest in physical chemistry was that some of its leading proponents strove to develop it into an anti-atomistic direction. By the turn of the century Ostwald questioned the existence of atoms and he was followed by Duhem in France and a few other prominent physical chemists including Georg Helm, a mathematical chemist from Dresden. As they doubted the existence of atoms, and generally wanted to free science of all materialistic hypotheses, so they downplayed the significance of the periodic system. Thomsen, on the other hand, was convinced not only that atoms and molecules existed as real bodies but also that they were composites of still smaller particles (see Chapter 7). To his mind chemistry without atoms was close to a contradiction in terms.

Not only because of his conservatism but also because of his rather unfriendly character and preference for isolation, Thomsen had very little contact with the younger generation of chemists. To describe him as a "leader of an important Danish school of physical chemistry," as one historian has done, is doubly wrong.[48] Not only

---

46. Thomsen (1888b). Thomsen also admitted his error in the Danish and English summary volumes of *Untersuchungen*. See Thomsen (1908), p. 278, and also Barkan (1999), p. 43.

47. The letter is reproduced in Körber (1961), vol. 2, p. 41. Ostwald and Thomsen exchanged several letters in the period from 1885 to about 1890 (Royal Library, TSC).

48. Barkan (1999), p. 85, who further writes (on p. 145) that Thomsen preferred "a

was Thomsen not a physical chemist in the ordinary meaning of the term, he also created no school and had no intention of doing so. Physical chemistry came late to Denmark, delayed perhaps by Thomsen and his professor colleague S. M. Jørgensen.[49] From a university prize essay announced in 1889 we get a glimpse of how the two professors viewed the ideas of the new physical theory. The subject of the essay was "A survey of the methods which during the last decade have been proposed about and applied to measurements of the relative weights of molecules." The evaluation committee consisting of Thomsen, Jørgensen and the physics professor Christian Christiansen found one of the answers valuable because it dealt in detail with osmosis, vapour tension, freezing point depression, and similar colligative properties. But they also had critical comments:

> The author mentions the theories of van't Hoff and Arrhenius but he does not develop the line of reasoning which have led these scientists, as well as [Max] Planck, to those theories, and he also does not suggest criticism of them. It must be admitted that ... the ideas expressed in the new theories have in no way won general scientific acceptance. On the other hand, the electric conductivity of solutions of electrolytes could deserve a closer examination. Not only have the results of [Friedrich] Kohlrausch and Ostwald in this area supported the new theories, but in this way it is also possible to determine the molecular weights of substances which otherwise are difficult to find.[50]

Despite their objections the three professors recommended a gold medal for the essay. The mention of the famous theoretical physicist Max Planck is interesting. At the time he was deeply immersed in the theory of physical chemistry and published in 1887 an article with many similarities to Arrhenius' from the same year. The near

---

small number of disciples to large research groups." But Thomsen had no disciples at all.

49. For physical chemistry in Denmark, see Bak (1974), and Nielsen and Kragh (1997).

50. *Københavns Universitets Aarbog 1890-1891*, pp. 690-691. I have not been able to identify the author of the essay, but it may have been Emil Petersen.

simultaneous discovery resulted in a minor controversy.[51] While it is
unknown to what extent Thomsen had acquainted himself with
Planck's theory of physical chemistry we know that the German
physicist studied Thomsen's work on thermochemistry to which he
referred in his *Vorlesungen über Thermodynamik* from 1897 and at other
occasions.[52]

The first to introduce physical chemistry in Denmark was Thomsen's successor Emil Petersen, who after a stay with Ostwald in
Leipzig converted to the new interdisciplinary field and included it
in his lectures from 1891. In 1889 and 1893 he published papers in
Zeitschrift für physikalische Chemie. The first of these was squarely in the
thermochemical tradition of Thomsen as it concerned the heats of
neutralisation of fluorides. Indeed, it contained numerous references to *Untersuchungen* and made use of Thomsen's calorimeter as
well as his general methods of heat measurements. "The investigations," wrote Petersen, "have been made in the laboratory of Copenhagen University whose director, Professor J. Thomsen, has
kindly put his apparatuses to my disposal; I am grateful for his ad-
vices and suggestions."[53] The other paper dealt with the heat of dissociation of weak acids in relation to Arrhenius' theory and it too
relied on some of Thomsen's measurements. The Danish version
that Petersen submitted to the proceedings of the Royal Academy
of Sciences was positively reviewed by Thomsen and Odin Christensen.[54]

Petersen's early work in physical chemistry grew to some extent
out of Thomsen's thermochemistry. But neither Petersen nor other
Danish scientists around the turn of the century contributed significantly to physical chemistry in the style of Ostwald, van't Hoff, Arrhenius, Nernst and others. The field was only institutionalised in
Denmark in 1930, when a new Institute of Physical Chemistry was
inaugurated in Copenhagen with J. N. Brønsted as its director.

51. See Crawford (1996), pp. 88-91.
52. Planck (1897), pp. 63, 68 and 228.
53. Petersen (1889), p. 384.
54. Petersen (1893). *Aarbog for Københavns Universitet 1891-1893*, pp. 54-56.

6.4. Electricity and batteries

Electrochemistry in the usual sense, such as electrolysis and measurements of electrical conductivity, did not belong to the research areas cultivated by Thomsen.[55] But he was seriously interested in electrical cells and batteries, an interest which covered their scientific as well as their technological aspects. During a period of forty years he worked on and off on various branches of electrical energy, sometimes connected with his thermochemical research programme but in other cases for different reasons. While employed as an adjuster of weight and measures in Copenhagen he communicated an investigation on the relationship between heat and electromotive force to the 1856 meeting of Scandinavian scientists held in Christiania. Two years later his work was published in the proceedings of the Royal Danish Academy of Sciences. It was never translated into a foreign language and consequently remained unknown to the large majority of scientists.

Thomsen's experiments were in direct continuation of his earlier work on thermochemical affinity. "If one could first determine how much electricity is required to produce the chemical effects of the electric current," he wrote, "one would have a new method to determine the chemical forces."[56] He consequently found it important to determine the flow of electricity which in a certain period of time was produced in an element. To express the electromotive force in terms of his thermal measure of affinity he examined the Daniell cell much used at the time. This type of cell, invented by the British physicist John F. Daniell in 1836, consisted of one plate of copper and another of zinc immersed in solutions of $CuSO_4$ and $ZnSO_4$ respectively; the two liquids were kept separate by a porous wall. Its electromotive force was known to be 1.1 volt. The experiments Thomsen performed to determine the electrical intensity of the Daniell cell and relate it to the evolved heat and the loss of metallic zinc were complex and sophisticated (Figure 6.3). For example, to

---

55. See Ostwald (1980) for an almost complete history of electrochemistry up to about 1900.
56. Thomsen (1858), p. 155.

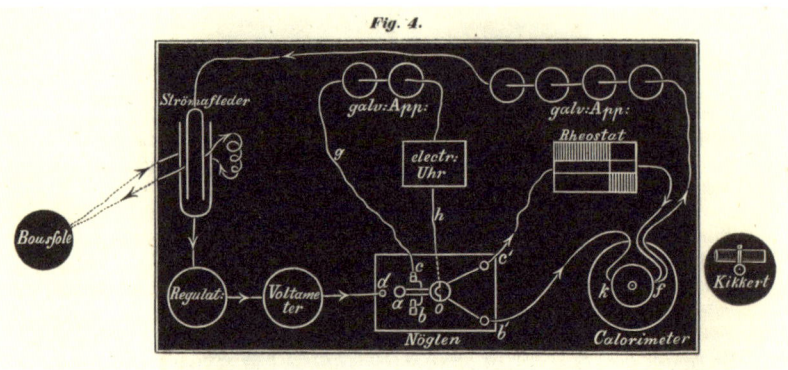

Figure 6.3. Thomsen's experimental setup for his measurements of the heat and the electric power of a battery. From Thomsen (1858).

measure the time he did not use a mechanical but an electric clock which he built into the circuitry.

Thomsen's main result was that the quantity of electricity in a Daniell cell corresponding to the dissolution of 16 g of zinc could be translated into an amount of heat equal to 3.329 kcal. The source of electricity was the chemical process

$$Zn + CuSO_4 \rightarrow Cu + ZnSO_4$$

From his earlier thermochemical measurements he inferred the chemical energy to be equal to the difference in heats of formation:

$$(Zn, H_2SO_4) - (Cu, H_2SO_4) = 6.660 - 3.460 = 3.200 \text{ kcal}$$

The small difference of 129 calories he explained away as experimental errors. There is no doubt, Thomsen concluded, that "the entire amount of force which under normal circumstances evolves as heat ... first appears in the state of electricity."[57] In other words, he thought, mistakenly as it turned out, that the electromotive force was a measure of the thermal affinity, or that all the energy gener-

---

57. Thomsen (1858), p. 165, and also Thomsen (1863e). As pointed out by Bjerrum (1909) and Brønsted (1909), Thomsen's result was fortuitous.

ated by the chemical processes was transformed into energy represented by the electric current. Retrospectively his conclusion can be explained by the fact that the change of entropy ΔS in a Daniell cell is very small. For this reason and because of the relation $G = H - TS$, the evolved heat is approximately equal to the free energy, $\Delta H \cong \Delta G$. Had Thomsen used other cells the disagreement between heat and electromotive force would have been more obvious.

In 1863 the French physicist François-Marie Raoult made more elaborate and precise comparisons between the energy developed in a cell and the energy that could be transformed by way of electricity in the form of heat. Raoult's numerical values were of the same order as Thomsen's, but the Frenchman did not subscribe to the general conclusion of the Danish chemist.[58] Thomsen did not follow up on his early attempt to explain electrochemical cells in terms of thermal effects except that he restated it in a more concise form in a paper of 1861. It followed from Thomsen's assumption of the electromotive force being proportional to the amount of chemical work that any conductive compound could be decomposed by a galvanic battery consisting of a large enough number of cells.[59] A proper thermodynamic theory of galvanic cells and batteries was only given much later, first by Helmholtz in 1882 and more fully by Nernst in important papers of 1888-1889.[60]

Thomsen briefly returned to the subject in a paper of 1880 in which he examined the chemical energy and electromotive forces of various cells.[61] From a letter written by Ludvig Lorenz two years later we learn that Thomsen at the time worked with the temperature coefficients of metals, the resistance of electric conductors, and the mechanical equivalent of heat measured electrically. Lorenz had for years been engaged in precision measurements of the ohm unit of resistance by means of a clever electromagnetic induction method, and in the letter to Thomsen he reported on his latest results.

---

58. On Raoult's work, see Ostwald (1980), pp. 778-784. Thomsen (1863e) noted with some regret that his Danish paper of 1858 was unknown abroad. He apparently thought that priority to some of the results published by Raoult belonged to him.

59. Thomsen (1861), pp. 110-111.

60. See Laidler (1993), pp. 220-227.

61. Thomsen (1880g).

The subject was evidently of interest to both of the Danish scientists. According to Lorenz, the mechanical equivalent of heat measured by electric means was somewhat greater than the value found by Thomsen.[62] Lorenz reported a best value of 436.1 gram-meter or 4.27 joule.

At the end of his 1858 paper Thomsen suggested that further investigations of the same kind might "have a decisive influence on the controversy between the contact theory and the chemical theory." As he made clear in a paper of 1863, he was convinced that the source of electricity in a cell was the chemical action and that contact between two different metals was itself unable to generate electric energy. "For nothing comes of nothing," as he wrote.[63] The so-called voltaic controversy that Thomsen referred to went back to Alessandro Volta's first construction of the pile in 1800 and continued throughout the nineteenth century. The most ardent champion of the contact theory during the first half of the century was the Danish-German natural philosopher Christoph Heinrich Pfaff, a professor at the University of Kiel, whereas Ørsted was more in favour of the chemical theory. The law of energy conservation was generally taken as support of the chemical theory, such as Thomsen and most chemists did. On the other hand, the contact theory was far from dead. In 1862, at the very time that Thomsen claimed victory for the chemical theory, the British physicist William Thomson – better known as Lord Kelvin – suggested a new version of the contact theory according to which the metallic contact force was responsible for the generation of current in a cell.[64]

In the mid-nineteenth century a variety of electric cells and batteries were in use, some for scientific purposes and others applied in the rapidly growing industrial sector that included telegraphy and galvanic electroplating. In a note of 1860 Thomsen announced one more type of cell, this time consisting of carbon and copper. He

---

62. Lorenz to Thomsen, 17 December 1882 (Uppsala University Library). The problem of an "absolute definition" of the ohm unit was important at the time. Lorenz's method was highly regarded by Lord Rayleigh and other British physicists.
63. Thomsen (1863e), p. 323.
64. The century-long, complex and confusing controversy is analysed in Kragh (2000).

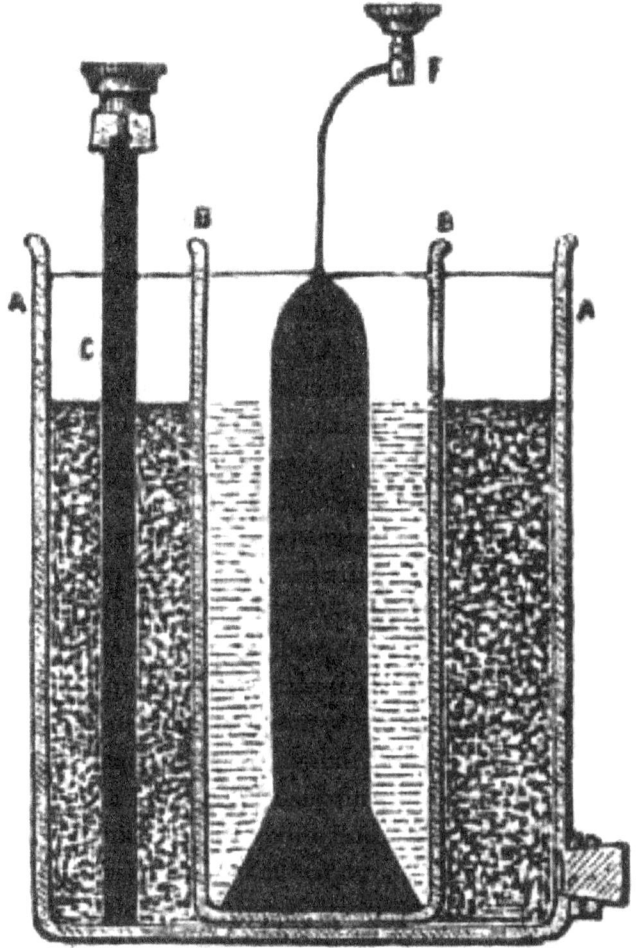

Figure 6.4. Thomsen's copper-carbon battery. Source: Bottone (1902), p. 245.

placed the copper plate in dilute sulphuric acid, contained in a porous cell; outside it was a negative carbon plate in a solution of pot-
ash, sulphuric acid, and water (Figure 6.4). The electromotive force
of the new copper-carbon cell was about 1.0 volt, a little less than
the one of the Daniell cell. Thomsen found the cell useful because
the copper was not attacked by the acid in the open circuit and also
because the sulphuric acid could be used for months without being

replenished. From a theoretical point of view he found the cell to be of no less interest: "As copper cannot decompose dilute sulphuric acid, the copper-carbon element is an example of a powerful apparatus in which chemical action and the disengagement of electricity are quite inseparable."[65] Thomsen's copper-carbon battery at first attracted some attention but it seems not to have been widely used in either the laboratory or for commercial purposes. It merely entered the large arsenal of different cells that could be used for special purposes.[66]

The most important result of Thomsen's interest in sources of electrical energy was the "polarisation battery" that he first described in his and his brother's Danish journal *Tidsskrift* in 1864. "I congratulate you with your nice invention of the polarisation battery which I came to know about in the last issue of your journal," the Swedish physicist Erik Edlund wrote. "I took the liberty to give an account of your invention at the recent meeting of the [Swedish] Academy of Science."[67] Thomsen's apparatus had a practical rather than scientific purpose in so far that it was designed to meet a need in the telegraph companies and other businesses relying on the current from batteries. The traditional way to build up a relatively high voltage was to combine a large number of galvanic cells, such as Daniell cells, into large batteries. Thomsen found a more efficient way to produce a constant current at a high voltage by an arrangement in which several plates of platinum were polarised by means of the current from a Daniell or Grove cell; the latter kind of cell consisted of zinc placed in sulphuric acid and platinum in nitric acid. Thomsen's polarisation battery of fifty cells (Figure 6.5) had an electromotive force of approximately 75 volts and was able to deliver a current of high intensity.

---

65. Thomsen (1860b), with English translations in *Philosophical Magazine* **21** (1861): 80, and *Journal of the Franklin Institute* **71** (1861): 336-337.

66. Together with many other cells, Thomsen's constant copper-carbon cell was described in Bottone (1902), p. 245.

67. For the polarisation battery see Thomsen (1864) and Thomsen (1865c). Edlund to Thomsen, 18 December 1864 (Royal Library, TSC). Edlund, who was a consultant to the Swedish Telegraph Office, was interested in Thomsen's invention for both scientific and technological reasons.

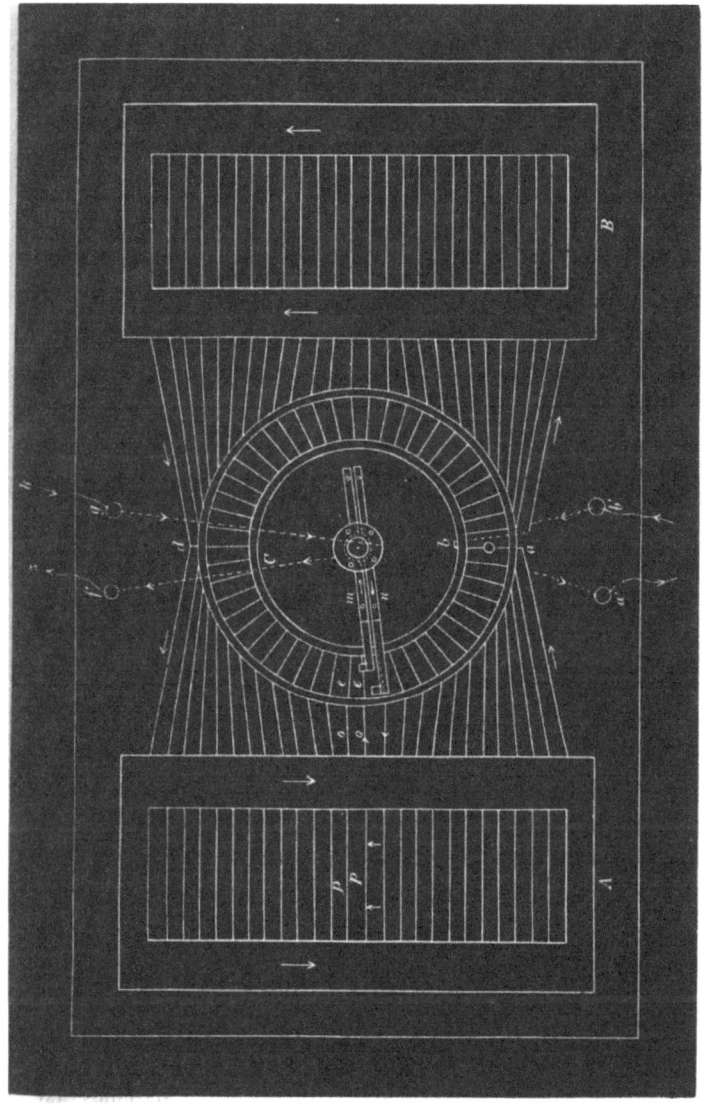

Figure 6.5. Thomsen's polarisation battery with its platinum plates above and below the rotating contact system. Source: Thomsen (1864).

When he published his invention in *Annalen der Physik und Chemie*, the journal's editor, the reputed German physicist Johann Christian Poggendorff, added a note in which he objected that there was little new in Thomsen's apparatus. It built on principles which he,

Poggendorff, had described more than twenty years earlier. Thomsen replied that although Poggendorff deserved credit for his work, his own apparatus was quite different in design and much more useful. Poggendorff's earlier apparatus was complicated and gave a discontinuous current, whereas the new battery was simple and generated a continuous current. "The polarisation battery is the only apparatus constructed so far which has succeeded in producing a continuous electric current at high voltage and constant intensity by means of a single galvanic cell."[68] The exchange of opinions did not, in this case, evolve into a controversy.

The polarisation battery attracted much interest as it promised an answer to a technological need. Thomsen was keen to market his invention, which he did by publishing a detailed account of it in the form of a German brochure which was as much a sales brochure as a scientific memoir. His polarisation battery was patented in many European countries and awarded prizes at the Art and Industry Exhibition in Stockholm in 1866 and also at the large Paris World Fair exhibition in 1867.[69] An American report from the Paris exhibition gives an impression of Thomsen's invention:

A remarkable battery, called by its inventor, Professor Jules [sic] Thomsen, of Copenhagen, a polarization battery, was exhibited in the Danish section. Fifty-two plates of platinum are immersed in dilute sulphuric acid, and these are successively brought into contact, by pairs, with the poles of a single cell of Daniell. ... This polarization gives rise to a powerful current in the platinum combination; and this is maintained nearly constant when the contacts succeed each other rapidly and regularly. The electrodes of the exciting battery are kept in rotation by means of an electro-magnetic motor. Professor Thomsen states that the fifty cells which correspond to the fifty-two platinum plates produce a current equal in intensity to that of seventy-three elements of Daniell.[70]

---

68. Thomsen (1865d), p. 165.
69. According to Bjerrum (1909) and Holst (1908).
70. Blake (1870), p. 559. Online as https://archive.org/details/reportsunitedst 02unkngoog.

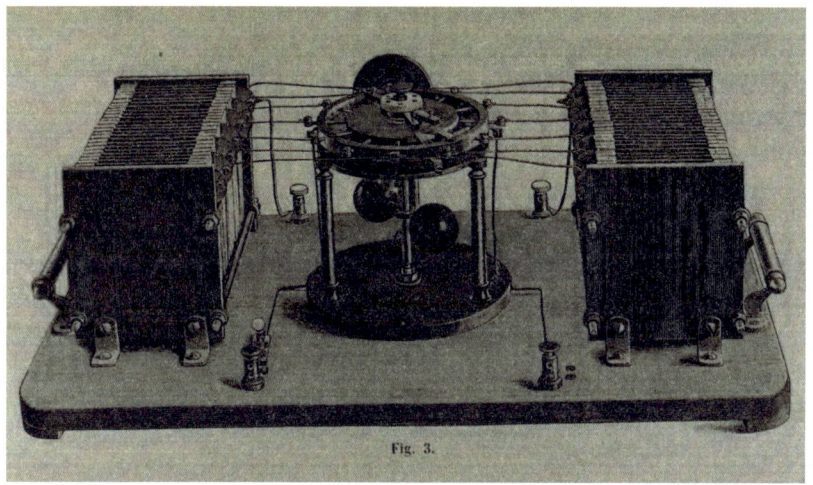

Figure 6.6. Thomsen's direct-current transformer designed to transform 110 volts to 8 volts. Source: Thomsen (1898b).

Already in 1865 the Danish State Telegraph bought the apparatus to use it in the main telegraph station in Copenhagen and it may also have been used by some other telegraph companies which expressed great interest in it. For example, in 1867 the Submarine Telegraph Works, a British company, invited Thomsen to come to the meeting of the British Association in Dundee in order to demonstrate his battery.[71] As far as I know, Thomsen did not accept the invitation.

Another area in which Thomsen's battery attracted some attention was the medical world. According to Eugen Ibsen, a Danish engineer and medical student, it was well suited for electrotherapy and other kinds of medical use.[72] He suggested that, when constant currents of high voltage were needed, the polarisation battery was preferable to the induction coils or series of batteries traditionally used at the hospitals. However, whether for the purpose of telegraphy or electrotherapy the commercial success was limited. With the

---

71. Submarine Telegraph Works to Thomsen, 20 August 1867 (Royal Library, TSC).
72. Ibsen (1865), according to whom a version of the polarisation battery specially designed for medical use was being tested in Copenhagen.

Figure 6.7. Thomsen in his laboratory, ca. 1898, possibly at work with his transformer system. The apparatus in front of him is a Ruhmkorff induction coil for generation of pulses of high voltages, so named after its inventor, the German physicist Heinrich Ruhmkorff. Source: Vinding (1941), p. 37.

introduction of dynamos as electric energy sources in the 1880s Thomsen's battery became superfluous.

By the 1890s Copenhagen had become electrified by means of electricity of 110 volts supplied by central power stations and electric light was competing with gas lighting. The problem was no longer to increase the voltage of batteries but to decrease the 110 volts from the dynamos to, for example, the 5-10 volts often used in laboratories and hospitals. For this purpose 72-year-old Thomsen invented a transformer the construction of which had much in common with his polarisation battery of the 1860s. But it worked in reverse, using a storage battery or accumulator instead of an ordinary battery consisting of primary cells (Figures 6.6 and 6.7). In his presentation of his new invention to the Royal Danish Academy in 1898 Thomsen said that he had tested the apparatus for more than a year.[73] It worked perfectly and was much more economic in use than other methods of transforming the voltage. Thomsen probably hoped that the new transformer would enter the commercial market but apparently it found no or only very little application outside his own laboratory.

---

73. Thomsen (1898b).

CHAPTER 7

Chemical elements

Although best known as a brilliant and meticulous experimenter, Thomsen also had a deep interest in the more foundational issues of chemistry and science generally. During most of his career, his emphasis on the value of precise experimental data in thermo-chemistry and elsewhere in chemistry co-existed with a desire to unravel the secrets of nature. He wanted to understand what matter was, ultimately, and thus in a sense return to chemistry as natural philosophy. Thomsen would have hated being called a philosopher, and yet his ambitious goal of understanding matter was essentially philosophical and not, in some respects, far from the one hailed by Ørsted. Latest by the mid-1860s he turned into an ardent supporter of the age-old idea of the unity of matter, the view that all the chemical elements were manifestations of a single primitive substance. During the last two decades of his life, work related to this view, known as Prout's hypothesis, occupied him increasingly. It led to a series of publications of a somewhat daring and speculative nature, very different from those dealing with his thermochemical measurements although not incompatible with them.

Some of Thomsen's ideas were original but most of them reflected the spirit of time characteristic for science in the late-Victorian era. Not only did he subscribe to the hypothesis of the unity of matter, he also combined the hypothesis with vague evolutionary speculations of the kind known as inorganic Darwinism. Characteristically (and again time-typically), he sought to justify his speculations by means of very precise determinations of the atomic weight of oxygen relative to hydrogen. The search for unity and order inevitably turned him towards the periodic system, which he presented in his own version in the 1890s. This version he used to predict a number of new elements which, in the wake of the discovery of argon, included a whole group of inert gases. One of them was the solar element helium which he thought might be simpler than hydrogen. The "other Thomsen" is no less interesting a figure than Thomsen, the pioneer of experimental thermochemistry.

7.1. The composite chemical atom

Although Thomsen in general favoured an empiricist attitude to science he fully recognised the value of theories and hypotheses. He was not foreign to speculations if only these were recognised to be just that. In an article of 1871 he spelled out his philosophical view of science, emphasising in a quite sophisticated way the dialectical relationship between hypotheses and experimental data. The chemist, he said, is not satisfied with bare facts, for these alone give no access to the causes of the phenomena, to nature's inner constitution. "From the safe domain of facts he enters the nebulous world of hypotheses where only a lucky thought, an inspiration ... can bring him further." By their very nature hypotheses are incomplete and uncertain, and yet they are absolutely essential to the progress of science. He elaborated:

> The reason for the impact of theory on the progress of science is not to be found in the theory's concordance with reality or truth but rather in its ability to unify and account for in a clear and simple manner a large amount of facts; in this way it may lead to the discovery of new facts. Even though a theory is erroneous the errors will often be harmless. The wealth of new facts will indicate the weaknesses of the theory and thus cause no harm to science; on the contrary, the weak points will spur the scientists to discover the reason for the disagreements and thus in the end to replace the older theory with a new one that agrees better with observations.[1]

In other words, a wrong theory is better than no theory; what matters is that the theory is fruitful, not that it is true; errors generate progress and progress is the ultimate measure of scientific success. Thomsen illustrated his general methodological lesson by referring

---

1. Thomsen (1871a), pp. 3-4.

to two of the classical cases in the history of chemistry. One was the alchemical theory of metallic transmutation, which "strictly speaking cannot be said to be wrong;" the other case was the phlogiston theory of the eighteenth century, which he judged had "influenced science in a most fortunate way."

By the 1860s there was a great deal of interest in atomism as the foundation of chemistry and the question of whether atoms – if they existed – were the indivisible units of the elements that Dalton had originally proposed.[2] In a lengthy paper of 1865 in *Tidsskrift for Physik og Chemi*, Julius Thomsen joined the speculations concerning the nature of what he called the "so-called elements." Were they really elementary units of matter or were they perhaps composite bodies consisting of even smaller entities? Although Thomsen in most cases preferred to speak of elements rather than atoms, he no longer referred to atoms in the purely stoichiometric sense, such as he had done in his early textbook of 1850. He apparently considered them to be real material particles, if not necessarily indivisible. What induced Thomsen to his speculations seems to have been the recent discoveries of new elements, bringing the number up to 64. In the introduction to his paper he referred to the discoveries of the metals rubidium, caesium, thallium, and indium due to "a recently developed analytic method, namely spectral analysis."[3] The elements were identified between 1860 and 1863 by R. Kirchhoff and R. Bunsen (Rb, Cs), W. Crookes (Tl), and H. T. Richter and F. Reich (In).

"One cannot," Thomsen said, "refrain from considering the idea that what we now think of as chemical elements are probably complex bodies, compounds of a few different elements in various ratios or maybe just different states of condensation of the same element. ... Perhaps we are not far from the time when one of the most important of the elements will be separated in different parts and

---

2. Much has been written on this subject. See, for example, Farrar (1965), Kragh (1982), Brock (1985), and Leone and Robotti (2003).

3. Thomsen (1865a), p. 65. As usual, Thomsen did not refer to the discoverers of the elements or to any names of scientists at all. Although his article was in the tradition going back to William Prout – he even suggested that "hydrogen may be a constituent of the various elements" (p. 107) – he did not mention Prout or other scientists in this tradition.

thus turn out to be a composite substance." Spectroscopic evidence was one indication of this scenario of the future, but only one among many. Thomsen suggested that a variety of experimental methods, facts and phenomena all pointed in the same direction, namely that the commonly recognised elements "are composite and thus do not deserve to be called elements." He repeated: "It seems quite evident that we cannot claim with any confidence that what we now call chemical elements should not really be composite bodies."[4]

Thomsen was one of the first Danish scientists to take an interest in the spectroscope. But although he found it a fascinating instrument, as a chemist of the traditional school he was somewhat sceptical about its usefulness in chemical analysis. At a meeting of May 1862 in the Royal Danish Academy (of which he had become a member two years earlier) he presented the spectroscope in the version constructed by the Munich instrument maker C. A. Steinheil. It was the same kind of instrument which Kirchhoff and Bunsen had used in their pioneering experiments two years earlier. Spectral analysis was able to detect minute quantities of chemical elements, Thomsen admitted, but

> ... in ordinary analytical investigations it [the spectroscope] will presumably only play a very limited role, whereas it can be invaluable in certain special investigations. ... As a method of ordinary chemical analysis its use will be restricted to the alkali metals and the alkaline earths when these have been freed from other metallic compounds by ordinary methods.[5]

Thomsen evidently underrated the power of the spectroscope as an apparatus for chemical analysis. He was not the only chemist of his generation to do so.

Among the chemical phenomena that possibly indicated the composite nature of atoms, Thomsen mentioned in his paper of

---

4. Thomsen (1865a). Quotations from pp. 65-66, 97 and 106.

5. Thomsen (1862c). He pointed out that whereas sodium fluoride and calcium fluoride produced characteristic spectral lines in the spectroscope, when they were mixed only the characteristic lines of sodium turned up.

1865 the allotropic states of some elements such as phosphorus and carbon. He found oxygen to be particularly interesting. Apart from the ordinary form O_2 the "active oxygen" or ozone with formula O_3 had been known since 1839 when it was discovered by the German chemist Christian F. Schönbein. But recently, Thomsen said, it had been discovered that "active oxygen can appear in two different modifications, which have been named ozone and antozone and which ... can mutually cancel the other's activity such that, when they are mixed, they will turn into oxygen in its ordinary inactive state."[6] What was he referring to? Ever heard of antozone? This non-existing substance was suggested and named by Schönbein in 1858 who conceived ozone to be negatively charged O_3 and antozone to be its positively charged counterpart. When the two substances were mixed, they would react as

$$O_2O^- + O_2O^+ \rightarrow 3\,O_2$$

Schönbein further assumed that antozone, contrary to ozone, reacted with water to form ordinary oxygen:

$$H_2O + O_2O^+ \rightarrow H_2OO^+ + O_2$$

While Thomsen in 1865 clearly believed in the existence of antozone, the large majority of chemists ignored Schönbein's discovery claim. It is not known when Thomsen realised his mistake.

In his search for evidence for the composite atom Thomsen paid particular attention to the regularities of the atomic weights or what he still referred to as atom-numbers. His groupings of the chemical elements according to atomic weights counts as one of the many incomplete anticipations of the periodic system, of the same kind as the better known groupings of John Newlands, Alexandre de Chancourtois, William Odling, and others. As Thomsen pointed out, for

---

6. Thomsen (1865a), p. 79. For details about the curious story of antozone, see Rubin (2009). Schönbein did not originally realise that ozone was a triatomic modification of oxygen.

selected groups of chemical elements, their atomic weights on the H = 1 scale would approximately agree with the formula

$$A = 16\,n + b \qquad (n = 1, 2, 5, 8)$$

In this expression b is a number which is characteristic for each group of analogous elements. For example, for what we call group II elements, $b = 8$, and their atomic weights were written as

$$Mg\ (24) = 16 \times 1 + 8;\ Ca\ (40) = 16 \times 2 + 8;$$
$$Sr\ (88) = 16 \times 5 + 8;\ Ba\ (137) = 16 \times 8 + 8$$

According to Thomsen, one could regard the four elements as consisting of 1, 2, 5 and 8 "atoms of a common constituent of atom-number 16" in combination with an atom of weight 8. He repeated the exercise for group VI elements ($b = 0$):

$$O\ (16) = 16 \times 1;\ S\ (32) = 16 \times 2;\ Se\ (79) = 16 \times 5;\ Te\ (128) = 16 \times 8$$

Thomsen further suggested to group together beryllium, aluminium and zirconium. By number manipulation he reproduced their weights as follows:

Be (14)	$\frac{1}{2}(40 + 2) = 14$
Al (27.2)	$\frac{1}{2}(80 + 2) = 27.3$
Zr (67)	$\frac{1}{2}(200 + 2) = 67.3$

As if to give his argument more force, he pointed out that the ratio between the differences of the atomic weights in the triad ($67 - 27.2 = 39.8$ and $27.2 - 14 = 13.2$) was almost precisely 3 (namely $39.8/13.2 = 3.015$). For a group of trivalent elements from boron ($A = 11$) to bismuth he noticed that on the assumption of "a hypothetical substance of atom-number 3" the atomic weights could be neatly reproduced as, for example,

$$N(14) = 11 + 3;\ \ P(31) = 2 \times 11 + 3 \times 3;\ \ Sb(120) = 10 \times 11 + 3 \times 3$$

Thomsen found regularities of this sort to be interesting, but he refrained from concluding that the elements actually consisted of smaller units of atomic weights 3, 8, 11 or 16.

The sort of numerical relationships which Thomsen suggested in 1865, were not uncommon in the period when similar numerology played a role in fertilising the soil for what in 1869 appeared as the periodic system of the elements. One of the period's more extreme chemical numerologists was the ten-year-younger Danish-born American scientist Gustavus Detlef Hinrichs, who in 1866 proposed atomic weight relations for the members of what soon became known as the chemical groups.[7] For the groups I, II and VI, he represented the atomic weights as

$$
\begin{array}{lll}
\text{Group I} & A = 4^2\, n + 7 & (n = 0, 1, 2, 5, 8) \\
\text{Group II} & A = 2^2\, n & (n = 3, 5, 11, 17) \\
\text{Group I} & A = 4^2\, n & (n = 1, 2, 5, 8)
\end{array}
$$

The following year Hinrichs published a lithographic reproduction of a handwritten work entitled *Atomechanik*, in which he presented a natural system of the elements – an anticipation of the periodic table. He made sure that copies of *Atomechanik* arrived in Copenhagen. Although Thomsen was presumably aware of Hinrichs's publications, he never referred to them in his own writings. And yet the similarity between Hinrichs' and Thomsen's formulae is remarkable.

Hinrichs was born 1836 in Holstein, which until the Danish-German war of 1864 was a duchy ruled by the kings of Denmark, and in 1853 he began studies at the University of Copenhagen with Forchhammer as his teacher in chemistry. In 1861 he emigrated to the United States, ending up two years later as Professor of Natural Philosophy, Chemistry and Modern Languages (!) at the University

---

7. For Hinrichs as a speculative atomist and precursor of the periodic system, see Zappffe (1969), van Spronsen (1969), pp. 116-125, and Scerri (2007), pp. 86-92. Hinrichs's analogies between astronomical and chemical phenomena were extreme and generally dismissed, but he was not the only chemist in the period which looked toward the heaven to understand the nature of terrestrial matter. To mention but one other example, so did the London chemist Henry Wilde in a booklet of 1892.

of Iowa. Hinrichs stayed in his new country until his death in 1923. He visited Europe a few times but not, to my knowledge, Denmark. According to his own statement, as early as 1855 he presented to Forchhammer his speculations on the unity of matter, suggesting that all elements were polymers of a primary substance. In his later writings he named the primary substance "pantogen" – that of which all is formed – and ascribed to it an atomic weight less than hydrogen's.

In a memoir of 1894 Hinrichs recalled that Forchhammer "most kindly and attentively listened to me." Although probably inaccurate and coloured by the passage of time, Hinrichs's little known recollection is of some historical importance. It deserves to be quoted:

> I confessed to him [Forchhammer], that ... while deeply interested in his lectures and experiments, I was mortified at the apparent impossibility of my ever *understanding* that science [chemistry], while physics and mathematics gave me no trouble whatever. He seemed much interested at my blunt confession and explained, that it was not possible to treat chemistry like physics or mathematics, because it was an experimental science." But it must be possible, it ought to be *made* possible," I rejoined – for which outburst I was rewarded by one of his most benign smiles, and an apparently greatly increased interest in myself. Less than a year after that conversation, I presented to him, for his consideration, a sketch of the first scientific contribution I ever made. It had for a basis the *hypothesis* of a single primitive substance, all atoms of which, therefore, were exactly equal; and I made the attempt to deduce the varied properties of the different chemical elements from the differences in *weight and form* of their complex atoms, built up from the equal atoms of the primitive substance. This was in February, 1855.[8]

Given that Thomsen served as Forchhammer's assistant at the time, it is hard to believe that Hinrichs and Thomsen never met person-

---

8. Hinrichs (1894), p. 54. Hinrichs was pleased to note that Forchhammer "was not by birth a Dane, for he was born in my own native province, Schleswig-Holstein." All the same, Forchammer as well as Hinrichs were born as Danish citizens.

ally. Nevertheless, this is what Hinrichs explicitly stated in his memoir. If Thomsen was in some way influenced by Hinrichs' ideas, it would have been indirectly and probably through Forchhammer.

In December 1868 Hinrichs addressed Steenstrup, secretary of the Royal Danish Academy, with a request of publishing his theory of *Atomechanik* in a Danish version in one of the Academy's periodicals. Should the Academy not want to publish the work, he wrote, could Steenstrup please ask Julius Thomsen if it could be brought in *Tidsskrift for Physik og Chemie*? "If the treatise is too large Thomsen is welcome to bring only parts of it." Steenstrup discussed the request with Thomsen, who was unwilling to include Hinrichs as an author in his and his brother's journal. In a long letter of February 1869 Steenstrup politely told the Danish-American natural philosopher that there was no possibility of a Danish publication.[9] Hinrichs never forgot the country of his youth. As late as 1892, congratulating the Royal Danish Academy of Sciences with its 150-year anniversary, he sent a copy of the most recent update of *Atomechanik* to the Academy.[10]

While Hinrichs was led to his speculations concerning the complexity of atoms by considerations in a Pythagorean spirit that included a mixture of astronomy, spectroscopy and atomic weights, there is little doubt that the main source of Thomsen's somewhat similar ideas was thermochemistry and its promise of revealing the nature of chemical affinity. Not content with numerology Thomsen discussed in his 1865 paper the possibility of testing the hypothesis of composite elements experimentally. He considered extreme heating, electrical dissociation, and chemical processes, but only to conclude that these methods were insufficient as they only affected the arrangement of atoms as held together by the ordinary force of affinity. The fact that present experimental techniques were unable to produce atomic decomposition indicated that "the hypothetical constituents of the elements must be bound together by a strong affinity," much stronger than the chemical one operating in molecules

9. Hinrichs to Steenstrup, 2 December 1868; Steenstrup to Thomsen, 29 January 1869; Steenstrup to Hinrichs, 12 February 1869. All letters in Royal Danish Academy, Main Archive 1868-1872.
10. Letter of 16 October 1892. Royal Danish Academy, Main Archive 1892.

and other compounds. As an alternative he suggested, vaguely and speculatively, that the action of light over long periods of time might conceivably "transform the atoms from their unstable equilibrium positions to another stable equilibrium position, with the result that a new substance with new properties would be formed."[11]

In the latter part of the nineteenth century, the idea that elementary atoms would decompose if only exposed to forces of extraordinary strength, was not uncommon. It lay behind the interpretation of spectra, as suggested by William Crookes and Joseph Norman Lockyer in Britain, and also behind Crookes' view of cathode rays as a "fourth state of matter." In 1879 the amateur astronomer Lockyer even claimed that he had succeeded in decomposing sodium and other elements into hydrogen by means of strong electric discharges.[12] Lockyer soon withdrew his claim, but not his belief that the elements might be decomposed into smaller parts. This belief was shared by the German chemist Victor Meyer, a professor at the University of Heidelberg. Meyer and his pupils undertook a large-scale research programme in pyrometry in which they succeeded in breaking down some simple molecules into atoms at temperatures approaching 1700 °C. For example, at this temperature molecular iodine decomposed into atoms according to

$$\cdot \quad I_2 \rightarrow 2\,I$$

It was a clearly expressed goal of the Heidelberg programme to demonstrate the decomposition of atoms into their smaller parts at even higher temperatures. Meyer did not consider this a utopian perspective and was convinced that the complexity of atoms was a legitimate research area for experimental chemistry. In an address of 1895, he said: "The complex nature of the elements, though unproved at the present time, must today be counted as a well-founded hypothesis which we are justified to choose as starting point for further research."[13] Thomsen, thirty years earlier, much agreed.

---

11. Thomsen (1865a), p. 114.
12. See Brock (1985), pp. 180-194.
13. Meyer (1895), p. 110, who referred to "Professor Thomsen's recent calculations

7.2. On the unity of matter

Thomsen only resumed his atomic speculations after he had completed his systematic series of thermochemical investigations during the period 1865-1882. In a University Address from 1884 he praised the atomic-molecular theory as the true foundation of chemistry and the cause for the amazing progress this science had experienced in recent time. Since we cannot observe the motion of the smallest particles, he said, we have to proceed indirectly and infer the laws of the micro-cosmos from macroscopic phenomena and experiments. This was a recurrent theme in Thomsen's work. More than twenty years later he spelled it out in the Danish summary version of *Untersuchungen*:

> An almost impenetrable darkness hides from us the inner structure of molecules and the true nature of atoms. We know only the relative number of atoms within the molecule, their mass, and the existence of certain groups of atoms or radicles in the molecule, but with regard to the forces acting within the molecules and causing their formations or destruction our knowledge is still exceedingly limited.[14]

In his address of 1884 Thomsen emphasised that the scientist should aim at a mechanical chemistry based on knowledge of atomic and molecular forces and treatable by means of the general methods of physics. However, as these forces were still unknown, for the time being one had to do with hypotheses and analogies. While planets moved in accordance with Newton's simple law of gravity, the law that governed the motion of atoms must be more complicated but not, he thought, of an entirely different kind.

Thomsen speculated that Newton's law was only a special case of a more general force law that asymptotically passed into the inverse-square law for large distances. "It is possible that the complete form of the law contains terms with higher negative exponents than

on the deviation from the whole numbers which is exhibited in some precisely determined atomic weights" (p. 97).
14. Thomsen (1905b), p. 3.

the inverse square," he wrote. "The difference in mass would then be expressed by these terms." In the style of the eighteenth-century natural philosopher Roger Boscovich he may have had in mind a modified force law of the kind

$$F = \frac{A}{r^2} \pm \frac{B}{r^n} , \quad n > 2$$

Whereas the coefficient A would be a constant, B would depend on the atomic weight. Repeating an analogy popular at the time, Thomsen compared the state of chemistry with that of astronomy at the time of Kepler or Descartes: "Chemistry is at present in a stage which reminds one of that of astronomy in the period shortly before Newton proved how all the movements in the universe could be deduced from general attraction."[15]

Thomsen's allusion to Newtonian mechanics as potentially useful in chemistry was far from new. The dream of a "Newtonian chemistry" goes back to the late enlightenment period and became a respectable orthodoxy in the early nineteenth century. For example, it was part of Boscovich's dynamical atomic theory, and it was also expounded by leading French scientists such as Laplace and Berthollet.[16] As early as 1782, Lavoisier expressed his hope that "one day the precision of the data [of chemical affinity] might be brought to such a perfection that the mathematician in his study will be able to calculate any phenomenon of chemical combination in the same way, so to speak, as he calculates the movement of the celestial bodies."[17]

However, according to Immanuel Kant this hope would never become a reality. In his post-Newtonian *Metaphysische Anfangsgründe der Naturwissenschaft* dating from 1786, the German philosopher argued that chemistry could never be a genuine science because its subject matter was intractable to the methods of mathematics and systematic deduction from higher principles. The results of chemis-

---

15. Thomsen (1884), p. 4 and p. 26.

16. For aspects of the dream of a Newtonian chemistry, see Levere (1971), pp. 196-211 and Gregory (1984).

17. Quoted in Crosland (1978), p. 49.

try, he claimed, were contingent and did not follow by necessity from the fundamental laws of nature, such as did the results of the much admired mechanical physics. Kant's philosophically based verdict on the non-scientific nature of chemistry was this:

> I assert that in any special doctrine of nature there can be only as much *proper* science as there is *mathematics* therein. For, ... all proper natural science requires a pure part lying at the basis of the empirical part, and resting on a priori cognitions of natural things. Chemistry can be nothing more than a systematic art or experimental doctrine, but never a proper science, because its principles are merely empirical, and allow of no a priori presentation in intuition. Consequently, they do not in the least make the principles of chemical appearances conceivable with respect to their possibility, for they are not receptive to the application of mathematics.[18]

But the great philosopher from Königsberg failed to imagine the rapid progress in theoretical chemistry that occurred during the nineteenth century and accelerated in the first decades of the following century. Thomsen, for one, was convinced that his own work in thermochemistry was a step along the road that would lead to a complete understanding of chemical affinity on the basis of mechanical physics. He probably was unaware of Kant's verdict and in any case disagreed with it.

In his 1884 address Thomsen did not mention Mendeleev's periodic system, which at the time had been introduced by younger Danish chemists. On the other hand, he briefly alluded to the kind of evolutionary scenarios that he had earlier dealt with in a popular context. For example, he imagined that our elements might be low-temperature condensates derived from other and more primitive forms of matter that only existed in the hot stars. This was an old idea, vaguely suggested by Colding amongst other in the 1850s (Section 3.3). As an interesting possibility Thomsen mentioned that there might be elements hidden in chemical compounds that exist

---

18. Kant (2004), pp. 6-7. It is worth pointing out that Thomsen's teacher H. C. Ørsted was greatly influenced by Kant's book.

in free form in the Sun. By 1884 evolution meant Darwinian evolution, a theme that Thomsen briefly alluded to in regard to the different affinities of the elements. "The atoms attract some other atoms more strongly than others; this shows that to some extent the right of the strongest is [also] valid in chemical processes," he said.[19] Like in his 1865 paper, Thomsen confirmed his belief that atoms were unlikely to be the most primitive form of matter. The "so-called atom" was neither simple nor indivisible:

> It is possible that the future will show the different atoms to be decomposable into smaller parts carrying common properties. Although all attempts to decompose the elements into different constituents have so far yielded a negative result – it is not unlikely that a discovery of new modes of action for the force will provide such means the range of which cannot at present form any reasonable idea. ... The decomposition of the elements into more elementary constituents is only a matter of time.

Thomsen, equally at home in pure and applied chemistry, not only presented chemistry as a fundamental science but also called attention to some of its recent and most amazing industrial applications. He referred to the progress in organic synthesis which had led to the manufacture of "the well-known dyestuffs in the indigo plant and madder ... and a series of so-called aniline dyes which now threaten to completely supplant the older dyes based on plants."[20] And this

---

19. Thomsen (1884), p. 12. He most likely had read Darwin's *Origin of the Species*, possibly in the Danish translation by the novelist J. P. Jacobsen which appeared 1871-1872, but we do not know when or how he got acquainted with Darwin's ideas. It should be pointed out that the quotation of Thomsen does not necessarily imply an influence from Darwinism. As early as 1850, years before the publication of *Origin*, Thomsen referred to "the right of the strongest" in the molecular world. It is perhaps significant that he spoke of "the strongest" and not "the fittest." See Thomsen (1850a), p. 11 and similarly Thomsen (1852), p. 157.

20. Thomsen (1884), p. 25. Alizarin was first synthesised in 1868 and a few years later manufactured on an industrial scale, leading to a sharp decrease in the production of dyes from madder. While Adolf von Baeyer and his collaborators succeeded to make synthetic indigo in the early 1880s, it was on laboratory scale only; by 1884, when Thomsen wrote his essay, no industrial indigo had yet been produced. That only

was only the beginning, for Thomsen stated, confidently and optimistically, that eventually chemists would be able to synthesise *all* organic substances from their elementary constituents. He saw no reason why the medically useful compounds quinine and morphine would not one day be produced in the laboratory. On the other hand, he stopped short at living organisms. Recall that in 1853 Thomsen had doubted if it ever would become possible to synthesise vegetable substances (Section 3.2).

Thomsen did not explicitly refer, either in his 1884 address or in earlier publications, to the periodic system of the elements which Dmitrii Mendeleev and Lothar Meyer had independently and largely simultaneously introduced in 1869. Yet he was almost certainly aware of the system at an early date. For example, in a critical response to one of Thomsen's papers, dealing with the constitution of iodic acid, Meyer referred in 1873 to his own and Mendeleev's periodic classifications of the elements.[21] Thomsen undoubtedly studied Meyer's paper carefully.

It took until 1887 before Thomsen referred to the periodic table or system. The first Danish chemist to review the chemical elements within the context of the periodic system was Odin T. Christensen in a paper in *Popular Expositions* of 1880. At the time an assistant at the laboratory of the Polytechnic College (and later a professor at the Agricultural College), Christensen expressed great confidence in Mendeleev's system because of its successful predictions of the metallic elements gallium, scandium, and germanium.[22] Four years later John Sebelien, a young polytechnic chemist, wrote an extensive prize essay on the history of atomic weight determinations in which he referred to the periodic system.[23] The subject of the prize

---

happened in the late 1890s, when BASF (Badische Anilin und Soda Fabrikation) started commercial production.

21. Meyer (1873); Thomsen (1873b).

22. Christensen (1880). For this and other Danish contributions to the periodic system and its local dissemination until about 1910, see Kragh (2015). Christensen further pointed out that the recent determination of beryllium's atomic weight to 9.1 (rather than approximately 14) was in agreement with Mendeleev's system.

23. Sebelien (1884b). Sebelien specialised in dairy chemistry and in 1889 he went to Norway, where he became professor of chemistry at the Norwegian Agricultural

Bilag: Grundstoffernes Atomvægte (se Side 21).

11	9	7	5	3	1	2	4	6	8	10	12
							V 51,1	Nb 93,7	Di¹) 142,1	Ta 182	
							Cr 52,45	Mo 95,9		Wo 184,0	U 239,8
							Mn 54,88				
					H 1		Fe 55,88	Ru 103,3		Os 192	
							Co 58,6	Rh 104,1		Ir 192,5	
							Ni 58,6	Pd 106,2	¹) Neodym 140,5 / Praseodym 143,5	Pt 194,3	
		Cs 132,7	Rb 85,2	K 39,03	Li 7,01	Na 22,99	Cu 63,18	Ag 107,7		Au 196,2	
	Er 166?	Ba 136,9	Sr 87,3	Ca 39,91	Be 9,08	Mg 23,94	Zn 64,88	Cd 111,7		Hg 199,8	
Tb 228	Yb 172	La 138,5	Y 89,6	Sc 43,97	B 10,9	Al 27,04	Ga 69,9	In 113,4		Tl 203,7	
Th 232		Ce 140,2	Zr 90,4	Ti 48,01	C 11,97	Si 28,0	Ger 72,3	Sn 117,4		Pb 206,4	
					N 14,01	P 30,96	As 74,9	Sb 119,6		Bi 207,5	
					O 15,96	S 31,98	Se 78,87	Te 125,0			
					F 19,06	Cl 35,37	Br 79,76	I 126,5			

Figure 7.1. Thomsen's first version of the periodic system. Source: Thomsen (1887a).

contest announced by the University of Copenhagen in 1882 was "A historical and critical exposition of attempts to determine the atomic weights of the elements."[24] The answers were evaluated by C. Holten, S. M. Jørgensen and Thomsen, who found Sebelien's essay worthy of a gold medal. Sebelien also discussed in some detail Prout's hypothesis, if only to conclude that it was probably "illusory." The following year yet another young Danish chemist, 23-year-old Rudolph Koefoed, published a review in *Tidsskrift* of what he called the periodic law and in which he referred to the works of Mendeleev and Meyer. Koefoed suggested that now the chemists were on their way to establishing their science on a principle nearly as universal and reliable as Newton's law of gravity was for the astronomers.[25]

In a pamphlet of 1887 written on the occasion of the king's birthday and entitled *Om Materiens Enhed* (On the Unity of Matter), Thom-

---

College.

24. *Københavns Universitets Aarbog 1882-1883*, pp. 197-198. One suspects that the subject of the prize contest was due to Thomsen.

25. Koefoed (1885), who paid particular attention to Meyer's system and the periodicity shown by the variation of the elements' atomic volumes.

sen finally addressed the periodic system and its relevance to the question of the complexity of chemical atoms with which he had dealt earlier. His two publications of 1884 and 1887 were published in Danish only and consequently not widely known outside Scandinavia. However, Hinrichs referred to them in his booklet of 1894 and so did the American chemist Francis Venable two years later.

Compared to his earlier writings, in his more bold account of 1887 Thomsen went further by not only suggesting that the elements were complex, but also that they consisted of various combinations of hydrogen atoms. Moreover, he discussed the periodic system (without using the term, though). As far as Thomsen's version of the periodic system is concerned – he judged it to be "rather satisfactory" – it did not differ substantially from Mendeleev's version (Figure 7.1). Open places in Thomsen's system indicated two unknown elements "with atomic weights approximately 100 and 188, and with properties analogous to manganese." The prediction of two manganese-like elements was not original, as Mendeleev and other chemists had already made the same observation.[26]

Thomsen found the idea that the atoms of all elements consist of hydrogen atoms to be highly appealing. The idea is known as Prout's hypothesis, a name referring to the British physician and chemist William Prout who suggested it as early as 1815.[27] From the hypothesis follows a testable relation for the atomic weight of any element A_X, namely that it has to be a whole multiple of hydrogen's weight:

$$A_x = nA_H \quad \text{with } n \text{ being an integer in the range 1 to ca. 240.}$$

Prout's hypothesis or "law" in its original form became increasingly difficult to defend as atomic weight determinations showed that it was not supported by experiment. Berzelius, for one, rejected it for

---

26. Mendeleev estimated the atomic weights of the two unknown metals to be 100 and 190, see Scerri (2007). Technetium with atomic weight 99 and rhenium with weight 186 were discovered in 1940 and 1925, respectively, the first synthesised in nuclear reactions.

27. See the detailed account in Brock (1985) and also Hamerla (2003). An older but still useful survey is given in Freund (1904), pp. 593-624.

this reason. With the very precise atomic weight determinations that the Belgian chemist Jean Servais Stas published in 1860, Prout's law greatly lost in credibility. Stas's measurements proved beyond any reasonable doubt that the atomic weights of the elements were not in general multiples of that of hydrogen, and that consequently Prout's law was incorrect. This conclusion was generally accepted.

However, it was easily possible to modify the Proutean idea of atomic unity of matter into versions that were not refuted, or even not refutable, by experiment. These possibilities were eagerly explored, and much ingenuity was applied in saving the philosophically attractive idea. After all, atomic weights of most elements did approach whole numbers so closely that it could scarcely be accidental; many physicists and chemists in the Victorian era felt it natural to conclude that there must be some underlying regularity in the structure of the elements, some modification of Prout's hypothesis. Julius Thomsen was one of them, and he was not the only Danish chemist who was fascinated by the hypothesis. Emil Petersen, who in 1901 was appointed professor at the University of Copenhagen, wrote in 1890 a popular article in which he supported the basic unity of matter and its relation to the periodic system.[28]

Although Thomsen supported the Proutean view of the unity of matter based on a common primitive constituent, he realised that Prout's law could not be maintained in its original form. To escape refutation of the law he mentioned two possibilities. First, the deviations of the atomic weights (based on H = 1) from whole numbers could have its cause in traces in the elements of some other, chemically related element. But this possibility he did not find entirely satisfactory. For many elements, and chlorine with its atomic weight of 35.5 in particular, it was hard to believe that their weights could be so distorted by pollution with other elements. No, Thomsen favoured another alternative, namely that "the atoms of our so-called elements are generated by combination of the uniform, minimal atoms of a primeval substance," this substance being lighter than hydrogen. A conjecture of this kind had been proposed by

28. Petersen (1890), which in spirit and content was closely similar to Thomsen's publications from the period. See also Kragh (2015).

Prout as early as 1831 and in a paper of 1873 F. W. Clarke argued that the primeval substance might have an atomic weight equal to one-half of hydrogen's. In that case Prout's law would be revised to

$$A_X = \frac{1}{2}nA_H \qquad \text{with } n \text{ being an integer in the range 2 to ca. 480.}$$

In this form the law would fit with the half-integral value of chlorine's atomic weight.

Thomsen thought that the primeval atoms were perhaps identical to those of the element "helium" inferred from solar spectroscopy. As he argued, the more simple an element is, the fewer lines there are in its emission spectrum. This assumption he saw confirmed in the case of hydrogen "whose atomic weight is the lowest and whose spectrum is so simple that it consists of only three lines." In fact, in the mid-1880s five emission lines of hydrogen had been detected but apparently without Thomsen being aware of it.[29] Since helium's spectrum consisted of just a single line, Thomsen continued his line of reasoning: "This hypothetical substance, the atomic weight of which must be assumed to be smaller than that of hydrogen because of the simplicity of its spectrum, has been named helium, and in this substance we might have matter in its primeval state from which our atoms and molecules have evolved by cooling and condensations."[30] The solar spectral line due to helium was first observed by Lockyer in 1868 but at the time without assigning it to a new element (Figure 7.7). The name "helium" first appeared in print in 1871. Thomsen was not the first chemist to speculate that helium's atomic weight was 0.5, for this is what Clarke suggested in his paper more than a decade earlier.[31]

---

29. A. J. Ångström identified three hydrogen lines in 1862 and by 1885 two more were detected. In the same year Johann Balmer proposed his famous relation which implied many more hydrogen lines outside the visible spectrum. Thomsen seems not to have followed the development in spectroscopy closely.

30. Thomsen (1887a), p. 32. More about helium and the inert gases follows in Section 7.6.

31. See Kragh (2009), p. 170 and Nath (2013), p. 210. It is unlikely that Thomsen knew of Clarke's paper published in *Popular Science Monthly*.

Why, Thomsen asked, do the elements only occur with certain atomic weights while others are missing, perhaps even prohibited? Why do the elements in the first period have atomic weights 7, 9, 11, 12, 16 and 19, while there are none with weights 8, 10, 13 and so forth? Amplifying on his comments of 1884, Thomsen formulated the question in quasi-Darwinian terms. With regard to the question of the origin of the elements, the chemist was "in a position similar to the one of the biologist with regard to the constancy of the species," he wrote. As Darwin and other biologists had arrived at the evolution theory without being able to conduct experiments on the evolution of species, so the chemist had to rely on indirect arguments in the case of inorganic evolution:

> Just as the biologist supposes that the right of the fittest has manifested itself in the evolution of the species … the chemist has shown that the atomic weights of the elements do not form a successive series of numbers. Many numbers are missing among the known elements, and he is tempted to seek the answer in the right of the fittest which has manifested itself and only allowed the formation of atoms of a structure firm enough for a continuous existence. … The biologist believes that evolution from one species to another occurs through a series of generations … and the chemist must presumably adopt a similar hypothesis if he is to suggest the mechanism by means of which the transformation or development of an element leads to another element.[32]

One may assume that the sources of Thomsen's quasi-Darwinian analogies did not only come from the local debate concerning Darwin's theory, but also from the international scientific literature. Since the early 1870s several chemists, physicists, astronomers and philosophers had suggested inorganic extensions of Darwinian evolution. The British-Australian physicist Morris Pell wrote in 1872 about an original "warfare among the molecules" governed by the principle of natural selection. According to Pell, the fittest of the molecules would conquer those who were too weak to survive in the

---

32. Thomsen (1887a), p. 37.

struggle for existence.[33] Other scientists who explicitly invoked Darwin's theory in more than just a metaphorical sense included the British-German physiologist William Preyer and the Austrian physical chemist Leopold Pfaundler, professor at the University of Innsbruck. The latter worked in areas close to Thomsen's, such as chemical dissociation, thermochemistry and equilibrium theory, and in 1876 he concluded that "Darwin's principles are valid also in the molecular world."[34] According to Pfaundler, equilibrium processes could be understood in analogy with Darwin's doctrine of the survival of the fittest in the struggle for existence.

Of more direct inspiration for Thomsen's 1887 essay was probably a much-discussed address which the prominent British chemist William Crookes gave to the British Association for the Advancement of Science in 1886. As indicated by the many similarities between the two works, Thomsen had almost certainly read the address. According to Crookes, it was natural to view "existing elements not as primordial but as the gradual outcome of a process of development, possibly even of a 'struggle for existence'."[35] Crookes further referred to the still hypothetical helium as a possibly a simpler element than hydrogen and "with atomic weight half that of hydrogen, [as] required by Mr. Clarke as the basis of Prout's law." Thomsen may have got his idea from the British chemist.

Compared to Crookes's address there was little new in Thomsen's 1887 essay except that the Danish chemist suggested that the "fitness" of the known elements might be associated with the degree of symmetry exhibited by the minimal constituents of the composite atoms. By arranging hypothetical material particles in symmetric patterns, he found the patterns suggestive of an explanation of the known atomic weights and to an explanation of the periodic table (Figure 7.2). In these atomic models, as they may be regarded, he imagined the material parts to be arranged in space, rotating around

33. Pell (1872), p. 185. See Kragh (2009b) for a survey of inorganic Darwinism.

34. Pfaundler's Darwin-inspired chemistry is described in Snelders (1977), which includes references to the literature.

35. Crookes (1886), p. 561, and Brock (1985), pp. 195-197. Whereas Thomsen (1887a) did not refer to Crookes' ideas, Petersen (1890) did. On the other hand, Petersen did not mention helium.

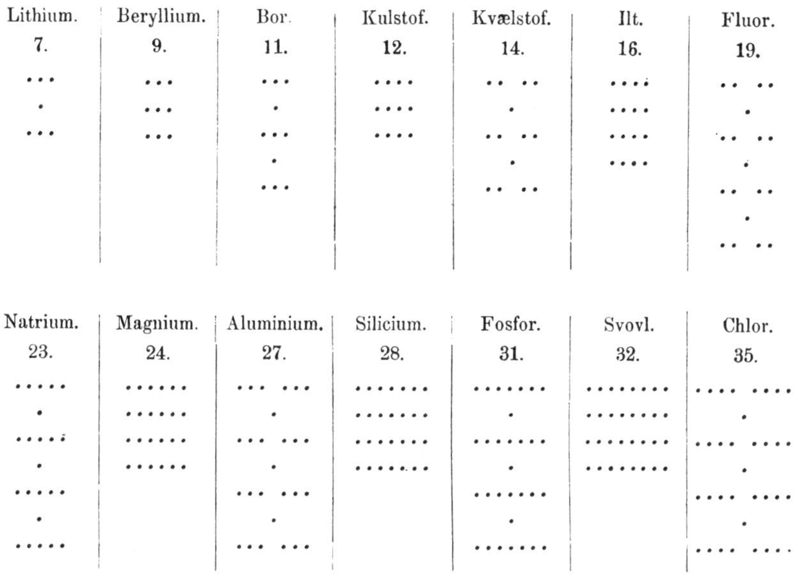

Lithium. 7.	Beryllium. 9.	Bor. 11.	Kulstof. 12.	Kvælstof. 14.	Ilt. 16.	Fluor. 19.
• • •	• • •	• • •	• • • •	• •　• •	• • • •	• •　• •
•	• •	•	• • • •	•	• • • •	•
• • •	• • •	• • •	• • • •	• •　• •	• • • •	• •　• •
		•		•	• • • •	•
		• • •		• •　• •		• •　• •
						•
						• •　• •

Natrium. 23.	Magnium. 24.	Aluminium. 27.	Silicium. 28.	Fosfor. 31.	Svovl. 32.	Chlor. 35.
• • • • •	• • • • • • •	• • •　• • •	• • • • • • • •	• • • • • • • •	• • • • • • • • •	• • • •　• • • •
•	• • • • • • •	•	• • • • • • •	•	• • • • • • • • •	•
• • • • •	• • • • • • •	• • •　• • •	• • • • • • •	• • • • • • • •	• • • • • • • • •	• • • •　• • • •
•	• • • • • • •	•	• • • • • • •	•	• • • • • • • • •	•
• • • • •		• • •　• • •		• • • • • • • •		• • • •　• • • •
•		•		•		•
• • • • •		• • •　• • •		• • • • • • • •		• • • •　• • • •

Figure 7.2. Thomsen's schematic illustration of the possible constitution of the atoms of chemical elements. Source: Thomsen (1887a).

an axis. He emphasised, however, that his pictorial atoms "should not be regarded as an expression of the true constitution, but merely as a vague indication that the cause of the elements having exactly the observed atomic weights might be sought in the symmetric structure of the so-called atoms."[36]

7.3. Prout's hypothesis revived

In the speculations mentioned so far, Thomsen had not directly faced the weighty arguments which Stas and others had launched against Prout's hypothesis. This he did in 1894, in a memoir which was published by the Royal Academy of Sciences in Danish language and also, with a slightly extended text, in French. It was read to the Academy on 14 December 1894. In this memoir he expressed as his hope that he would "succeed in breathing new life into this

---

36. Thomsen (1887a), p. 29.

hypothesis [Prout's] which has been several times sentenced to death."[37] His elaborate attempt at revival consisted in considering, not the atomic weights themselves, but their deviations from integral values. As Thomsen's work can only be understood within the discussions which in the late nineteenth century took place as to a possible reconciliation of Prout's hypothesis with experimental data, a brief survey of this discussion will be useful.

The existence of an all-pervading world ether was generally accepted in the late nineteenth century, when it served as the very foundation of physicists' theories of electrodynamics and optics. Although the concept of the "luminiferous" ether was primarily of interest to the physicists it also played a role in the way that many chemists thought about the secrets of matter. Thomsen was one of the chemists who took the ether seriously. According to the "ether condensation hypothesis," the deviations from integral atomic weights might be due to amounts of the ethereal substance which condensed upon the surface of material atoms. As early as 1872, Lothar Meyer imagined that "apart from the particles of this primary matter [hydrogen] there may be included in the constitution of the atom larger or smaller amounts of that substance, perhaps not being completely weightless, which fills up the universe and we use to call the light ether."[38] He repeated the suggestion in later editions of his influential textbook *Die Modernen Theorie der Chemie*. Several other chemists made use of the ether for similar purposes, including Isidor Traube and Julius Quaglio in Germany, Thomas Carnelley in Britain, and William Livermore in the United States.

The ether condensation hypothesis was tested experimentally in the early 1890s based on the reasonable assumption that the quantity of condensed ether would change during a process between two different substances. In other words, an exact determination of weights before and after a simple chemical reaction, such as

$$Hg + Br_2 \rightarrow HgBr_2,$$

---

37. Thomsen (1894a), p. 321. The French version was Thomsen (1894b).
38. Meyer (1872), p. 293. For chemical uses of the ether at the end of the nineteenth century, see Kragh (1989).

might provide an answer. D. Kreichgauer in 1891, Hans Heinrich Landolt in 1893 and Adolf Heydweiller in 1894 performed weight determinations of unsurpassed accuracy only to conclude that the relative change of weight was less than 2×10^{-8} and thus unable to account for the deviations from whole numbers. Landolt's conclusion that "the last resource which was still open for Prout's hypothesis, is now blocked," was generally accepted.[39] However, although admitting the accuracy of the measurements of the German chemists, Thomsen was not ready to accept Landolt's conclusion. Perhaps a change in weight would turn up in processes where only one sort of atom was involved?

In careful experiments conducted over several months Thomsen investigated the allotropic and exothermic transformation of white into red phosphorus, which can be written as

$$P_4 \rightarrow 4\,P + energy$$

Contrary to what should be expected from the ether condensation hypothesis, the result was that "loss of potential energy and its transformation into heat ... does not affect the weight of the body."[40] Although Thomsen was now forced to abandon the ether condensation hypothesis, his belief in Prout's law remained undiminished. The solution, he thought, might lie in a reconsideration of the atomic weights.

During the last two decades of the nineteenth century a dispute went on concerning the proper unit for atomic weights – basically H = 1 versus O = 16 – and how the choice of unit would affect the evaluation of Prout's hypothesis.[41] A central issue in the dispute was the experimentally determined atomic weight ratio between oxygen and hydrogen. Did the ratio deviate significantly from 16? Attempts to answer the question had one very important if completely unex-

39. Landolt (1893), p. 34. For references to the experiments, see Ehrenfeld (1906) and Kragh (1982).

40. Thomsen (1894a), p. 312. Thomsen never published his data, apparently considering the experiments to be a failure and thus of no interest.

41. See, for example, Meyer and Seubert (1885). Hinrichs (1894) was part of the dispute.

pected consequence: It was Lord Rayleigh's measurements of 1888 (with the result $O : H = 15.880$) that became the starting point of the process that six years later led to his and William Ramsay's momentous discovery of argon.[42]

Thomsen had for long taken an interest in the issue, if not always in relation to Prout's hypothesis. As early as 1870 he found on the basis of Avogadro's law that the $O : H$ ratio was somewhat smaller than 16.[43] By means of even more precise atomic weight determinations Thomsen hoped to prove that the atomic weight of oxygen relative to hydrogen was 16 exactly (Figure 7.3). In late 1893 he performed very precise measurements of the $O : H$ ratio by means of a new indirect method based on the process

$$NH_3 + HCl \rightarrow NH_4Cl$$

Denoting the ratio between the molecular weights of NH_3 and HCl by x, it follows from mass conservation that

$$A_H = \frac{xA_{Cl} - A_N}{3 - x}$$

Thomsen found $x = 0.467433$ and in this way a value for $O : H$ very close to 16. "The result of this investigation shows without any doubt," he concluded, "that the atomic weight of hydrogen relative to that of oxygen is as close to 1:16 as can be expected if the unavoidable uncertainties of the atomic weights [of N and Cl] are taken into account."[44] However, Thomsen was able to reach this conclusion only by using atomic weights for N and Cl that fitted his purpose but were not accepted by the majority of chemists.

---

42. See Rayleigh (1892) and also Scott (1896), p. 209. Argon and its consequences are considered in Section 7.6.

43. Thomsen (1870a).

44. Thomsen (1894c), p. 406. Thomsen's conclusion was criticized by Meyer and Seubert (1894), who thought it was unfounded.

Fig. 3.

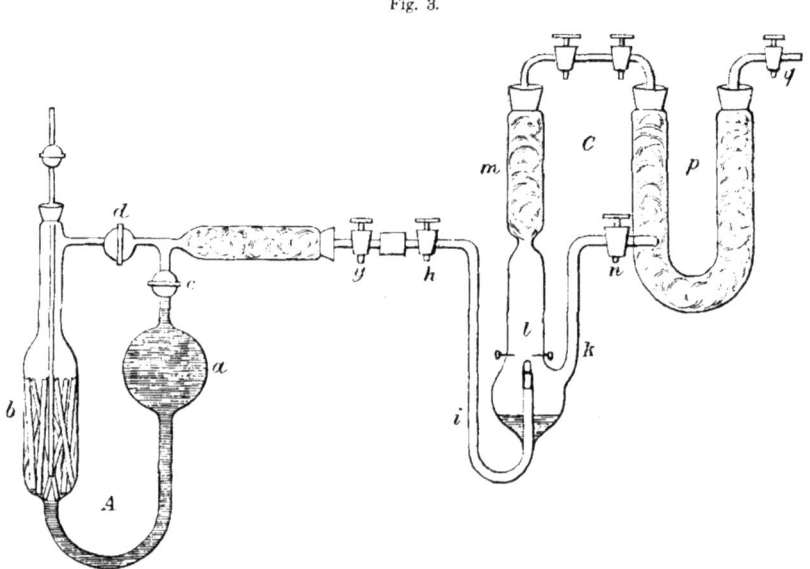

$^1/_5$ natürlicher Größe.

Figure 7.3. Part of Thomsen's experiments to measure the atomic weight ratio of oxygen and hydrogen. Source: Thomsen (1895a).

His further efforts did not result in what he had hoped for.[45] Works between 1894 and 1896 gave a value less than 16 (in terms of H = 1) and in good agreement with the O : H ratio found by other chemists. At about the same time the American chemist Edward Morley, at the Case Western Reserve University, completed his own and no less meticulous study on the subject. He arrived at a result that differed only slightly from Thomsen's:[46]

---

45. Thomsen (1895a); Thomsen (1896). The two papers were dated 24 September and 19 November 1895, respectively.

46. See Clarke (1896), pp. 197-201, who also referred to Thomsen's value. Details are given in Hamerla (2003). Morley is perhaps best known for his collaboration with Albert Michelson in the famous Michelson-Morley experiment aimed at determining the motion of the Earth relative to the ether. This experiment is often, if not quite correctly, seen as the origin of Einstein's theory of relativity.

Morley $O : H = 15.879 \pm 0.00032$

Thomsen $O : H = 15.8690 \pm 0.0022$

Comparing the two sets of measurements, William Noyes at the Rose Polytechnic Institute in Indiana admitted that Thomsen was "an experimenter of unusual ability" but nonetheless argued that Morley's value was definitive.[47] The question concerned more than just experimental accuracy and minutiae, for behind it lurked the ghost of Prout. As Morley said in 1896, referring to the legendary bed of Procrustes:

> This fable was really a prophetic vision; the bed is Prout's hypothesis; our friends who admire it want to stretch the most unyielding quantities, and to lop off numbers which have been determined with the greatest precision. The chapter has come to an end. Prout's hypothesis cannot be proved by experiment.[48]

While Morley's colleague in Copenhagen may have agreed by 1896 that Prout's hypothesis was untenable, two years earlier he was not so sure.

Thomsen's new idea was to base the atomic weights on a "rational" scale, instead of the one in which the weight of oxygen was arbitrarily fixed to 16. The new scale was supposed to show the regularity of the deviations from whole numbers which was not exposed by the $O = 16$ scale. Fixing the weight of oxygen to $16(1 + q)$ instead of 16, the increase for an element of atomic weight m would then be mq. Thomsen asked if there exists a value of q for which the deviations from whole numbers show a remarkable regularity, and found that with $q = 0.00075$ the wanted regularity turned up. For this value the atomic weight of oxygen becomes

$$A_O = 16(1 + 0.00075) = 16.0122$$

---

47. Noyes (1896). The British chemist Alexander Scott, at the time a demonstrator in Cambridge, agreed. He saw Thomsen's new results, if taken together with those of Morley, as proof that Prout's law in its original form was wrong. See Scott (1896).
48. Quoted in Hamerla (2003), p. 371.

"I then tried," Thomsen wrote, "to find the result, all the mentioned atomic weights being ... expressed in terms of an atomic weight of 16.0122; much to my surprise, it turned out that the deviations from whole numbers for all atomic weights had to be multiples of approximately 0.0120."[49] With Z and n being integers, Thomsen's formula was

$$M = Z \pm n \times a, \quad \text{with } a = 0.012$$

He found that if the rational atomic weights written in this manner were divided by 1.0075, and thus transformed to the standard O = 16 scale, the resulting atomic weights were very close to those found experimentally. Table 5 gives Thomsen's data and his recalculated atomic weights compared to Stas's values.

Element	Rational atomic weights ($a = 0.012$)	Empirical atomic weights	Stas's values (O = 16)	Thomsen's values (O = 16)
Ag	$108 + a = 108.012$	107.930	107.930	107.9299
Cl	$35.5 - 2a = 35.476$	35.449	35.457	35.4494
Br	$80 + a = 80.012$	79.951	79.952	79.9510
I	$127 - 4a = 126.952$	126.856	126.850	126.8556
S	$32 + 7a = 32.084$	32.060	32.0742	32.0606
Pb	$207 + 5a = 207.060$	206.904	206.034	206.9042
K	$39 + 15a = 39.180$	39.150	39.1425	39.1507
Na	$23 + 6a = 23.072$	23.055	23.0455	23.0543
Li	$7 + 3a = 7.036$	7.031	7.022	7.0307
N	$14 + 4a = 14.048$	14.038	14.055	14.0396

Table 5. Thomsen's atomic weights for select elements.

According to Thomsen, the formula $M = Z \pm n \times a$ was not fortuitous, but exposed the regularity of the inner structure of the atoms. The

---

49. Thomsen (1894a), p. 317. See also Thomsen (1894d) for his recalculations of Stas' atomic weights. His attempt to rescue Prout's hypothesis was mentioned in Ehrenfeld (1906) together with other similar attempts.

atoms of the elements, he wrote, "have evolved out of combinations of particles of a common basic substance; but to the masses of the atoms something unknown is also joined, an amount of energy or something which materially influences the chemical character of the atoms and their apparent ways in such a way that the empirical atomic weights do not exactly express the real masses of the atoms." As regards the nature of the mysterious "something unknown" Thomsen had to fall back on pure imagination, possibly inspired by the no less imaginative ideas of contemporary chemists and physicists.[50]

As an alternative to the ether condensation hypothesis Thomsen suggested that the unknown quasi-substance might be due to electrical forces. If one assumes that the primitive atoms are inseparably associated with a quantum of electrical charge, positive or negative, then the globe would be a huge, charged body, attracting or repelling the individual atoms. And then the weight of a chemical substance found empirically, would be the result of two forces, a gravitational and an electrical one, acting in the same or in the opposite direction:

> In the first case the apparent weight of the substance would be higher than the true mass of the substance, and in the second it would be lower. If we imagine the atoms to be charged with positive or negative electricity, then the first mentioned atoms would appear with a higher weight, the last mentioned with a lower weight ... than the value which corresponds to the true mass of the atom and which finds its expression in the whole numbers.[51]

Clearly, Thomsen's attempt to give a physical explanation of the observed atomic weights in agreement with Prout's hypothesis was

---

50. According to one version of the so-called perturbation hypothesis, inter-atomic forces might act as perturbing forces masquerading as a mass; as a result, the weight of a compound atom would not be exactly the sum of the weights of its elementary atoms. Somewhat similar ideas were carried over in the new atomic and relativistic physics of the early twentieth century in which Einstein's energy-mass equivalence as given by $E = mc^2$ was taken into account. See Farrar (1965) and Siegel (1978).

51. Thomsen (1894a), p. 323.

vague and speculative. Hinrichs criticized it, not unreasonably, for being "a sort of reduction ad absurdum of the whole system of Stasian atomic weights."[52] Thomsen may have received inspiration from the classical theory of electrons and the idea of an electromagnetic mass which at that time – even before experiments revealed the reality of the electron – was much discussed by physicists. However, this is nothing but a conjecture. Thomsen never referred to the physicists' discussions and also did not refer to the electron discovered by J. J. Thomson and others in 1897.[53]

That Thomsen might nevertheless have been aware of these discussions is suggested by a pamphlet published in 1897 by Peer Sophus Wedell-Wedellsborg, a young Danish nobleman. Wedell-Wedellsborg was trained in mathematics and defended in 1894 a doctoral dissertation on the three-body problem in celestial dynamics, but his real interest seems to have been foundational theories of physics and chemistry. Inspired by Thomsen's ideas of the relative weights of atoms – and after having discussed the subject with him – Wedell-Wedellsborg proposed a new unitary theory of matter, energy and electromagnetism, indeed of everything.[54] Referring to recent work by Ostwald, J. J. Thomson, John Poynting and others he sketched a speculative theory that combined chemical forces with gravitation, light pressure and much more. As to the interatomic force between two atoms of mass m_1 and m_2 he suggested that it was due to a gravitational potential of a form somewhat similar to what Thomsen had referred to in 1884, namely

$$U = G \frac{m_1 m_2}{r} + \sum_{n=1}^{\infty} \frac{a_n}{r^{2n+1}}$$

Describing atoms in terms of electro-gravitational forces, he stated that "each atom permeates the entire universe." It is unknown if

52. Hinrichs (1894), p. 60.

53. Thomson was a neo-Proutean no less than Thomsen was. His discovery of the electron and first atomic models based on the elementary particle led him to sketch an explanation of the periodic system. In this work he was much inspired by the chemists' view of a common primitive substance. See, for example, Kragh (2002).

54. Wedell-Wedellsborg (1897). The author thanked Thomsen for conversations with him and for having inspired him to take up his work.

Thomsen took Wedell-Wedellsborg's over-ambitious speculations seriously or that he encouraged him to develop them. The author of the pamphlet published a few papers on theoretical physics in recognised journals such as *Annalen der Physik* and *Zeitschrift für physikalische Chemie* but then seems to have left science.[55] Thomsen never referred to him.

7.4. Periodic systems

As mentioned, the establishment of the periodic system of the elements gave much impetus to the ideas of complex atoms and the unity of matter. "The periodic law, together with the revelations of spectrum analysis," Mendeleev said in 1889, "have contributed to again revive an old but remarkably long-lived hope – that of discovering ... the primary matter."[56] Lothar Meyer, one of the originators of the system, was convinced that the regularity of the atomic weights was the key to understanding the unity of matter and the atoms as consisting of compounds of even smaller bodies. Dozens of other chemists agreed that there was a profound connection between the new periodic system and the constitution of atoms, and yet the belief was far from generally accepted.[57] Many chemists and physicists of a positivist inclination considered explanations of the periodic system in terms of atoms or subatomic entities as pure metaphysics, threatening the entire scientific character of chemistry. Although the periodic system was recognised to be a useful means of classification and one of considerable predictive power, they felt that atomistic, evolutionary or numerological speculations were not only premature but fundamentally wrong. Such speculations constituted a reactionary movement, a return to the non-scientific thinking of the old alchemy.

Positivists and phenomenalists could draw on no less an author-

---

55. These papers were on the theory of terrestrial magnetism and on Poynting's theorem in electromagnetism. P. S. Wedell-Wedellsborg was born in 1864 and died 1951.

56. Mendeleev (1889), p. 643.

57. Venable (1896) provides a comprehensive overview of the many periodic systems from 1869 to 1896 and attempts to interpret them theoretically.

Figure 7.4. D.I. Mendeleev, the founder of the periodic system.

ity than Mendeleev, according to whom neo-Proutean attempts in explaining the periodic system were not only "utopian" but also based on "prejudices." In his 1889 Faraday lecture – given the same year as he was elected a corresponding member of the Royal Danish Academy – he made his point clear:

> The periodic law, based as it is on the solid and wholesome ground of experimental research, has been evolved independently of any conception as to the nature of the elements; it does not in the least originate in the idea of a unique matter ... it affords no more indication of the unity of matter or of the compound character of the elements, than the law of Avogadro, or the law of specific heats, or even the conclusions of spectrum analysis.[58]

58. Mendeleev (1889), p. 647. The Faraday lecture is reprinted in Jensen (2002), pp. 162-188.

Mendeleev was not foreign to the idea of the periodic table some-how expressing the nature of the elements, but he thought that such ideas were premature and not worth engaging in at the present stage of scientific knowledge.

To the extent that Thomsen subscribed to a philosophy of sci-ence it can with some qualification be characterised as positivist and empiricist (but see Section 7.1). It was not far from the one defend-ed by Mendeleev in his Faraday lecture. And yet, as we have seen, he was at the same time a believer in the unity of matter and the composite nature of the chemical elements. These ideas predated the periodic system and it took him some time before he recognised in the system a reflection of the unity of matter that he had held for more general reasons. He first spelled out the connection in his 1887 paper, where he presented his own Mendeleev-like version of the periodic system. Eight years later he suggested a more original ver-sion of the system, an alternative to the standard version with hori-zontal periods and vertical groups (Figure 7.5).

In Thomsen's system of 1895, the groups were arranged horizon-tally and the periods vertically, which resulted in an "easily grasped relationship between the elements." He explained: "The lines con-necting the related elements proceed from hydrogen in two ways; on the one hand to the electropositive lithium and on the other to the electronegative fluorine. Between these members are grouped in the usual way the other members of the first row."[59] In a comment to a paper by the American scientist Matthew Carey Lea, Thomsen pointed out that the colours of the elementary ions varied with the atomic weight in a manner that appeared clearly in his new periodic table.[60] Thomsen's system was well known in the *fin de siècle* period, but it was only one system out of many and did not attract particu-lar attention. Walther Nernst referred to it in his widely read text-book, mentioning that Thomsen "lays stress chiefly on the electro-

---

59. Thomsen (1895b), p. 132. The memoir was read to the Royal Danish Academy on 22 March. It appeared in a slightly different German version as "Systematische Gruppierung der chemischen Elemente" in *Zeitschrift für anorganische Chemie* 9 (1895): 190-193, and was also translated into English in *Chemical News* 71 (1895): 89-91.
60. Thomsen (1895d).

positive and negative character of the elements, and brings this out more clearly in his arrangement."[61]

Thomsen's use of horizontal groups and vertical periods was not entirely original, as a somewhat similar system had been proposed by Thomas Bayley in 1882 and later also by Thomas Carnelley and Henry Bassett (in 1886 and 1892, respectively). However, at the time Thomsen was unaware of these earlier writers.[62] His new system gave a clear survey of the physical and chemical relations between the elements and was an improvement not least with regard to the problematic rare earth metals. It was immediately suggestive of an atomic or evolutionary explanation as it indicated a common structure among the elements based upon the unit of hydrogen. In Thomsen's published version of 1895, thorium and uranium were curiously omitted, possibly because he was uncertain of where to place the two elements. However, he suggested that the long 31-element row starting with caesium would probably be followed by an analogous row to which thorium and uranium belonged. In an unpublished version of 1898, Thomsen placed the two radioactive metals (as they were then known to be) at the end of the caesium row, but without any lines connecting them to other elements.[63]

One of the merits of Mendeleev's periodic system which impressed even its opponents, was its success in predicting new elements, as dramatically demonstrated with the discovery of gallium – or "eka-aluminium" – in 1875. On the basis of his 1895 system, Thomsen pointed to a missing element of atomic weight 180 or 181 just outside the rare earth group: "One sees from the table how the full lines pass from zirconium to two opposite directions, one to cerium with the atomic weight 140, the other to a still undetermined

---

61. Nernst (1904), p. 186.

62. For references to the tables of the three British scientists, see Kragh (1982) and van Spronsen (1969). In a letter to Thomsen of 27 September 1895 the American chemist Francis Venable pointed out the similarity between his system and those of Bayley and Carnelley (Royal Library, TSC). In his letter of response Thomsen wrote that "the work of these authors was entirely unknown to him." See Venable (1896), p. 272.

63. George von Hevesy found this version of the periodic system in archival material belonging to Thomsen. It is reproduced in Hevesy (1927), p. 5. I have been unable to locate the Thomsen *Nachlass* mentioned by Hevesy.

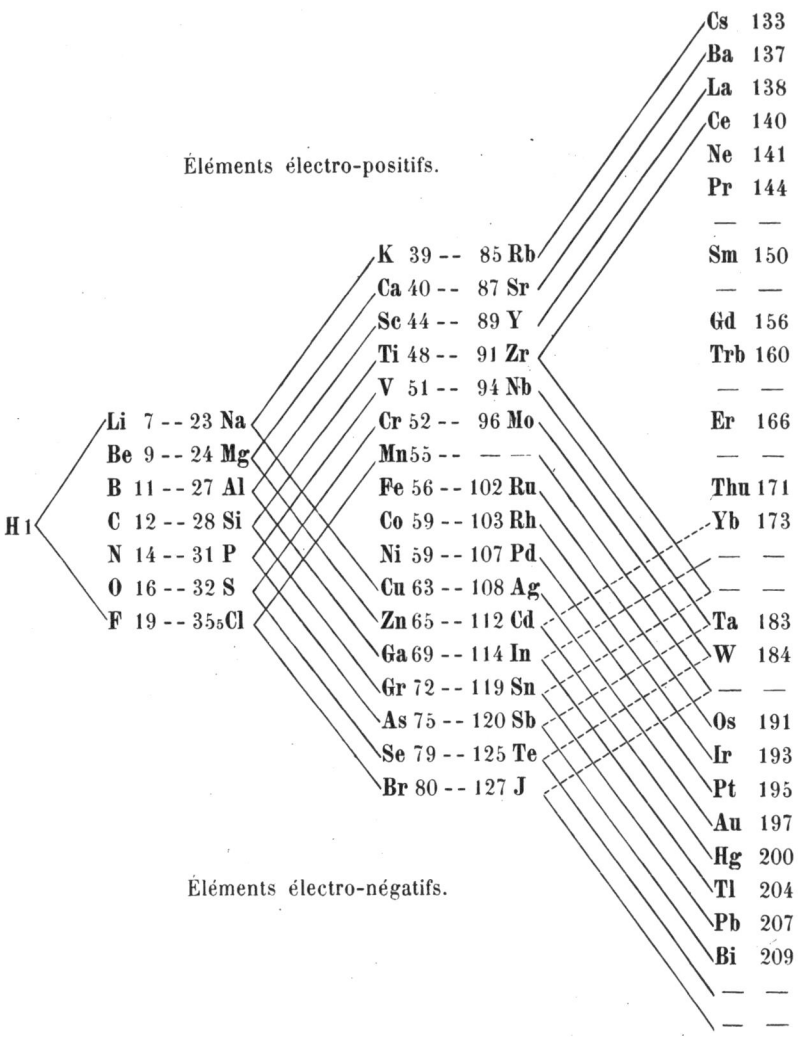

Figure 7.5. The periodic system as Thomsen presented it in Thomsen (1895b).

element the atomic weight of which should be 180. In between these two extremes there are a large number of rare-earth elements."[64] The element in question was later identified as hafnium. In 1895 the idea of a new element related to zirconium and titanium was not quite new and the supposed element had in fact been claimed at several occasions since the 1840s. Mendeleev, in his 1869 account of the periodic system, had a row of "Ti = 50, Zr = 90, ? = 180" and thus may be said to have predicted hafnium in about the same sense that he predicted gallium and germanium. Yet Thomsen was the first one who explicitly referred to the missing element and placed it correctly; that is, outside the group of the rare earths. As the distinguished Hungarian chemist and future Nobel Prize laureate George von Hevesy suggested many years later, he may even have undertaken a search for the new element in zirconium minerals.[65]

It was a novel feature of Thomsen's periodic table that the rare-earth elements did not bear any relationship with the elements of the preceding period from rubidium to iodine. The rare earths did not belong to any of the eight groups, but were fitted in between group IV and V as an intergroup.[66] The classification of the rare earth metals as a separate group was also present in Bassett's system from 1892, but while Bassett assumed the group to contain 18 metals, Thomsen was the first to define the rare earths as those lying between cerium and the unknown element of atomic weight approximately equal to 180. He must thus have counted 14 rare earth elements. When the missing element of A = ca. 180 was discovered in late 1922 by Hevesy and Dirk Coster, working in Copenhagen, the two discoverers did not fail to point out that it had been anticipated by Thomsen, professor in the same city.[67] Apart from the un-

---

64. Thomsen (1895b), p. 135. In the German version the element appeared with atomic weight 181.

65. Hevesy (1927), pp. 117-118. See also Hevesy and Thal Jantzen (1924).

66. See Thyssen and Binnemans (2011), pp. 67-71, where Thomsen's system is called "a particularly interesting classification."

67. Thomsen prepared the compounds K_2ZrF_6 and $NiZrF_6$, possibly with the purpose of examining them for content of the new element of A = 180. In later investigations of the compounds, provided by Thomsen's successor at the University of Copenhagen, Einar Biilman, Hevesy found a small amount of hafnium. See Hevesy

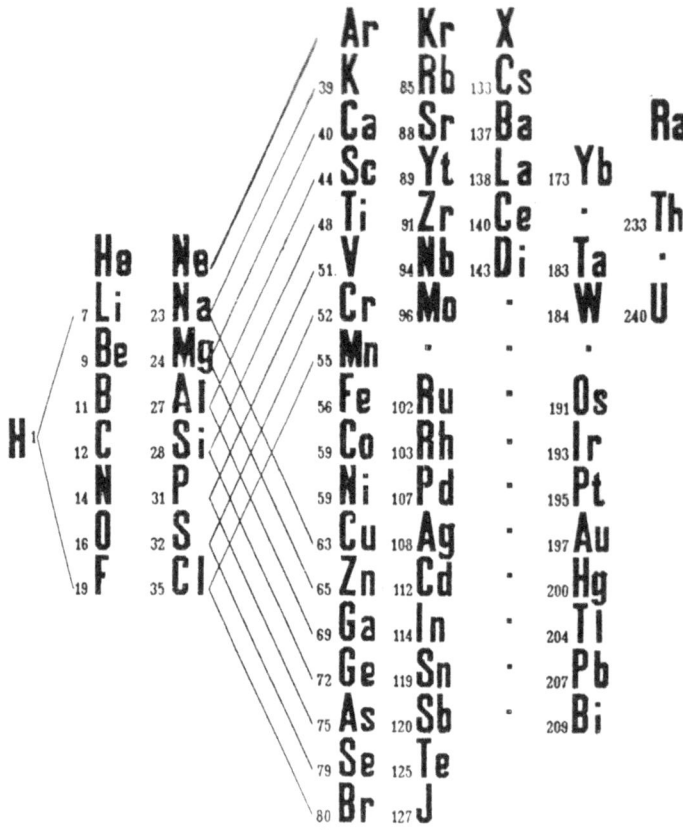

Figure 7.6. Thomsen's unpublished version of the periodic table, as he used it from about 1898 in his chemistry lectures at the Polytechnic College. Source: Kragh (1982).

known zirconium homologue, there was room for a couple of other missing elements in Thomsen's table. He left spaces vacant between praseodymium and samarium, terbium and erbium, and erbium and thulium, and two spaces after ytterbium. These vacant spaces were eventually filled with the elements promethium, europium, dysprosium, holmium, lutetium, and hafnium.

At the end of his brief 1895 paper on the classification of the elements, Thomsen pointed out what he called a "curious detail,"

and Thal Jantzen (1924).

namely that the number of elements in the periods amounted to 1, 7, 17 and 31. As he observed, these numbers can be written in the symmetric form

$$1$$
$$3 + 1 + 3$$
$$5 + 3 + 1 + 3 + 5$$
$$7 + 5 + 3 + 1 + 3 + 5 + 7$$

Row number n would thus have

$$N = 2n^2 - 1$$

as its sum. The way he commented on this numerological relationship differed in the two versions of his paper. "Is this relation more than a coincidence?" he asked in the French version, answering: "Only the future will show; but I have nevertheless wished to expose the possibility of a more profound cause." What this profound cause might be, he did not intimate, but he probably had in mind a symmetric arrangement of the proto-atoms of which he suspected the elements to be composed and which he had suggested in his 1887 paper. In his comment in the German version Thomsen pointed out that 1, 3, 5 and 7 were prime numbers, but said that this was probably coincidental. He did not refer to any cause, profound or not.

Thomsen may have had second thoughts concerning his version of the periodic table and the associated relationship between the numbers of elements in the vertical rows. In connection with his introductory lectures on inorganic and general chemistry he had a large table made which was placed in the chemical lecture room at the University of Copenhagen. It hung there for at least a decade. At some time after 1898 Thomsen completed the table by adding the newly discovered radium and also the inert gases from helium to xenon (Figure 7.6). This system differed from the published one in several respects. The long period of 31 elements had disappeared and was replaced by three new periods, the last of which now contained thorium and uranium, indicated as belonging to the Ti-Zr group and the Cr-Mo group, respectively. In the demonstration version the group of rare earths was broken and much of the suggestive symmetry lost. Had Thomsen grouped helium together with hydrogen and placed the other inert gases at the end of the periods, he would have arrived at

$$N = 2n^2$$

instead of $N = 2n^2 - 1$. But he never took this step.

7.5. Digressions: Thomsen, Mendeleev, Bohr

Not only was Thomsen thoroughly aquainted with the scientific work of Mendeleev, he also had met the famous Russian chemist when the two stayed in Uppsala in 1884.[68] Five years later, on 5 April 1889, Mendeleev was elected a foreign member of Royal Danish Academy, most likely on the initiative of Thomsen, who at the time served as president of the Academy. The letter of motivation was written by Thomsen and signed jointly by him and S. M. Jørgensen, the other leader of Danish chemistry. However, Jørgensen did not appreciate the periodic system. The system was absent from both the first and the second editions of his textbook on inorganic chemistry, published in 1888 and 1896, respectively, and it also did not appear in his widely read 1902 textbook on general chemistry.[69] One can reasonably assume that the proposal to elect Mendeleev was actually due to Thomsen. The main part of the motivation formally signed by the two chemists was this:

> During many years, Prof. Mendeleev has conducted a great number of excellent investigations, in part of a general chemical nature and in part of a physico-chemical nature, and they have all been character-

---

68. Mendeleev to Thomsen, 9 March 1885 (Royal Library, TSC). The subject of Mendeleev's letter was a request that one of his former students came to Copenhagen to work in Thomsen's laboratory.
69. For S. M. Jørgensen's systematic neglect of the periodic system, see Kragh (2015), which is also the source for the connection between Mendeleev and the Royal Danish Academy.

ized by a superior mind. ... Mendeleev's name has become generally known by his brilliant work on the theory of how the chemical and physical properties of the elements depend on their atomic weights – the so-called periodic system. In this way he has opened a wide field for philosophical discussion of the most important chemical phenomena; his theories have several times been remarkably confirmed by the discovery of elements whose existence and most important properties he had predicted as a consequence of the system. Objections can indeed be raised against the full justification of the system, such as can be done against many other theories; but the system has, to a very high degree, advanced chemistry as a science, and for this reason Mendeleev's name will forever be inscribed among the first in the history of chemistry.[70]

Mendeleev quickly responded to the invitation, expressing in a letter of 14 April how great an honour it was for him to become a foreign member of the Royal Danish Academy. He was pleased to accept this sign of "the scientific brotherhood of the peoples," which he considered a manifestation of "the sympathy which unites the Danes and the Russians." He undoubtedly had in mind the close family relations between the Danish king Christian IX and the Russian tsar Alexander III.

At the time of his election to the Royal Danish Academy, Mendeleev was well acquainted with Thomsen's work in thermochemistry and other branches of general and inorganic chemistry. In his massive textbook *Principles of Chemistry* Mendeleev referred numerous times to the Danish chemist. In fact, there were more references to Thomsen than to any other chemist apart from Mendeleev himself; while Thomsen appeared 22 times in the name index of the German translation, S. M. Jørgensen appeared twice (and Thomsen's rival Berthelot 21 times).[71]

When Niels Bohr enrolled at the University of Copenhagen in

---

70. Letter of 25 February 1889 in Thomsen's handwriting, addressed to Hieronymus G. Zeuthen, secretary of the Royal Danish Academy. Mendeleev was elected a foreign member of the prestigious Royal Society in 1892, and ten years later the same honour was bestowed on Thomsen.

71. Mendeleev (1892).

1903, Thomsen had retired.[72] As far as I know, Bohr never personally met the old chemist, but he probably knew about him through his father Christian Bohr, a well-connected professor of physiology and thus a colleague of Thomsen. Incidentally, Niels Bohr and Julius Thomsen, two of Denmark's great scientists, came to enter into a family relationship, if only through their offspring and many years after Thomsen had passed away.[73] There was in fact some slight connection between the two great scientists separated in time by nearly sixty years. In 1905 the Royal Academy of Sciences announced a prize essay on the theory of vibrating liquid jets for which young Bohr was awarded the Academy's gold medal. In early 1907 he received a brief but welcome letter signed by Thomsen and H. G. Zeuthen on behalf of the Academy: "Upon recommendation of the Section for Natural Sciences and Mathematics, the Royal Danish Academy of Sciences and Letters at its meeting on February 22 has awarded you its gold medal for your solution of the Physics Prize Problem set in 1905. In sending you the medal, the Academy extends to you its congratulations."[74]

Bohr's education included not only physics, but also solid courses in astronomy, chemistry and mathematics. In 1905 his teacher in inorganic analytical chemistry was Niels Bjerrum, who had graduated a few years earlier under S. M. Jørgensen and was on his way to a brilliant career in national and international chemistry. Familiar with the attempts to explain the periodicity of the elements in subatomic terms, in 1907 Bjerrum suggested that the periodic system "can hardly be explained without assuming an internal constitution

---

72. Abraham Pais (1991, p. 209) refers to Thomsen as "Bohr's distinguished teacher," which is clearly incorrect.

73. This came about as follows. Julius Thomsen's daughter Ellen (1865-1958) married in 1890 Andreas B. Richter (1861-1939). They had the son Einar D. A. Richter (1898-1980), who with Ruth A. A. Ammentorp (1897-1965) got the daughter Else Richter (1924-2011). In 1950 Else married Ernest David Bohr (b. 1924), a son of Niels and Margrethe Bohr. Julius Thomsen was thus a great-grandparent to Niels Bohr's daughter in law.

74. Quoted in Rud Nielsen (1972), p. 7. The Academy decided to award a gold medal also to the engineer P. O. Pedersen, to whom we shall refer in Section 8.3.

of the atom."[75] Bohr most likely read Bjerrum's article. During his studies 1898-1902 Bjerrum had only limited contact with Thomsen, whom he recalled "was not involved in teaching at all except that he gave an introductory course to all students, including the medical students."[76] Bjerrum did not follow this course, but for a time he worked as an assistant for Thomsen's lectures.[77]

From the lectures in the chemistry auditorium Bohr must have spent many hours in front of Thomsen's periodic table, which in his later career came to play some role for him. As part of extending his quantum atomic theory from one-electron atoms to all the elements, in the years 1921-1923 Bohr developed an ambitious theory of atomic structure which he summarised in a modified version of Thomsen's old periodic system, but now based on the atomic number rather than the atomic weight. The periodic system used by Bohr is sometimes known as the Thomsen-Bohr system. Bohr, who had evidently studied Thomsen's periodic table of 1895, found this version suggestive of atomic interpretations.[78] "Compared with usual representations of the periodic system," he said in a paper of 1921, "this method, proposed more than twenty years ago by Julius Thomsen, of indicating the periodic variations in the properties of the elements is more suited for comparison with theories of atomic constitution." Likewise, when Bohr gave his Nobel lecture in Stockholm on 11 December 1922, he used the occasion to pay tribute to the Danish chemist. As he pointed out with regard to the element of atomic number 72 (hafnium), Thomsen had predicted this element as a homologue of titanium and zirconium. The discovery of the element in 1923 caused a heated priority controversy in which the long dead Thomsen became involved.[79]

The numerical law which Thomsen tentatively suggested in 1895, namely, that the number of elements in the periods can be

75. Bjerrum (1907), p. 77.

76. From Bjerrum's autobiography written about 1942 and published in Nielsen (2004), pp. 7-100. Quotation on p. 25. See also Chapter 8.

77. See Bjerrum (1909), p. 4986, which includes comments on Thomsen's personality.

78. For Bohr's theory of the periodic system and his references to Thomsen, see Kragh (2012) and the literature cited therein.

79. See for example Paneth (1923) and Hevesy (1927).

written as $N = 2n^2 - 1$, or $N - 2n^2$ if the inert gases are included, came to be known as Rydberg's rule. The name refers to the Swedish physicist Janne Rydberg, who proposed the rule in 1906 and earlier, in 1897, had come close to it. Apparently Rydberg was unaware of Thomsen's speculation of 1895, just as were the atomic physicists in the later tradition of Bohr's quantum theory. Arnold Sommerfeld and other physicists in this tradition were greatly fascinated by the relationship, which they suspected was a key to understanding the periodic system in terms of atomic physics. It is still not generally recognised that Thomsen suggested the rule, or a slightly more limited version of it, several years before Rydberg.[80]

The numerical rule was eventually deduced from quantum theory, first by the British physicist Edmund Stoner in 1924 and the next year by his Austrian colleague Wolfgang Pauli. Pauli's version was carried over into modern quantum mechanics which explains the Thomsen-Rydberg rule by means of Pauli's exclusion principle and the allowed quantum numbers for a given principal quantum number n. With ℓ denoting the azimuthal quantum number and taking into regard the two spin states of the electron, the result becomes

$$N = 2 \sum_{l=0}^{n-1} (2\ell + 1) = 2n^2$$

Thomsen would presumably have been pleased.

7.6. The inert gases

As mentioned earlier in this chapter, as early as 1887 Thomsen was willing to consider the hypothetical "helium" as a real substance

---

80. See Kragh (2015), p. 189. On the other hand, in 1897 Rydberg suggested that the periodicity of the elements should be based on a whole ordinal number different from the atomic weight, thus foreshadowing the atomic number. Thomsen remained committed to the standard view that the periodic system expressed variations in atomic weights.

with atomic weight smaller than hydrogen's. Of course, at that time he could not know that helium actually exists as the lightest member of a group of inert-gas elements. It was the sensational discovery of argon, and not helium, which first indicated that the periodic system was seriously incomplete. When Ramsay and Rayleigh announced their detection of a new, chemically inactive gas in the atmosphere, the discovery aroused intense interest. The two Britons reported the element to be mono-atomic and with atomic weight 39.9 – greater than that of potassium – but how could such an element be fitted into the periodic system? Mendeleev was one among many chemists who protested that "If we admit that the molecule of argon contains but one atom, there is no room for it in the periodic system."[81] He consequently suggested that perhaps Ramsay's and Rayleigh's element was actually N_3, an allotropic form of nitrogen with molecular weight about 42. The suggestion was supported also by Berthelot and a few other chemists of prominence. It was only with the subsequent discovery of helium that it became clear that an eighth group had to be added to the periodic system.

The discussions which followed in the wake of the discovery of argon induced Thomsen "to publish some ideas, with which I have been occupied for years, but which I have wished not to publish until now in order not to encumber science with hypotheses that cannot easily be tested by experiment."[82] He read the paper to the Royal Danish Academy on 19 April 1895, less than a month after his paper on the periodic system and at a time when Ramsay had not yet published his discovery of terrestrial helium (he communicated his discovery to the Royal Society on 27 March). Thomsen's considerations were based on his old idea that the chemical properties of the elements should be described by well-behaved functions of the atomic weights. If this was the case the discontinuity in electrochemical character, when passing from the halogen family to the alkali metals, would seem to demand the existence of a group of intermediate elements. As he expressed it, "The transition from one

81. Mendeleev (1895). For the argon controversy, see Giunta (2001).
82. Thomsen (1895c), which was a German translation of the paper published in the Royal Danish Academy's *Oversigt*, pp. 137-143. See also Venable (1896), pp. 273-276.

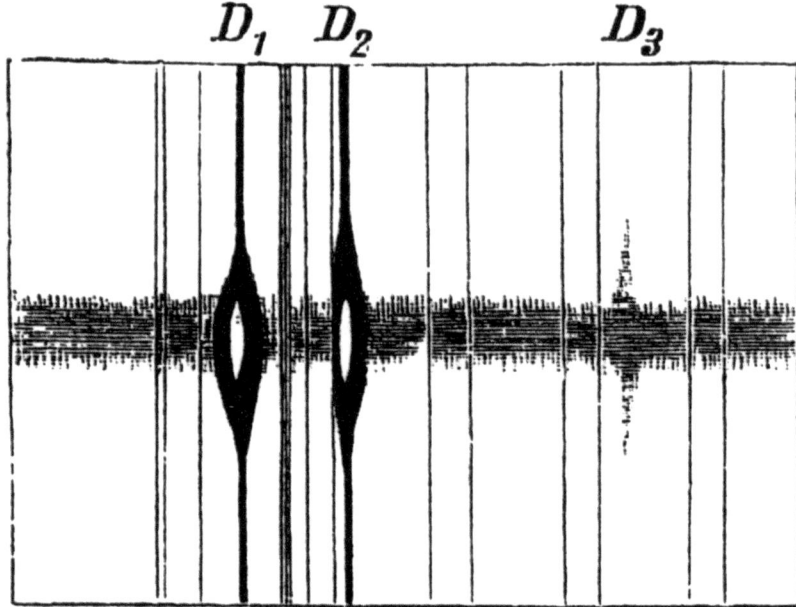

Figure 7.7. Emission spectrum of D lines from sunspot, showing the D_3 or helium line. The two stronger lines to the left are the sodium doublet lines. Source: H. Kayser, *Lehrbuch der Spektralanalyse* (Berlin: Springer, 1883), p. 188.

row [period] to the next is formed by an element whose electro-chemical character is $\pm \infty$, meaning that it is electrochemically indifferent." Thomsen argued that these elements had to be of zero-valence and thus be unable to form chemical compounds.

Based on his recently published periodic system, Thomsen suggested that "there exist the mentioned inactive elements with atomic weights 4, 20, 36, 84, 132, 212 and 292, the electrochemical character of which is indifference and the valence of which is zero." Somewhat remarkably he did not mention helium as the candidate element for atomic weight 4. In the parallel period to the long one of 31 elements, in which only thorium and uranium were known, he now added the inactive element of atomic weight 292 as its end point. Thomsen's prediction was surprisingly good, as shown by a comparison with the presently known elements in group VIII:

Symbol	He	Ne	Ar	Kr	Xe	Rn	Uuo
Atomic number	2	10	18	36	54	86	118
Atomic weight, Thomsen	4	20	36	84	132	212	292
Atomic weight, modern	4.0	20.2	39.9	83.8	131.3	222	(294)

Table 6. The noble gases and Thomsen's predictions.

The last of the predicted elements calls for a comment, as it refers to what is presently the heaviest known element. A few atomic nuclei of the element Z = 118, provisionally named ununoctium, was manufactured in nuclear reactions in 2006; ten years later the element received official recognition as oganesson instead of ununoctium. Based upon his interpretation of the periodic system, in 1922 Bohr predicted the electron configuration of the hypothetical element, stating that it would be an inert gas with chemical properties similar to radon. He merely presented the structure of Z = 118 as an illustration of the power of his atomic theory and did not, in this case, refer to Thomsen's earlier prediction.[83]

Thomsen considered his new periodic table well suited for expressing in a mathematical way the relations between the physical-chemical properties of the elements and their atomic weights. Attempts of this kind had been made by several other chemists using a variety of mathematical formulae, but none of the attempts were even approximately successful. In his 1889 Faraday lecture Mendeleev explicitly criticized the idea of representing the periodicity of the elements with some mathematical curve, something he considered fundamentally flawed. The periods of the elements, he maintained, have "a character very different from those which are so simply presented by geometers."[84] Undeterred, Thomsen thought that trigonometric or elliptic functions offered appropriate measures for

---

83. On the prediction of Z = 118 and other transuranic elements, see Kragh (2013). The name oganesson derives from the Russian nuclear physicist Yuri Oganessian.

84. Mendeleev (1889), p. 633. See Freund (1904) for various suggestions of mathematical formulae representing the periodicity of the elements. In the same lecture Mendeleev dismissed Lockyer's hypothesis of solar helium, suggesting that the spectral line belonged to a known element placed under such extreme conditions that the line had not yet been observed in the laboratory.

a mathematical relationship of the kind he desired. For the elements between lithium and chlorine, he suggested the following formula as a rough expression of the electrochemical character E as a function of the atomic weight A:

$$E = \cot\frac{A-4}{16}\pi$$

Similarly he suggested an expression for the dependence of the valence v on the atomic weight. With F being a trigonometric function, his expression was

$$v = 4F\left(\sin\frac{A-14}{16}\pi\right)^2$$

Trigonometric functions of a similar kind had been considered eight years earlier by the Russian chemist Flavian Flavitsky at the Kazan University. His aim was the same as Thomsen, to give a mathematical representation of the periodic system. Flavitsky's paper of 1887 was published in Russian only, but when he became aware of Thomsen's paper he felt obliged to claim priority in the *Zeitschrift für anorganische Chemie* and restate his original results.[85] Thomsen had independently arrived at the same mathematical expressions, which Flavitsky found most remarkable but without suggesting that Thomsen had actually plagiarised him.

According to Thomsen, his classification of the elements was significant with regard to the hypothesis of a common constitution of the atoms. Relating to his earlier speculations of 1885, he wrote:

> If one assumes the hypothesis of the unity of matter which, despite all attacks, cannot be displaced from the minds of scientists, one finds that the hypothetical inactive elements must have been formed by a regular and closed arrangement of the prime atoms, in such a way that the equilibrium of the molecules thereby formed have no disturbing points of action for chemically active substances; therefore they would not be able to form stable compounds but would only follow the general laws of gravity.[86]

85. Flavitsky (1896). See also Venable (1896).
86. Thomsen (1895c), p. 285.

Thomsen's prediction of a whole group of inert elements, and his attempt in interpreting them in accordance with neo-Proutean views, were not the only one at the time. For example, the French chemist Paul-Émile Lecoq de Boisbaudran, the discoverer of gallium, claimed in 1895 to have predicted several of the inert elements. Although his atomic weights were more accurate than Thomsen's (in the sense that they contained more decimals), they were based on numerical speculations even more obscure than those of the Danish chemist.[87] Without specifying his reasons Boisbaudran gave the atomic weights 20.0945, 36.40 ± 0.08, 84.01 ± 0.20, and 132.71 ± 0.15. C. J. Reed, an American amateur chemist, had entertained the idea of several inert and monatomic elements since 1885; ten years later he suggested that their atomic weights were 4, 20, 36, 84, 132 and 196. In a letter to Thomsen he wrote: "Believing that this theory would not meet with much favor ... I have never advocated it any further. It is gratifying to me, however, to find one so eminently qualified as yourself has been led by a method less empirical to conclusions which are substantially the same."[88]

A few other Danish chemists followed with interest the amazing discovery of argon and the consequences it might have for the well-established periodic system. In a paper of June 1895, Emil Petersen supported the claim of Ramsay and Rayleigh, confidently pointing out that there was no fundamental disagreement between argon and Mendeleev's system. Later the same year, the delicate question was reviewed by Søren P. L. Sørensen, who at the time worked at the chemical laboratory of the Polytechnic College and a few years later moved to the Carlsberg Laboratory. Sørensen expressed support of Thomsen's view of the periodic system and its "convincing argument for the existence of a group of elements of an inactive character."[89] Neither Petersen nor Sørensen mentioned helium.

---

87. Boisbaudran (1895).

88. Reed to Thomsen, 31 March 1898 (Royal Library, TSC). Reed (1895) appeared on 3 May 1895 and was thus independent of Thomsen's prediction.

89. Petersen (1895); Sørensen (1896). S. P. L. Sørensen, a student of S. M. Jørgensen, specialised in protein chemistry and is best known for his 1909 introduction of the pH scale for acidity. In 1921 he was nominated for the chemistry Nobel Prize, without success.

At the fifteenth Scandinavian Meeting of Scientists held in Stockholm in 1899, Thomsen gave a report on the state of the art concerning the inert gases, which was a part of a broader survey of recent progress in the chemical and physical sciences. Thomsen referred to advances in low-temperature physics made by James Dewar and others, and also to the French chemist Henri Moissan's experiments with carbon and other substances subjected to extreme heat. In this way it had become possible to manufacture calcium carbide as an industrial commodity, which was destined "to play an important role as a source of light and heat in the future." He was referring to the burning of acetylene produced by

$$CaC_2 + 2 H_2O \rightarrow Ca(OH)_2 + C_2H_2,$$

a process which was already used widely for domestic lighting, bicycle lamps and other purposes. Nor did Thomsen forget to highlight Guglielmo Marconi's system of wireless telegraphy and the sensational discovery of X-rays made by "the Austrian" Wilhelm Röntgen (who was in fact German). But he dwelled in particular on the discoveries of the inert gases, not only argon and helium but also the very new krypton announced by Ramsay and his assistant Morris Travers in June 1898. According to Thomsen, the serendipitous discovery of argon was a beautiful example of how progress was made in the empirical sciences. He said:

> It is the study of anomalies that leads to new truths. By accident and quite unexpectedly the scientist makes some observations which do not agree with theory or what was known previously. He infers that either must the theory be wrong or the observations erroneous; further investigations will then lead to an understanding of the cause of the phenomenon and thus to either a discovery or just a disappointment.[90]

Although Thomsen chose to illustrate his methodological point by means of argon, he could just as well have chosen Röntgen's discov-

---

90. Thomsen (1899), p. 68.

ery as an example of serendipity. Thomsen thought that the inert gases had once been part of the primeval Earth and that the present scarcity of helium was due to its very small molecular weight which had caused it to evaporate into space. As to the new krypton gas he suggested that it might be responsible for the spectral lines of the aurora borealis and thus "provide a likely explanation of this enigmatic phenomenon." The so-called green line of wavelength 5570 Å was at the time a mystery and Thomsen was among the first to suggest the krypton hypothesis which remained popular for about a decade.[91]

Thomsen kept an interest in the new class of inert gases until his last years. In 1898 he took up again the chemical analysis of a red-brown fluorite mineral from Greenland which he had first investigated "about 20 years ago" in connection with the cryolite deposits. Thomsen had found that the mineral, which apart from fluorspar also contained small amounts of rare earth metals, developed light and heat when pulverized. Assuming that it was a kind of fluorescence phenomenon, he laid the experiment aside without communicating it in the form of a paper. Only after the discovery of helium in terrestrial sources did he return to it, suspecting that a compound of the new gas might be involved. As Thomsen reported, after having removed carbon dioxide and water he examined the spectrum of the residual gas, finding "a strong helium spectrum and also the spectra of hydrogen, nitrogen and carbon; the light at the negative electrode showed a strong and nearly pure spectrum of helium mixed with only traces of hydrogen."[92] He suspected the gas to exist in the mineral in the form of an unstable compound with other elements. The compound, which he speculated might be conditioned by the presence of rare earths, would dissociate and helium being released in an endothermic process.

Suggestions of unstable helium compounds similar to Thom-

---

91. Thomsen (1899), p. 70. The mystery of the green auroral line was only solved in 1925 when it turned out that the line was the result of a quantum transition between two metastable states of oxygen. See Kragh (2010) for chemical speculations concerning the origin of the aurora.
92. Thomsen (1898a), p. 113.

sen's were made also by a few other chemists, but none of them were confirmed. Indeed, to this date no helium compounds are known, making the element unique in this respect. In a later investigation, one of his last, 78-year-old Thomsen determined the quantity of helium liberated on heating from the mineral to be 24-27 cm^3 kg^{-1}. He tried in various ways to "decompose the possible compound of helium," but with no luck; from this he concluded that the fluorite mineral contained helium in the form of a very firm compound.[93] His attempts to detect the new gas in other fluorite minerals containing rare earths also failed.

Ramsay found Thomsen's investigation to be of great interest. In a letter of early 1898 he wrote: "I am very glad that you have taken up the subject, and look forward with interest to your thermal measurements, which are certain to be more trustworthy than mine. ... I strongly suspect that we are dealing with an endothermic compound of helium." And at the end of his letter: "I dare say you know that I can read Danish fluently, though I am too lazy to write it, knowing that you of course understand English. So please, in answering this, take the way of 'least friction'."[94] The investigations of Thomsen and Ramsay attracted considerable attention. As a British chemist remarked in the summer of 1898, "Prof. Julius Thomsen and Prof. Ramsay have examined two minerals which are endothermic; when gently heated they glow red-hot, and at the same time they evolve helium."[95]

As mentioned in Section 5.2, Ramsay had known Thomsen since the mid 1880s. He followed Thomsen's work closely and frequently referred to his thermochemical results. For example, in 1898, the same year that he addressed Thomsen with respect to possible helium compounds, Ramsay published an investigation of the occlusion of hydrogen and oxygen by palladium.[96] Thomsen had in 1877

93. Thomsen (1904). Thomsen's work attracted international attention. See, for example, *Science* **20** (1904): 442.

94. Ramsay to Thomsen, 9 February 1898 (Royal Library, TSC).

95. Hartley (1898), p. 725.

96. The occlusion of hydrogen by palladium was discovered by Thomas Graham in 1867, who believed that the phenomenon could be explained by conceiving hydrogen as a kind of volatile metal. He suggested the name "hydrogenium" for the hypothetical

studied the thermochemical properties of platinum and palladium compounds and determined the heat of formation of palladium hydroxides to be

$$Pd + O + H_2O \rightarrow Pd(OH)_3 + 22.71 \text{ kcal},$$

and

$$Pd + O_2 + 2\ H_2O \rightarrow Pd(OH)_4 + 30.43 \text{ kcal}$$

Ramsay and his collaborators now used their own measurements in combination with Thomsen's data to suggest that the absorption of oxygen by palladium was a true phenomenon of oxidation and not merely a physical process.[97]

To return to helium, much of the interest in the gas was at the time associated with radioactive substances and the evidence that helium is closely connected with alpha radioactivity, indeed that alpha particles are made up of helium ions. The English physicist and meteorologist Robert Strutt, the son of Lord Rayleigh, was active in the exciting field of radioactive research and in contact with Thomsen, who supplied him with a sample of the Greenlandic fluorite mineral. Strutt's analysis revealed no radium, but heating of a solution of the mineral gave abundant "thorium emanation" (radon), from which he concluded that thorium was a substantial constituent of the mineral.[98] The chemistry professor Odin Christensen, who knew Thomsen well, recalled:

> Just a few months before his death he was much occupied with Ramsay's investigations concerning the action of radium emanation on certain elements and their subsequent "degradation." He thought

---

metallic state.

97. Mond, Ramsay and Shields (1898); Thomsen (1877); Thomsen (1882-1886), vol. 3, p. 436.

98. Strutt (1907), who referred to Thomsen's papers of 1898 and 1904. Robert Strutt became the fourth baron of Rayleigh in 1919, whereas his father John William Strutt, generally known as Rayleigh, was the third baron. The mineral examined by Thomsen was later shown to be radioactive, but with a thorium content of only ca. 0.3 per cent. See Pauly (1986).

that Ramsay's investigations might produce such results that they led
to a whole new era in chemistry and a total change of our presently
valid chemical theories.[99]

Ramsay, Nobel laureate and president of the Chemical Society of
London, had in 1907 stunned the scientific community by announc-
ing that radon – at the time commonly known as "radium emanation" – in contact with copper sulphate produced argon. Moreover,
he found evidence that "the copper, acted upon by the emanation,
is 'degraded' to the first member of its group, namely lithium."[100]
Most likely, Thomsen was inspired by Ramsay's sensational investigations which seemed to confirm his old hypothesis of chemical elements as compounds bodies built up from some *prima materia*.

Whereas there is no evidence that Thomsen did research on radioactivity, we know that he was occupied with the inert gases at the
very end of his life. According to Bjerrum, he was led to believe that
argon was not truly elementary after all; he tried to demonstrate its
compound nature by synthesising it and thought to have obtained
some positive results. Perhaps realising that the evidence was after
all shaky, Thomsen did not publish his work.[101] But he did write a
draft of an article and from this we know what kept him busy during
his very last years. The following is a slightly abridged version of
Thomsen's draft.

Is Argon a New Element? Experimental Investigations

For about one and a half year I have been occupied with answering
the question of whether or not argon i a new element. Its elementary
nature has earlier been doubted, but mainly on a theoretical basis;
experimental tests have rarely been performed. ... A decision regarding the nature of argon can of course only be reached experimentally;

---

99. Christensen (1910), p. 169.
100. Ramsay (1907). For Ramsay's experiments and alchemical speculations based
on radioactivity, see Trenn (1974) and Morrisson (2007), pp. 115-121. Ramsay
believed as late as 1913 that his experiments proved controlled element transmutation.
101. Bjerrum (1909), p. 4986.

two routes have been explored, either attempts to dissociate it or to form it from known substances. W. Ramsay largely followed the first route and his experiments have shown that argon cannot be dissociated by the means presently known. But the question of whether argon is an element or not is not answered decisively in this way, for the investigation has only shown that its dissociation does not follow by the applied means. For this reason I pursue the other route and have tried to synthesise the mentioned substance from known elements.

Such a work must for the time being rely on hypotheses. My first assumption was that argon might be a carbon-nitrogen compound, namely CN_2 with molecular weight 40. Perhaps such a compound would be formed in small amounts by the dissociation of carbon and nitrogen compounds. I initially focused my attention on the cyanogen compounds such as potassium ferrocyanide, potassium ferricyanide, sodium nitroprusside, silver cyanide, and other metallic cyanides, and next on silver rhodanide, silver cyanamide, cyanuric acid, hexamethyleneamide, biuret, etc. Substances of this kind were heated to dissociation either directly or by the assistance of weak oxidisers such as nickel oxide, lead oxide, nitrogen oxide and the like. The gases and vapours produced in this way were subsequently oxidised by copper oxide and collected over mercury. The gas, which mostly consisted of nitrogen, was then oxidised by means of oxygen and induction sparks. Oxidation products such as nitrogen oxides and some carbon dioxide were absorbed in a concentrated solution of potassium hydroxide. After the oxidation the gases (mostly oxygen and a little mercury) were absorbed on read-heated copper. The small rest mass was finally transferred to a Geissler tube and examined spectroscopically. ... The amount of nitrogen was in each of the experiments 100-120 cm^3.

The oxygen used for the oxidation was produced galvanically ... and led directly to the oxidation vessel so that entrance of atmospheric air was ruled out. Even the provisional experiments gave the result that, after oxidation and absorption, nitrogen formed in this way contained a small rest in which argon was identified spectroscopically. This observation caused me to re-examine my method in order to locate any errors. During the last year I have performed a couple of hundred experiments of this kind. The amount of argon produced in each experiment was about 0.05 per 100 cm^3 nitrogen; however, in many of the experiments no argon turned up. For this reason I examined other reactions such as those between gaseous substances, for example carbon dioxide and nitrogen dioxide, cyanogen and carbon dioxide, carbon disulphide and nitrogen, etc. It was my intention to find a reaction more energetic than those examined, but this work was in vain. The amount of argon remained very small.

The produced argon was apparently independent of the nature of the chemical reactions, which I found to be remarkable. It led me to think of the possibility that argon was not formed by the chemical processes but only by the electro-oxidation of the nitrogen. So I turned my investigation toward the nature of the nitrogen itself and was soon able to show that nitrogen produced by ammonium chloride and potassium nitrate, after it had been oxidised as described, also caused a formation of argon. This I found most remarkable, given that W. Ramsay had made aware of the possible formation of argon from ... .[102]

The small amounts of argon that Thomsen thought to have detected by means of his spectroscope were possibly due to traces of atmospheric air. Another possible explanation is that the spectral lines he attributed to argon were in reality due to other elements or compounds. Thomsen had only limited experience with spectroscopic measurements and he may easily have been led astray such as were many other non-experts in spectroscopy.

102. Handwritten manuscript in German, 4 pp. (Royal Library, TSC). The manuscript ends abruptly, suggesting that one or more pages are missing.

CHAPTER 8

Science, society, and technology

First and foremost a brilliant chemist, Thomsen spent countless hours alone in his laboratory, deeply immersed in precise chemical experiments. The outcome of these countless hours, his thermo-chemical data bank, is what he is primarily remembered for and what has been described in earlier chapters. But chemical research was only part of his long and active life. In a Danish context he was a well-known and respected public figure who contributed heavily to the nation's system of science, technology, and higher education. Not only did he serve as Rector of the University in two periods, he was also an efficient director of the Polytechnic College and for many years president of the Royal Danish Academy. Thomsen was never a member or supporter of any political party but as an indi-vidual he contributed to municipal politics through his long-time membership of Copenhagen's City Council. In this capacity he worked in particular with gas lighting, city planning, sanitation, and other technical issues.

Scientifically a loner, Thomsen formed no school and had but few students. At several occasions he was addressed by young chemists from abroad who wanted to work with him, but he al-ways declined and that even when the request came from a col-league as distinguished as Mendeleev (Section 7.5). All the same, his impact on the younger Danish generation was considerable and in part the result of the lectures he gave on general and inor-ganic chemistry over many years. These lectures were followed by a large number of students some of whom found them inspiring and valued Thomsen as a teacher. But to most students he was a stern and distant authority, not one they could feel any bond of affection to. Cold and unsympathetic Thomsen might appear, yet as a private person he was no less human than most people. Only very few sources cast light on his personality and private life, but what little there is gives a glimpse of "another Thomsen," a caring family man with interests far beyond science. This chapter aims at characterising Thomsen as a human being and a major player in Danish nineteenth-century science and society. The latter task is easier than the former.

8.1. A leader of Danish science

Apart from his much-acclaimed contributions to chemical research Thomsen undertook a series of administrative duties which brought him to the very top of the Danish system of science and higher education. As a skilled negotiator with experience in administration, business affairs and local politics he was well suited for these duties. Thomsen became the front figure in at least three of the most important institutions of Danish science, namely, the University of Copenhagen, the Polytechnic College, and the Royal Academy of Sciences. But first a few words about a very important institution of the late nineteenth century with which Thomsen was *not* or only peripherally involved.

The Carlsberg Foundation and its associated Carlsberg Laboratory was established by the wealthy brewer Jacob C. Jacobsen, the founder of the Carlsberg Brewery.[1] With these institutions Thomsen had no direct connection although he was indirectly involved with them through his years as president of the Royal Danish Academy which appointed the Foundation's board of directors. But it were other members of the Academy who played important roles in Jacobsen's scientific empire, in particular the zoologist Japetus Steenstrup and the chemist Christen Barfoed who were directly involved in the formulation of the statutes of the Foundation in 1876. The two sat on the first board of directors of the Carlsberg Foundation together with the classical philologist Johan N. Madvig – who was Thomsen's predecessor as president of the Royal Academy – the historian Edvard Holm, and the physiologist Peter L. Panum. From 1885 and until his death in 1914 S. M. Jørgensen, professor of chemistry and Thomsen's close colleague, served on the board of both the Carlsberg Foundation and the Carlsberg Laboratory. He

---

1. See Glammann (2003).

was chairman of the Carlsberg Foundation 1909-1913. Julius Thomsen was absent from the Carlsberg empire.

As university professor of chemistry and member of the Senate Thomsen exerted much influence on the teaching of chemistry at the University of Copenhagen and also on the more general questions discussed within the Faculty of Science. One way of promoting a particular branch of science was to announce a prize essay, a tradition which went back to the eighteenth century. Thomsen occasionally evaluated essays of a chemical nature and most likely was also responsible for some. For example, in 1872 the Faculty announced a prize essay on the subject "A historical and critical survey of investigations which until the year of 1868 have been made on the heat phenomena accompanying chemical processes."[2] One may safely assume that this was Thomsen's idea but apparently no one submitted an answer. Two other prize essays probably reflect Thomsen's later interest in atomic weights and the unity of matter. One of them, dating from 1882, has been mentioned in Section 7.2, and in 1890 the University announced a prize contest on new methods applied to the determination of the relative weights of molecules. The latter subject, which essentially concerned physical-chemical methods, has been described in Section 6.3.

In the academic year 1886-1887 Thomsen served as Rector for the University, to be followed by another period in 1891-1892. During the first period he was involved in the negotiations which on 1 April 1887 led to the promotion of S. M. Jørgensen from lecturer (or associate professor) to full professor. The preparations for the celebration of the royal couple's golden wedding in 1892 were taken very seriously by the University and Thomsen played a central role in them during his second tenure as Rector (see below).

The University of Copenhagen had grown considerably in the preceding decades and continued to grow, the influx of matriculated students per year increasing from about 130 in the mid-nineteenth century to about 350 at the turn of the century.[3] Still at the time when

---

2. *Aarbog for Københavns Universitet 1872-1873*, p. 233.

3. Petersen (1993), p. 434. The total student population grew in the same period from about 1,000 to nearly 5,000.

Thomsen served as Rector the three traditional faculties – theology, medicine and law – dominated the university with approximately 85 per cent of the students. The number of students in chemistry and the other fields of science was increasing but remained small. In 1892, when a new law of university teachers' salary was passed, the number of ordinary professors was 49 of which 10 belonged to the Faculty of Science. During Thomsen's periods as Rector there were constant problems with inadequate funding from the government and also, on a quite different level, with an increasing ideological divide within the corps of teachers. While traditionally the professors and the University Senate had been anchored in conservative-national values, at the time they were challenged by teachers who propagated liberal, positivistic, and radical views. Despite Thomsen's positivistic leanings in matters of science, he had no sympathy for the latter group.

When Julius Thomsen was appointed director of the Polytechnic College on 1 August 1883 the 54-year-old institution faced several problems left over from his predecessor, the physicist Carl V. Holten. For one thing, the rules and structure of the College were archaic and in need of serious reform. These were implemented by Thomsen during the first year of his directorship.[4] His insistence on strict order and work discipline led him on collision course with Julius F. Wilkens who since 1849 had served as professor of mechanical technology. Wilkens had in the early 1850s been a member of a commission to reform the Polytechnic College and thus been part of the same debate in which young Thomsen was involved (see Section 3.1). He had then argued that the primary purpose of the school should be to serve industry and trade. During the following decades Wilkens continued to express his strong dislike of the scientific orientation of the Polytechnic College, complaining that he was insufficiently appreciated and that Thomsen actively opposed and even harassed him.[5] Thomsen chose to ignore the difficult and somewhat paranoid professor who retired in 1886.

4. Lundbye (1929), pp, 191-196. *Aarbog for Københavns Universitet 1883-1884*.
5. In his later years Wilkens reported lectures without actually having given them, something which Thomsen found totally unacceptable. See H. I. Hannover's memorial article on Wilkens in *Ingeniøren* no. 18 (1912): 163-168.

As part of the reform process a new board of the College was formed, consisting of the director and a council of teachers. The College was also in need of an extension of its staff of teachers; negotiations with the Ministry concerning this problem occupied much of Thomsen's time. He responded in part to a general dissatisfaction with the engineering education which had mounted during the 1880s and culminated in the early 1890s. Thomsen stressed in particular the need for an extension of courses in technical subjects, including electrical and chemical technology. Since the introduction of civil engineering in 1857 the field had been covered by a single teacher only, namely professor Ludvig F. Holmberg. As Thomsen pointed out in a letter to the Ministry of 1891, during this period civil engineering and its number of students had drastically increased.[6] Funds for more teachers and improved facilities were urgently needed. After a lengthy debate a major reform was finally implemented in 1894. Although the seeds for a modern engineering education in Denmark were thus planted under Thomsen, it was only after Hagemann had replaced him as director in 1903 that they turned into flowers.

The political circumstances of course played a major role in the slow process of reforming the institutions of science and technology in Denmark. During the period 1885-1894 the country was effectively governed by the right-wing Council President Jacob Estrup by means of provisional laws which made it particularly difficult to obtain State funding for higher education and other purposes. The dictatorial Estrup was forced to resign in 1894, but the political situation only changed markedly in 1901 with the introduction of parliamentarism under a new liberal government (in Danish history this is known as "systemskiftet," meaning the change of the system). It was only then that the effective power over Denmark was exerted by the Parliament and not by the king and his advisors. One reason why Hagemann succeeded where Thomsen did not was simply that the former worked under more favourable political and economic circumstances.

6. See Harnow (1998), p. 143 and, for Danish engineering education, Harnow (1993).

Figure 8.1. The impressive new building of the Polytechnic College at about 1900. To the left the University's astronomical observatory is visible.

The most important, and most visible, change of the Polytechnic College during Thomsen's directorship was the establishment of new buildings outside the crowded inner city. The costly plans for a new Polytechnic College went back to shortly after Thomsen's appointment as director, but due to a combination of political opposition and financial problems they were delayed. Construction finally began in 1888 and two years later the new building complex at the corner of the city's Botanical Garden and close to the University Observatory was completed (Figure 8.1). The new Polytechnic College was a neighbour to the Mineralogical Museum, but due to Thomsen's initiative the two institutions remained separate. The inauguration on 1 September 1890 was a great day for Thomsen who gave the main speech in front of the king and a large number of the country's most distinguished persons. As part of the celebration a marble bust of Thomsen and also a "Professor Julius Thomsen grant" of 10,000 kroner were donated to the Polytechnic College. The donator was anonymous, but it is known that he was G. A. Hagemann.

During the busy years of 1852-1853 young Thomsen served as chairman of the Polytechnic Association (Polyteknisk Forening) which had been founded in 1846 as a loose group of students and other people involved in science, technology and mathematics. In its early days the association hosted lectures by prominent scientists such as Ørsted and Forchhhammer, but it was as much a social club for teachers and students as it was a professional organization for science and engineering. Thomsen appreciated the Polytechnic Association and remained a life-long member of it. He was also a member of the larger Technical Association (Teknisk Forening) dating from 1877, a group that included many of Denmark's important engineers and businessmen, several of them with experience from the City Council.[7] There was no restriction on membership in this organisation. Among the members of the Technical Association were Hagemann, Copenhagen's city engineer Charles Ambt (the successor of L. A. Colding), and Carl Jacobsen from the New Carlsberg Brewery.

However, when the far more influential and politically active Association of Danish Engineers (Dansk Ingeniørforening) was founded in 1892 Thomsen stood outside. The twenty-one founders were mostly prominent engineers, amongst them the ubiquitous Hagemann. The new association criticized from the very beginning the level and content of the education at the Polytechnic College, something Thomsen perceived as an attack on his role as director. Nonetheless, he was forced to compromise and arrange a meeting with the critical engineers. The reform of 1894 was to a large extent based on the recommendations of the working group established by Thomsen and the engineers. Yet Thomsen had no sympathy at all for the Engineering Association and in particular not for its ambitions of interfering in the business of the College and in the political process generally. His dislike of the new organization only increased when it recommended that the Polytechnic College should establish a testing laboratory oriented towards concrete and other building materials.[8] Thomsen ignored the recommendation and instead

---

7. On the two associations, see Harnow (1998) and Nielsen (2000), pp. 199-202. The Technical Association published its own journal, *Den Tekniske Forenings Tidsskrift* (1878-1941).

8. Nielsen and Wistoft (1996), p. 99. For the early history of the State Testing

suggested that the Engineering Association created itself the laboratory it desired. After negotiations with the ministry this is what happened. Statsprøveanstalten (the State Testing Institution) was founded in 1896 as a commercial project under the Engineering Association but with some public support. It was taken over by the Danish State in 1909.

Thomsen's attitude towards the Association of Danish Engineers was shaped by his experience with the manufacture of cryolite soda at a time when an engineer was an individual succeeding only because of his own talent and determination. The Association of Danish Engineers was not only a professional body but also an organization serving the economic and labour interests of the engineers and in this regard a kind of a trade union, something which Thomsen resented. Whether consisting of workers or engineers, he disliked trade unions and what they stood for.

Despite his polytechnic background Thomsen identified himself as a scientist rather than an engineer. This may have been an additional reason for staying outside the Association of Danish Engineers, but it cannot have been the whole reason for he also decided to stay outside the Danish Chemical Society founded in 1879 as the first Scandinavian society of its kind. The initiative came principally from three chemists of the younger generation, Odin T. Christensen, Christian Steenbuch, and Thomas Thomsen, Julius' youngest brother. The object of the society was "through lectures, discussions, minor communications and social meetings to establish a tie between men who have an interest in chemistry."[9] Julius Thomsen – at the time Denmark's only chemist of international repute – could not possibly be against such an activity and at one occasion, in 1884, he gave a lecture to the society. But he never became a member. On the other hand, Thomsen's colleague S. M. Jørgensen took a most active interest in the Chemical Society. Not only did he serve as its first chairman, he also remained in this position for more than twenty years. At the time of Julius Thomsen's death in 1909 membership

Institution, see *Ingeniøren* no. 19 (8 May 1909): 163-168.

9. Quoted in Nielsen (2008), p. 75. This source and Nielsen (2000) provide details about the early history of the Danish Chemical Society.

in the Chemical Society had grown to about 140 of which more than a quarter were pharmacists.

No less than the University and the Polytechnic College, the Royal Danish Academy of Sciences and Letters was a permanent base for Thomsen's activities. For nearly half a century he rarely missed a meeting. He presented a large number of communications and often evaluated papers submitted to the Academy's two periodicals, *Skrifter* (Proceedings) and *Oversigter* (Transactions). On 18 May 1888 Thomsen was elected the eleventh president of the Academy, receiving 16 votes out of 23. He would stay in the office until his death in 1909, during the whole period assisted by the mathematician Hieronymus G. Zeuthen as secretary of the Academy.[10] The election period was five years, and Thomsen was thus re-elected several times if not always without opposition. In the Academy as elsewhere his temper and authoritarian style of leadership invited confrontation. A minority group of members preferred another president and yet as late as 1908 the octogenarian Thomsen was re-elected for another five-year period with the votes 25 for and 3 against. He presided over his last meeting on 8 January 1909, about a month before he passed away on 13 February. In March another Thomsen was elected president, this time the eminent philologist Vilhelm Thomsen. Julius Thomsen was buried together with his wife at the present Solbjerg Parkkirkegård in Frederiksberg.

At the time of Thomsen's election the Royal Academy was in an economically troubled situation which caused him to propose that it applied for government support. However, because of fear that such support might endanger the independence of the Academy the proposal was shelved. On Thomsen's instigation the crown prince Frederik, the later king Frederik VIII, was in 1894 elected honorary member of the Academy, a political move to increase its status (and Thomsen's as well). As Julius Thomsen, the director of the Polytechnic College was faced with the problem of new buildings, so Thomsen, the president of the Royal Academy, was faced with the problem of finding new premises for the Academy. Happily the

---

10. See Lomholt (1942-1973), vol 1, pp. 472-480 and Pedersen (1992), pp. 233-240, for Thomsen and the Royal Danish Academy.

problem turned into a non-problem when the Carlsberg Foundation in 1893 generously offered to house the Academy in a new mansion to be built in central Copenhagen. The mansion, which housed both the Academy and the Carlsberg Foundation and thus signalled the close connection between the two institutions, was completed in 1899. It was and still is located just opposite another Carlsberg monument, the New Carlsberg Glypthotek, the museum housing the exquisite art collection which Carl Jacobsen (the son of J. C. Jacobsen) had donated to the public a few years earlier.

At a few occasions Thomsen proposed new members to the Academy, always jointly with S. M. Jørgensen. We have already referred to their nomination in 1889 of Mendeleev (see Section 7.5). In this year they also nominated the Swedish chemist Lars Fredrik Nilson, "one of the most reputed chemists of our time whose work in the areas of analytic, general and theoretical chemistry has won common and well-deserved recognition." They highlighted Nilson's discovery of the two new elements scandium and ytterbium as particularly important. Nilson's scandium survived but his ytterbium did not (although the name did).[11] In 1890 Thomsen and Jørgensen nominated yet another foreign member, the German chemist Hermann Kopp whom they praised as an eminent and versatile scientist as well as a pioneering historian of chemistry.[12] Finally, the same year they nominated two domestic members, namely Odin T. Christensen and Johan Kjeldahl. All the nominees were accepted by the Academy.

---

11. Letter of 5 April 1889, Royal Danish Academy, Main Archive. In 1880 Thomsen had listened to Nilson and his colleague Per Theodor Cleve reporting on their research of the chemical elements in rare earth minerals. See Thomsen (1880f). The discovery of ytterbium and Nilson's isolation of scandium were considered a triumph of Mendeleev's periodic system which predicted an element ("eka-boron") with properties similar to those of scandium. By 1890 Nilson's ytterbium was believed to be elementary but within a decade or two it turned out that it was a composite substance. For the complex history, see Kragh (1996).

12. Letter of 16 February 1890, Royal Danish Academy, Main Archive. Kopp died in 1892. He was the author of the important *Geschichte der Chemie* in four volumes (1843-1847).

Figure 8.2. Thomsen at the height of his career. Wikimedia Commons.

Apart from what has been mentioned Thomsen was also involved, if much more indirectly, in the establishment in 1892 of the
privately funded Pharmaceutical College. He had since 1866 acted
as chairman of the Commission for the Pharmaceutical Exam and

his chemistry lectures covered large doses of pharmacy. Thomsen's primary interest in a reform of the pharmaceutical education may have been "to get rid of the pharmacists" at the Polytechnic College, as Emil Koefoed cynically expressed it.[13] At any rate, in 1888 Thomsen became chairman of a commission to reform the pharmaceutical education. In the report written by Thomsen the commission proposed to establish a new school specifically aimed at pharmacists. The school should be independent of the University and the Polytechnic College. Although the government did not follow up on the report, a pharmaceutical school or college was established a few years later by the wealthy apothecary Christian D. A. Hansen. As director of the Polytechnic College, Thomsen participated in the negotiations concerning the new college and its relations to his own institution. Of course, when the Pharmaceutical College was inaugurated on 1 November 1892 Thomsen was invited as a guest of honour. In his obligatory speech of celebration he recalled the great impact of pharmacists in the Danish history of chemistry and also that he was the first chemistry professor without a background in pharmacy.[14]

8.2. Public life and public science

As one of Denmark's most admired scientists and industrial entrepreneurs, and at times Rector of the University and the Polytechnic College, Thomsen gave several public addresses for the consumption of the educated bourgeoisie. Most of these addresses dealt with science in one form or other, but he also referred to subjects outside science and technology. We have already mentioned the lecture of 1884 on the theory of atoms and molecules that he gave to the annual banquet of the University (see Section 7.2). It is worth notic-

---

13. Koefoed (1974), p. 53. Thomsen, he wrote, disliked the pharmacists, "probably because they ... competed with the polytechnic candidates for many positions." The report of 1888 is reproduced in the same source on pp. 79-87. Koefoed served as chemistry assistant at the Polytechnic College between 1882 and 1892, after which he moved to the Pharmaceutical College.

14. See Kruse and Kofod (1992), p. 44, where Thomsen is described as a key person in the planning of the Pharmaceutical College.

ing that the occasion was a commemoration of the Reformation of 1536 in which the Lutheran-Protestant church became established in Denmark. And yet Thomsen did not mention the Reformation with a single word and he also did not refer to other aspects of theology and church history. What he did mention, and that at length and with much enthusiasm, was science and the many wondrous inventions that followed in its wake. "Our century is the age of science," he asserted. Steamships, railways, telegraphs and telephones, electric light, artificial fertilizers, spectroscopy, rational agriculture, and the direct use of solar heat – "All of these results and many others are almost exclusively the products of science."[15] Scientific research held a truly amazing potential, Thomsen said, adding that the sheer number of wonders flowing from the laboratories might cause people to take scientific and technological progress for granted. He worried that attention was turning into indifference and that discoveries and inventions no longer invoked the admiration they had enjoyed just a few years ago.

Thomsen deeply believed that science was an unmitigated benefit for society, but he was neither naïve nor uncritical in his praise of science. At the 1886 meeting of Scandinavian scientists in Christiania he gave a plenary lecture on "The Fundamental Hypotheses of Natural Science." These fundamental hypotheses included the two conservation laws of energy and matter but also, he said, the *impossibility* of creating life out of lifeless matter.[16] As mentioned, four years later, at the inauguration in 1890 of the new buildings of the Polytechnic College its director, Julius Thomsen, spoke to a large audience including the king, the queen, the crown princess, and other members of the royal family. He expressed his view of science and its future in terms that indicated a certain reservation. Although he was confident that science would continue making progress, he did not believe that a complete understanding of nature would ever be possible. Despite the enormous advance in scientific knowledge, most likely there were limits which could not be transgressed. Some

---

15. Thomsen (1884), pp. 1-2. See also Kragh et al. (2008), pp. 243-245.

16. Thomsen's lecture is briefly summarized in *Tidsskrift for Physik og Chemi* **7** (1886): 289-290.

of these limits were given by the fundamental laws of physics while others were of a more practical, technological nature:

> Man is able to form and transform matter, but not to increase or re-
> duce the amount of it; he is able to transform one activity into an-
> other, but never to generate an activity from nothing or annihilate an
> amount of work. ... No one has the capacity to form a cell that is able
> to live, or a seed that is able to sprout. And I think this ought to be
> kept in mind; for in our day and age, science is presenting so many
> wonderful findings, in the field of physics and that of chemistry, that
> it easily kindles the greatest aspirations and leads to an over-estima-
> tion of the reach of science; to an idolatry against which one must
> caution fully as much as H. C. Ørsted once zealously denounced the
> superstition of his times at the founding of the Polytechnic College.[17]

Thomsen referred to Ørsted's speech on 5 November 1829 in the presence of the king, Frederik VI, which dealt with "The Intellectual Influence Exercised by Natural Science in its Practical Applications." Thomsen might also, on the occasion in 1890, have mentioned his predecessor as director of the Polytechnic College, J. G. Forchhammer, who in a lecture of 1858 said: "No chemist, physiologist or geognostic has, as yet, been able to explain to us the manner in which inorganic matter could be transmuted into an organic being."[18] Thomsen went on with a mathematical analogy:

> In mathematical terms, the relationship of science to the enigma of
> existence is a hyperbolic function; for just as the hyperbole on its
> rule-bound path moves ever closer and closer to its asymptote with-
> out ever being able to reach it, so the development of science brings
> us ever closer to the recognition of truth, but without it will ever suc-
> ceed to fully reach its source.

17. Lundbye (1929), p. 223. The first lines refer to the laws of conservation of matter and energy, respectively. With the discovery of radioactivity in the late 1890s it turned out that matter conservation is not an absolute law of nature.
18. Quoted in Kragh et al. (2008), p. 136.

Views of the same kind would later be argued by several twentieth-century philosophers of science.

As a final example of Thomsen's public addresses consider a university lecture he gave as part of the celebration in 1892 of the golden wedding of the royal couple, king Christian IX and queen Louise.[19] As one might expect at an occasion like this, the lecture was a panegyric praise of the king and the great progress Denmark had experienced during his reign. Mixing panegyric and patriotism Thomsen traced Denmark's much needed reform period to the democratic constitution of 1849, a turning point which caused "a sudden change in the character of the people." But he also complained that a large part of the population was still not ready to take on the responsibility associated with democratic reforms. Thomsen referred to the catastrophic war with Germany of 1864 which had caused the government to focus on domestic progress rather than power in the international arena. What was needed, he said, was a determined development toward "an energetic, entrepreneurial and civilized state." In this great national project the University was an indispensable partner: "The influx of students has never been as high as during the latter years; never before have all classes of the people ... gathered around the university, recognizing that knowledge is power." Yet Thomsen well knew that the University was an elite institution and not really one for "all classes of the people."

In 1861 Thomsen was elected a member of Borgerrepræsentationen (Copenhagen City Council), an organisation founded in 1840 and consisting of 36 citizens of repute.[20] As the highest authority in the capital the City Council was a powerful institution responsible for the appointment of mayors and the development of the city. The members, who were unsalaried, sat for a period of six years with the possibility of re-election. They met about thirty times a

19. Thomsen (1892b), published separately and also in the newspaper *Berlingske Tidende*. Christian IX (1818-1906) was the first Danish king since 1660 which did not hold absolute power. Thomsen spoke in front of a most distinguished audience including not only the royal couple but also the Tsar and Tsarina of Russia as well as many other members of the European royal houses.

20. Information about the City Council during Thomsen's time can be found in Lange (2006), pp. 135-141, and Vinding (1942), pp. 210-245.

year. At the time membership of the City Council was in practice reserved the economic and commercial elite such as bankers, factory owners, and merchants; also a few engineers, professors, and architects entered what has been characterised as "a self-constituted welfare council."[21] Although they were formally elected, in reality they were chosen by the tightly-knit group of existing members. There were no political parties represented in the City Council but in a general sense it was a bastion of conservatism until the end of the century.

In 1893 the bourgeoisie running the city was challenged when several liberals and social democrats were elected, much to the regret of Thomsen who regarded them as "insolent disturbers of the peace."[22] The entrance of liberal and even socialist ideas much annoyed him and may have added to his decision to leave the City Council in 1894. It was not his only reason, though. There was at the time a major discussion of whether the city should invest in a new system of sewers or in a new and much-needed Town Hall building. Thomsen argued for the first solution but was overruled by a majority in the City Council, which greatly dissatisfied him.[23] He thought that the expensive Town Hall project could wait until the problem of the city's sewers had been solved. The magnificent Town Hall building, designed by the architect Martin Nyrop, was completed in 1898.

Thomsen had close connections to several of the city councillors and used the platform to establish further connections and extend his social network. His former business associate C. F. Tietgen also entered the council in 1861 but left it again after only four years. Thomsen valued his connections to Tietgen who, on his side, found

21. Vinding (1942), p. 211.

22. According to Marstrand (1928), p. 118. Jacob Marstrand, Mayor of Copenhagen 1904-1917, was one of the new "insolent disturbers of the peace" which were elected to the City Council in 1893. He recalled that Hagemann's attitude was much more sympathetic than Thomsen's.

23. See *Illustreret Tidende* of 23 February 1896, pp. 313-314, where Thomsen is portrayed as one of the most influential members of the City Council: "There was never established a significant committee without Thomsen as a member, and nor was there any serious debate without Thomsen taking part in it."

Thomsen's technological expertise useful and occasionally employed him as a kind of technical consultant. For example, when the Great Northern Telegraph Company planned to lay a cable between England and Denmark, Tietgen requested Thomsen's advice regarding the best isolation material for the cable. The traditional material for submarine cables was gutta-percha, a natural thermoplastic material, but Tietgen was aware of other proposals and asked Thomsen to evaluate the possibilities for an alternative.[24]

Another of Thomsen's friends from the days of the cryolite adventure, G. A. Hagemann, was a member of the City Council from 1882 to 1902, and he was no less active than Thomsen. Like him, he mostly dealt with technical matters, infrastructure, and building projects; also like Thomsen, he emphasised that the City Council was – or ought to be – non-political in the sense that the members did not represent the political parties. Thomsen took his work as a City Councillor seriously, as he did with all work in which he got involved, and he spent much time in the political negotiations with his fellow councillors. His focus was on problems of a technical and economic nature, whether transportation, energy supply, sanitation, or building constructions. In this work he established a fruitful collaboration with one of the mayors, Edvard D. Ehlers, a polytechnic candidate from 1835 who was responsible for the city's technical infrastructure. His contacts with Ehlers were professional as well as social. True to his habits Thomsen was frequently involved in minor disputes with, for example, his old collaborator L. A. Colding who until 1886 served as City Engineer.[25]

In addition to his work in the City Council, Thomsen was in 1863 appointed member of a commission to reform the country's system of measures and coins, and during 1882-1892 he served as chairman of the Harbour Council. In 1870-1873 he worked in a commission whose aim it was to improve the medical education and to coordinate it with the needs of the hospitals. A major result of the first commission was the coin reform of 1873 where the modern system of 1 krone = 100 øre was introduced. According to Hagemann,

24. Tietgen to Thomsen, 9 January 1868 (Royal Library, TSC).
25. Vinding (1941), p. 39.

for many years the technical expertise in the City Council was al-most identical with Thomsen. "The gasworks, the waterworks, and the sewage system are his municipal memorials," he exaggerated.[26] In 1903, at the age of 77, Thomsen was appointed chairman for a commission which should oversee the move of the University's ob-servatory at Østervold.[27] However, nothing came out of the commis-sion's plan and it took until 1946 before the inadequate observatory in Copenhagen was replaced by another one far away from the city.

Gas works and lines for lighting and other purposes were intro-duced in Denmark in the mid-nineteenth century, first in Odense in 1853, which at the time was the country's second-largest city. Co-penhagen followed suit four years later when more than 2,000 gas-lights were installed to replace the city's traditional oil lamps.[28] As mentioned in Section 2.3, the leading planning engineer in the early phase of the Copenhagen coal gas system was the polytechnic can-didate Georg Howitz with whom Thomsen had close connections and for a time entered a business partnership to turn the invention of the cryolite soda method into a manufactory. In 1858 Howitz became the first manager of the Copenhagen Gas Works (Vestre Gasværk). In Denmark as elsewhere the new technology was seen as a revelation; it attracted massive scientific, technological, and economic interest. One indication of the interest is the large num-ber of articles on the subject in *Tidsskrift*, the journal established by the Thomsen brothers in 1862. Most of the articles and notes deal-ing with gas technology were written by either Julius or August Thomsen.

Julius Thomsen's interest in the matter was primarily rooted in technological and economic considerations rather than in scientific curiosity. After having become a member of the City Council he went to Paris to study the energy economy of various kinds of gas burners and their possible uses in Copenhagen. Upon his return he made a series of systematic measurements with the purpose of find-

26. Hagemann (1909), p. 130.

27. *Aarbog for Københavns Universitet 1911-1912*, pp. 707-727.

28. See Hyldtoft (1994) for a detailed history of the Danish gas system during the nineteenth century.

ing the type of burner with the highest luminosity and least waste of gas energy in the form of heat. He estimated that as much as 30 per cent of the city's consumption of gas was wasted and that consumers would save up to 50 per cent if the efficiency of their burners was maximized. In his photometric experiments at the Copenhagen Gas Works he determined the gas consumption at varying luminosities for Argand burners and different designs of slit burners.[29]

These practically oriented experiments resulted in recommendations of which burners to use and how to regulate them in order to minimise the cost. Thomsen's general recommendation to the consumer was to use either an Argand burner or a carefully regulated slit burner. "The most important rule for the economy of gas illumination is this: use few and large flames."[30] As he pointed out, not only would this save money for the consumer it would also result in a cleaner, less polluted air. It was more than just recommendations, for in 1879 Thomsen, dissatisfied with the high prizes of consumer gas, brought up the question before the City Council with the result that the prize was actually reduced.

Danish scientific contributions to gas technology were few and relatively modest but not without local impact. The most important was undoubtedly Howitz's method of using bog iron to purify the coal gas (see Section 2.3), but also the engineer J. Charles Krayenbühl's invention of a coal tar centrifuge in the 1880s deserves mention.[31] As a member of the City Council, Thomsen was involved in the implementation of several of the innovations related to gas technology. For example, in 1881 he criticized the Copenhagen gas for its smell of sulphurous gases, especially hydrogen sulphide, arguing that this was in part the result of using bog iron as purifier; he suggested that the method should be supplemented with lime as purifier.[32] In 1879 Thomsen invented a new type of manometer which was

---

29. The much used Argand lamp or burner was originally an oil lamp invented by the Swiss scientist Aimé Argand (1750-1803) in 1780. The principle of Argand's invention was later adapted to gas burners.

30. Thomsen (1863c), p. 81.

31. For a brief description of Krayenbühl's invention see Kragh and Petersen (1995), pp. 287-288.

32. Hyldtoft (1994), p. 142. According to August Thomsen (1877), the gas in

more sensitive than existing instruments and particularly suited for measuring small differences in gas pressure. He suggested that his "sinus manometer" was not only of scientific but also of technological importance as it could be used to regulate the flow of air in ventilation shafts or gas lines.[33] Thomsen emphasised that the instrument would be useful in maintaining a constant gas pressure in the gas lines of large cities.

As Thomsen was keenly interested in gas technology, so he took an interest in the electric power system which was a subject of much discussion in the City Council in the late 1880s. But he was far from enthusiastic and thought for a while that public electricity supply was a luxury for the few that the city could not afford. After all, there was a well-functioning gas system and should the city now invest in a system that would compete with it and in all likelihood produce energy at a higher cost than previously? Besides, he considered electric power plants and the distribution of electricity to be scientifically complex tasks which required competences of quite a different kind than the gas system. Thomsen warned the City Council that "The person in charge of an electric lighting plant must have expert knowledge in mathematics and physics; for electric illumination is completely based on a scientific foundation and rests on calculations performed with the utmost accuracy. The operation of gas works, on the other hand, is an industry whose management does not to the same extent require mathematical knowledge."[34] After much discussion the City Council approved in 1889 to establish a municipally owned electric power station which opened three years later. Thomsen was initially somewhat sceptical but decided to follow the majority.

Shortly before leaving the City Council in 1894, Thomsen proposed a plan with the purpose of improving the city's sanitary conditions caused by the inadequate removal of night soil. His idea was

---

Copenhagen contained about 0.001 g sulphur per litre, which was substantially more than in London, Leipzig and Berlin.

33. Thomsen (1879a). I am not aware if Thomsen's manometer was actually used for these or other purposes. It was briefly mentioned in *Nature* **20** (1879): 46.

34. Quoted in Rode (1942), p. 39.

to establish a central station for night soil on the island of Amager. Some of the night soil should be sold to local farmers as fertilizer and the rest, together with waste water from the sewer lines, should be pumped into the Øresund waters by means of a pumping station driven by a powerful steam engine. Thomsen argued that there would be almost no pollution effects: "The mass of solid night soil which each day is discharged and mixed with salt water ... will be diluted in a ratio of 1 to 5 million or just one drop in four barrels of water. Its effect will at most be an insignificant change in the colour of the water; but at such a dilution a chemist will hardly be able to detect any amount of night soil at all."[35] The proposal was well received by the medical authorities. After protracted negotiations a pumping station for night soil, largely following Thomsen's plan, was established in 1901.

8.3. Personality and impact

Thomsen never wrote a textbook for the higher education; he made his chemical research experiments himself and his long list of publications in chemistry does not include a single one with a co-author; in his teaching he had but few assistants and even fewer graduate students; with one possible exception (Emil Petersen) he had no pupils. This and his generally unapproachable nature would suggest that he had very little impact on the course of chemistry in Denmark. Yet, although Thomsen did not form a scientific school or anything just remotely like it, on an individual basis his influence on the younger generation was considerable.

Among Thomsen's assistants was Christian D. A. Hansen who worked with him at the chemical laboratory of the University between 1870 and 1872. Trained as a pharmacist and without having matriculated to the University, Hansen won in 1872 its gold medal, the first time ever that the honour was awarded a non-academic. His essay on "the relationship between the crystal form of a substance and its chemical composition" was evaluated by Thomsen and the geology professor J. F. Johnstrup. While an assistant under Thom-

---

35. See Lindegaard (2001), p. 224, where the case is discussed in detail.

sen, and with his support, Hansen got interested in dairy chemistry and the problem of manufacturing animal rennet for cheese-making on a scientific basis. The result was a successful company which eventually grew into an international corporation in biotechnology.[36] Thomsen was impressed by the entrepreneurial spirit of young Hansen who, on his side, considered Thomsen his mentor. In 1891 the now wealthy businessman and privy councillor Hansen donated a large sum of money to establish the Pharmaceutical College (see also Section 8.1). The idea may have come from Thomsen, who approached Hansen one or two years earlier, suggesting that he used his wealth for the noble cause.[37]

In addition to Hansen, Thomsen had other assistants at his disposal which helped him with the lectures and the experimental demonstrations which were often parts of them. We have already mentioned his brother Thomas (Section 1.1) and also Emil Petersen (Section 6.3) who may have been the one closest to Thomsen and the only one to work, if only briefly, in his thermochemical research area. Erich C. W. Steenbuch, a pharmaceutical candidate and later an apothecary, worked as assistant at the University's chemical laboratory from 1872 to 1889, in part as responsible for Thomsen's lecture experiments. According to the recollections of Emil Koefoed, later a professor of chemistry at the Pharmaceutical College, "His [Thomsen's] lecture experiments were regarded as the very best of what could be performed on a lectern." But, he continued: "Woe to assistant Steenbuch if something went wrong; he would then be rebuked in front of the audience, which we felt was embarrassing. Thomsen's lectures were good but more fitted for the polytechnic than the pharmaceutical students. ... The tone in the laboratory was not very pleasant, and especially not under Steenbuch. We had the impression that as he had been bullied by Jul. Thomsen, so he would take revenge on us."[38]

---

36. For Hansen's essay see *Aarbog for Københavns Universitet 1871-1873*, p. 238. See also Kruse and Kofod (1992) and Kragh and Petersen (1995), pp. 201-202. Today Chr. Hansen A/S is a global company producing natural ingredients for the food, beverage and agricultural industry with more than 2,300 employees.

37. According to Koefoed (1974), p. 53.

38. Koefoed (1974), p. 47 and p. 50. His impression is confirmed in Bergsøe (1945), p.

Haldor F. A. Topsøe served as chemical assistant under Thomsen from 1867 to 1873 during which period he completed a doctoral dissertation. In 1876, after the retirement of Jesper Bahnson, he was appointed chemistry teacher at the Army's Military Academy, since 1868 the successor institution of the Military High School; this was the same position which Thomsen had unsuccessfully applied for back in 1859 (Section 1.1). A specialist in crystallography, Topsøe was elected a member of the Royal Danish Academy in 1877. As a member of the Academy and also as General Secretary of the meetings of Scandinavian scientists in 1892 (Copenhagen) and 1898 (Stockholm), he must have been in close contact with Thomsen, who served as President of the 1892 meeting (Figure 8.3). The two chemists shared an interest in Greenland and its cryolite deposit, and after Thomsen's death Topsøe became director for the Cryolite Mining and Trading Company (Section 2.5). Johan Kjeldahl, one of Denmark's most important chemists, was not an assistant to Thomsen but as a student at the Polytechnic College he followed his lectures. He later became the first director of the Carlsberg Laboratory. Kjeldahl appreciated Thomsen with whom he mixed not only professionally but also socially.[39]

Numerous chemistry students at either the Polytechnic College or the University listened to Thomsen's lectures, and a few of them later made important contributions to science, technology or culture. Ejnar Hertzsprung has a name in the history of astronomy as a pioneer of stellar classification, witness the Hertzsprung-Russell (HR) diagram relating the luminosity of stars to their colours; but he was trained as a chemical engineer and not as an astronomer. According to one source, he became interested in chemistry when reading a book by Thomsen.[40] Having received his degree as a poly-

---

110, and also in Christensen (1910), p. 168: "Only very rarely did one of his [Thomsen's] experiments fail; he demanded that they were carefully prepared and could be rather rough [with his assistant] if he thought that the unfortunate result was due to inadequate preparation." Jensen (1984) suspects that Steenbuch prepared most of the solutions and other substances that Thomsen used in his thermochemical investigations.
39. According to Jerslev (1963), p. 121.
40. See Strand (1968) who refers to "a small book on this subject [chemistry] by the famous Danish chemist, Julius Thomsen." The book must have been Thomsen

Figure 8.3. The fourteenth meeting of Scandinavian scientists, Copenhagen 1892. Thomsen, who served as President of the meeting, is in the lower right part of the composite photograph; H. Topsøe, the meeting's General Secretary, is placed second to his right. Source: Peter C. Kjærgaard, ed., *Dansk Naturvidenskabs Historie*, vol. 3 (Aarhus: Aarhus Universitetsforlag, 2006), p. 137.

technic candidate in 1898 Hertzsprung worked as an engineer for a German acetylene company in St. Petersburg. In 1901 he went to Leipzig to study photochemistry under Ostwald and prepare for a doctoral dissertation. Private circumstances caused him to return to Denmark and it was only then that he changed, with great success, from chemistry to astronomy. However, as an astronomer Hertzsprung did almost all his work in Germany and the Netherlands.

From his studies at the Polytechnic College Hertzsprung may have known another student of chemical engineering, the four

---

(1853b). Kai Aage Strand was a student of Hertzsprung, who told him about his youth in a conversation of 1961.

months older Paul Bergsøe who graduated in 1899. Bergsøe has a name in Danish history as a chemical engineer, founder of a metallurgical company, and an accomplished author and science populariser. He followed for a period Thomsen's lectures for polytechnic and medical students, about which he recalled: "As he stood there behind his lectern in the University's chemical auditorium in Østervoldgade, there was something distant as well as awe-inspiring about him. Busy assistants had covered the long table with flasks and other glassware; the lecture was constantly accompanied by experiments which from time to time resulted in fireworks with smoke and vapours. Thomsen, who was dressed in a frockcoat, took himself and his lectures very seriously."[41] When Bergsøe was a young man he discovered an electrolytic method of removing tin from tin clippings. His efforts to establish a manufacture based on the method received support from Hagemann and Thomsen. According to Bergsøe's memoirs, Thomsen confirmed that the method agreed with the relevant thermochemical data.

The most famous pair of Danish inventors is arguably Peder Oluf Pedersen and Valdemar Poulsen who were both, each in his own way, indebted to Thomsen. A 15-year-old farmer's boy from Jutland, Pedersen thought in 1889 that he had made an invention of a new, eternally working watering device and had the audacity to address the king, requesting support for his idea. His letter ended up on the desk of Thomsen, who requested the engineer Simon Borch to look at it. Although Borch quickly realised that Pedersen's machine, essentially a *perpetuum mobile*, would not work, he and Thomsen were impressed by the boy's mechanical talent. There is reason to believe, Thomsen wrote, "that the applicant would be able to make progress in the field of technology if he, by following a course at a technical institute, could attain the theoretical skills that are necessary for work of this kind."[42] Thomsen raised economic support for the youngster and arranged for him to come to Copenhagen to stay with Borch. In 1892 Pedersen entered the Polytechnic College from where he graduated as a construction engineer

41. Bergsøe (1945), p. 109.
42. Quoted in Larsen (1950), p. 210.

Figure 8.4. Thomsen in his private residence in Ny Vestergade 11. Photograph from ca. 1900, as reproduced in Vinding (1941), p. 45.

five years later – eventually to become director of the College in 1922.

In a radio broadcast of 1941 Pedersen spoke movingly of the "unique helpfulness and kindness" which Thomsen had showed him as a teenager. Thomsen evidently cared about the young man: "He arranged that I entered a gymnastics class; he paid the expenses for it and undoubtedly also controlled that I actually participated in the class. Many a Sunday evening I was a guest in his home … [and] during the summer these evening parties were sometimes extended to a visit in Tivoli."[43] After Pedersen married in 1899 he and his wife often visited the Thomsen family. The last time, in company with Poulsen, was a few weeks before Thomsen's death.

---

43. Pedersen (1941), based on broadcast lecture of 28 May 1941. Tivoli is the amusement park in the centre of Copenhagen.

In 1889 Poulsen entered the University of Copenhagen to study medicine but soon dropped out. A subsequent attempt to pass the entrance exam to the Polytechnic College failed because of poor grades in mathematics and instead he became employed as a technical assistant at the Copenhagen Telephone Company. This led in 1898 led to the invention of the "telegraphone," the world's first functional magnetic recorder. He soon teamed up with Pedersen with whom he developed an arc transmitter for wireless telegraphy. For a period this technology known as the "Poulsen arc" was a much-used alternative to other types of wireless telegraphy based on Marconi's system. Although Poulsen, an autodidact mechanical genius, never studied at the Polytechnic College, during his brief period as a student of medicine he listened to Thomsen's lectures in chemistry. They made a lasting impression on him and wetted his appetite for science and its applications:

> Never before nor later did I, by nature a scientifically and artistically inclined student, experience more enthralling lectures … marked by the famous chemist's dazzling experiments and his picturesque, authoritative appearance. Thomsen's rather grim and somewhat melancholic look disappeared on occasions, as when he advised the students of an upcoming experiment: "Now, my dear gentlemen, you will observe a rather lively reaction," and immediately thereafter the auditorium would resound with a thundering boom that more than woke up those sleeping unworthily on the back rows.[44]

In 1900, when the Association of Danish Engineers inaugurated its new premises in central Copenhagen, Hagemann decided to honour six young Danes for their inventiveness, including Poulsen, Pedersen, and Bergsøe. This he did by awarding them a bronze copy of the gold medal which had been presented to Thomsen on the occasion of his seventieth birthday four years earlier. The copies were made with Thomsen's permission and Hagemann pointed out

---

44. Poulsen (1930). According to Pedersen (1941), Poulsen was fascinated by Thomsen: "None other of Poulsen's teachers made as deep an impression on him or influenced his development as strongly as Thomsen did."

that they should remind the recipients of "the man whom we all look up to with such reverence and respect."[45]

Thomsen took pride in his lecture demonstrations, which he perfected over the years and in some cases developed himself. He could be a bit of a chemical showman, although of the serious kind. In 1870 he reported to the German Chemical Society about some particularly "simple and elegant" chemical reactions illustrating reversible combustions that were suitable for lecture experiments.[46] One of the spectacular experiments demonstrated how the burning of hydrogen in an oxygen atmosphere ($2 H_2 + O_2 \rightarrow 2 H_2O$) can as well be conceived a burning of oxygen in an atmosphere of hydrogen.

In the 1880s Thomsen lectured four times a week on inorganic and general chemistry for a large audience consisting mainly of medical and pharmaceutical students but also with some polytechnic and university students in the audience. Victor Klæbel, a pharmaceutical candidate of 1884, recalled Thomsen as an excellent speaker whose lecture experiments made quite an impression on the students. Thomsen preferred to leave the laboratory exercises to his assistants, but occasionally he turned up in his frockcoat, asking a few questions to the students. "We all had a great respect for him, I believe; there was no giddiness when he was around."[47]

Of course, not all students shared Poulsen's enthusiasm with regard to Thomsen's qualities as a teacher. Some found his lectures too elementary, more directed toward medical students than to serious students of chemistry. "Julius Thomsen hardly had much understanding of the views of the younger generation," an engineer reminisced; "it is doubtful if he could be called a beloved teacher."[48] The author of another obituary, the chemist Hans Jessen-Hansen,

---

45. Vinding (1942), p. 465; Bergsøe (1945), p. 159. The Julius Thomsen medal awarded in 1900 was Hagemann's donation and only given on this occasion. The other recipients were Aa. Kirchner, A. Theilgaard, and C. J. Kielberg. What is presently known as the Julius Thomsen Medal dates from 1936.

46. Thomsen (1870d); Bjerrum (1909), p. 4978.

47. Quoted in Kruse and Kofod (1992), p. 24. Klæbel found Jørgensen's lectures less satisfactory: "Contrary to Thomsen, he did not have a fortunate hand when he demonstrated experiments; they sometimes went wrong."

48. Christian Petri in a paper of 1909. Here quoted from Lundbye (1929), p. 376.

referred to Thomsen as "an elegant and bold experimenter ... [whose] experiments left an impression on all listeners, even though he did not win their affection to any extent."[49] Carl F. Jarl, an engineer who had followed Thomsen's lecture course in the 1890s, similarly highlighted the lecture demonstrations: "His lectures on introductory chemistry were at times somewhat monotonous, but he was an exceptionally good experimenter who impressed [the audience] by his well-prepared and boldly performed experiments."[50]

The late nineteenth century was the period in which female students first made their entry in the academic halls of science. In Denmark women were allowed to the university in 1875 and fourteen years later Sofie Rostrup, a student of plant pathology, became the first Danish woman to defend her master's thesis on a subject of natural science.[51] Kirstine Meyer, one of the first female graduates in physics and the aunt of Niels Bjerrum, recalled about her laboratory training in chemistry in about 1886 that she did not experience any problems because of her gender. S. M. Jørgensen outfitted a temporary changing cubicle for women near the old chemical laboratory. "Our famous chemist Julius Thomsen," on the other hand, "was not a friend of women entering academic studies; he preferred to ignore their existence." As director of the Polytechnic College and responsible for the furnishing of the new buildings Thomsen "opposed any special arrangements in the toilets and changing rooms that might anticipate the presence of women among the student body – better close the eyes for the possibility."[52]

But whether Thomsen wished it not, women did enter the study of chemistry at the new Polytechnic College. In 1897 the two first women graduated in chemical engineering. Thomsen was a personal friend of the distinguished Uppsala chemist Per T. Cleve, who was a supporter of women's equality. In 1898 Astrid Cleve, the

---

49. Jessen-Hansen (1910), p. 348. Jessen-Hansen, a polytechnic candidate of 1891, worked from 1892 at the Carlsberg Laboratory as an assistant of J. Kjeldahl and S. P. L. Sørensen.

50. Jarl (1909b), p. 53. Jarl served 1902-1911 as director of the Øresund Chemical Factories and wrote the first history of the company, Jarl (1909a).

51. Kragh et al. (2008), pp. 349-356; Nielsen (2000), p. 77.

52. See Meyer's untitled article in Jacobsen (1925), pp. 81-85.

daughter of Per Cleve and his wife Alma, became the first woman in Sweden to earn a doctorate in natural science; after she married the German-born biochemist Hans von Euler-Chelpin the couple entered a fruitful scientific collaboration in organic chemistry. Thomsen, by then retired, followed the development in Uppsala with surprise but also with genuine interest, perhaps even admiration. In a letter of 1908 to Alma Cleve, he wrote, "I was interested to hear how Astrid and her husband worked together in the scientific field and divided their time between science and their three children."[53]

Thomsen sharply separated his private life from his life as a scientist, an administrator, and a participant in public affairs. He did not consider his private life and views to be of interest to anyone but himself and his family, and consequently he never wrote about himself except in a professional, terse, and factual way. Contrary to many other scientists of fame, he did not write any kind of autobiography. In 1879, when receiving an honorary doctorate from the University of Copenhagen, Thomsen was obliged to write an autobiographical sketch but characteristically it was brief, kept in the third person, and limited to a factual account of the most important steps in his scientific career.[54] It was no more personal or emotional than his long tables of thermochemical data. But, as we have seen in previous chapters, Thomsen did have emotions and he did have a life outside the laboratory and the conference rooms. Indications of this life and Thomsen's personality generally appear from his extensive correspondence with members of his family. As historian of science Finn Aaserud points out, "the private sphere of scientists is far less documented that that of other groups of intellectuals, which is part of the reason that it has only received scant attention among historians of science and other interpreters of the documentation that scientists have left behind."[55] One might imagine that Thomsen's brothers, daughters or wife exerted some kind of indirect im-

---

53. Thomsen to A. Cleve, 24 January 1924, Uppsala University Library, as quoted in Espmark and Nordlund (2012), p. 86. In 1929 Hans von Euler-Chelpin was awarded the Nobel Prize in chemistry for his work on the role of enzymes in alcoholic fermentation. By that time he and Alma had divorced.

54. Thomsen (1879b).

55. Aaserud and Heilbron (2013), p. 3.

pact of how he approached science or thought about it in wider contexts, but I have found nothing in his personal correspondence or other sources that suggest such an impact.

Another source that illuminates Thomsen's personality is the dozen or so obituaries and memorial articles that were written on the occasion of his death, most of them in Danish but also a couple in other languages. Although obituaries are of course problematic as historical sources and have to be looked at in a critical light, if taken together they convey information that cannot be found elsewhere. Almost all obituaries and contemporary sources refer to Thomsen's fighting nature, his temper and strong opinions, and to his dislike of being contradicted.[56] The many scientific controversies in which he was involved speak their own language. "Thomsen was a fiery and crusty nature," wrote Bjerrum. "He could attack an opponent quite without self-command ... but shortly after the thunderstorm he would again be calm and friendly."[57]

As Thomsen saw it, life was one long struggle and he wanted to win every battle whether large or small. His self-confidence and stern authority derived from his fighting experiences which convinced him that in scientific as well as other matters he was usually right. If people found him reserved and cool, even unfriendly, he did not mind. Thorpe wrote in his obituary to the Chemical Society:

> The students respected and even feared him, but his cold and unsympathetic nature evoked no warmer feelings. It was said of him by one who knew him intimately that he never learned to draw the young to him, to create in them an interest for his work, to form a school. Thomsen was a homely man, but not even in his home, says the same authority, was it possible for him to change his active, earnest, strenuous disposition – what his friends called his fighting character.[58]

---

56. Apart from shorter notes, obituaries and memorial articles include Bjerrum (1909), Hagemann (1909), Jørgensen (1909), Muir (1909), Jarl (1909b), Christensen (1910), Thorpe (1910), Jessen-Hansen (1910), Brønsted (1909) and (1932), Christiansen (1929), Vinding (1941), and Pedersen (1941).

57. Bjerrum (1909), p. 4987.

58. Thorpe (1910), p. 165. The authority which Thorpe relied upon may have been Bjerrum.

Indeed, phrases like "more feared than beloved" appear in several of the obituaries, even the more praising of them. This was the case with Hagemann's memorial address given on 2 March 1909 at a solemn ceremony in the Polytechnic College. Hagemann, who had known Thomsen for half a century, was not blind to the weaknesses of the great chemist but he focused on his undeniable importance for Danish science and society. Ending his address he was moved to cite Henry Longfellow's "Psalm of Life":

> Life of great men all remind us,
> We can make our life sublime,
> And, departing, leave behind us,
> Footprints on the sands of time.[59]

Thomsen could certainly be difficult to work with and even more+ difficult to work under. His close colleague S. M. Jørgensen generally had a good relationship to Thomsen but on occasions found him difficult and annoying. Jørgensen was a life-long friend of the Swedish mineralogist and chemist Christian W. Blomstrand with whom he shared an interest in the structure of complex inorganic compounds. The two exchanged many letters and in a few of them Thomsen turns up. Thus, in a letter of 1886 Blomstrand referred to his critical view of some of the ideas of Thomsen, "who cannot stand any sort of criticism from older workers like myself."[60] Jørgensen responded that Thomsen was "unpredictable and erratic, even if he is getting much better now, as the years pass, than he ever was before." In a letter three years later Jørgensen mentioned that "Thomsen is director of the Polytechnic College and for many years has been impossible to work with."[61]

---

59. Hagemann (1909), p. 132.

60. This may have been a reference to Thomsen's investigations a few years earlier of complex platinum double salts, as mentioned in Section 4.3. Blomstrand thought that the compound Thomsen had found had a chain-like character, contrary to Thomsen's formula. See p. 88 in Kauffman (1976), where Blomstrand's paper of 1871 is translated into English.

61. The letters of 1886 and 1889 are summarised in Kauffman (1977). In a letter of 1896 Jørgensen referred to Thomsen's seventieth-year birthday, calling him a great

It was Jørgensen's task to commemorate Thomsen at a meeting in the Royal Danish Academy on 5 March 1909 attended by the king, Frederik VIII. Jørgensen dwelled upon his colleague's exceptional energy, intelligence and perseverance which had resulted in *Untersuchungen*, a work he called "the pride of any Danish chemist and the most important Danish contribution to the development of chemistry." Jørgensen further suggested that Thomsen belonged to the same pantheon as Tycho Brahe, Ole Rømer, and H. C. Ørsted. But he also added a word of reservation: "Thomsen was not a learned man; it was at times striking how ignorant he could be of what had been done earlier. On the other hand, he was undoubtedly one of the most brilliant experimenters ever. And chemistry, being an experimental science, is all about experiment."[62] Although Thomsen was not a learned man in the traditional sense, he was an avid reader of the scientific literature and collector of books. At about the time of his retirement he donated part of his private library, some 1,150 volumes, to the library of the University's Chemical Laboratory.[63]

Jørgensen's emphasis of Thomsen's unbound energy was appropriate; a main reason for Thomsen's successful career was simply his enormous work capacity. For him, work was the very essence of life and not something he had to do in order to earn a living for himself and his family. When he retired in early 1902 he just continued working in his laboratory as he had always done. When he received the Grand Cross of the Dannebrog Order (Storkorset af Dannebrog) in 1896 he had to choose a motto for his coat of armour. His choice was "Work Makes Life Valuable."

There was another side of Thomsen, one far away and well hidden from the ever-active and fierce laboratory workaholic. He was not after all a cold and brusque rationalist whose only concern was his scientific career. P. O. Pedersen wrote about Thomsen's "shell of

---

master in the art of experimentation. Thomsen met Blomstrand at the 1880 meeting of Scandinavian scientists and possibly at other occasions.

62. Jørgensen (1909), p. 30.

63. Jensen (1983), p. 504. Sadly, the valuable collection of books and periodicals, including a complete collection of *Annales de Chimie*, does not exist any longer. The books have been dispersed to various libraries or just been thrown away.

cool unapproachableness" but believed it was just that, a shell covering "a more gentle and emotional inner."[64] As mentioned in Section 1.1, as a young man Thomsen was much attached to poetry; he was social and extrovert, with many interests besides science, and when he got older he maintained an interest in music and art. He liked to go to the theatre and used to recreate, and sharpen his mind, by solving mathematical problems. This pastime led in one case to a paper on one of the classical (and unsolvable) problems of mathematics, the trisection of an angle by means only of a compass and an unmarked straightedge.[65]

After having settled as a professor Thomsen often went on travels either in Denmark or abroad. Remarkably, these travels were rarely work-related but vacations to places he wanted to experience just because of their scenic beauty or cultural history. He loved the mountains and on several occasions travelled to Switzerland, Austria, Norway or the Harz mountains in Germany to enjoy the beautiful nature. On these travels he was normally accompanied by a few friends or members of his family. But even when on vacation he wanted to be in charge and make his preparations with scientific precision. According Bjerrum, "He was an excellent tour leader which arranged everything and could tell about everything. He could cite the heights of the mountains and give lectures on their geology and mineralogy. On these travels he always had to be the one who made the decisions. ... Thomsen not only kept a diary during his travels, he also carried with him a sketch book in which he drew neat pictures of what caused his excitement."[66]

For example, in the summer of 1887 Thomsen went on an extensive journey with his daughters Ellen and Johanne. The company travelled over Berlin and Dresden to Prague, where they visited the city's impressive castle and the Tyn Church housing the tomb of the

---

64. Pedersen (1941).

65. See also Pedersen (1941). The paper, Thomsen (1900), is incorrectly listed in Thomsen's bibliography of his own works as dating from 1890. Thomsen did not claim to have solved the problem but just proposed a way in which an angle could be trisected by means of a hyperbolic construction.

66. Bjerrum (1909), p. 4977. Some of Thomsen's sketch books and travel diaries are kept at the Royal Library in Copenhagen.

famous Tycho Brahe. From Prague they went through Bavaria to Salzburg and from there to Switzerland, walking in the scenic mountains. The route back to Copenhagen was the direct one over Munich and Berlin. Throughout the journey Thomsen made many pencil drawings and also a few paintings in watercolour of what he saw, that being landscapes, houses, monuments or peasants in their traditional clothes (Figures 8.5 and 8.6). Thomsen's human side is further illustrated by episodes in his life such as his concern for the young P. O. Pedersen mentioned above. Moreover, for a year or so his nephew Thomas Marius Thomsen, an eighteen-year-old son of Sigismund Gotthelf Thomsen, lived in his home in Ny Vestergade. The young man had just passed his high school exam and under the influence of his uncle he enrolled as a chemistry student at the University. However, he soon changed to literary and historical studies, eventually to become a highly respected archaeologist and ethnographer at the Danish National Museum.[67]

The human dimension in Thomsen's life was emphasised in a more direct way by his daughter Johanne, who at the age of 59 devoted a booklet written in verses to her father and mother. There she gives a picture of Julius Thomsen as a loving father who cared for and used much time on his family.[68] She recounts the family's vacations on Langeland, where her mother came from, and how her father used to read books for her. He frequently went with his daughters to the nearby National Museum. She tells about her parents' devastation when their son died in 1883 and how her father immersed himself in work to escape thinking of the tragic loss. Johanne recalls how her father took her and her older sister Ellen with him to Switzerland, where they saw the large glaciers, and how they payed a visit to Tycho Brahe's tomb in Prague; and she describes her fascination by looking in silence and awe on her father working in his magic laboratory filled with strange devices and bottles with

---

67. For Thomas M. Thomsen (1870-1941), see *Dansk Biografisk Leksikon*. In his diary from the time he described several times his uncle as "patronising." Royal Library, Copenhagen, box NKS 4306-4°.

68. Drechsel (1926). The many personal letters kept at the Royal Library in Copenhagen confirm Thomsen's affection for his family and also the sharp distinction he drew between his private and public life.

Figure 8.5. Pencil drawing by Thomsen from 1887, showing the Prague Castle. Royal Library, Copenhagen.

colourful liquids. The Julius Thomsen she speaks about is very different from the one known in public or from the chemistry auditorium. And yet it was the very same person.

8.4. Thomsen's legacy

We have dealt in some detail with Thomsen's unique status in Danish science and culture, such as demonstrated by his role in the University of Copenhagen, the Polytechnic College, the Royal Danish Academy, and the Copenhagen City Council. As briefly mentioned in Section 1.1, this status appears in an artistic form from the two large paintings made by Peder Severin Krøyer, a celebrated artist who in his youth had made scientific illustrations for the zoological literature. The two paintings have different histories. One of them, depicting a meeting in the Royal Academy and completed in 1897, was a gift from the Carlsberg Foundation on the occasion of the inauguration of the Academy's new premises.[69] The speaker standing at the blackboard is the zoologist Japetus Steenstrup, who had a status within the Academy as high as that of Thomsen (Figure 8.7). Not only had Steenstrup served as secretary 1866-1878, he was also the Academy's oldest member in terms of age and seniority. Moreover, he was close to the Carlsberg Foundation. The other central figure is Julius Thomsen, sitting watchful and upright at the table. He is the president, in full command of the session and placed next to (on his right side) the crown prince. The second person to the left side of Thomsen is S. M. Jørgensen. Also the chemists O. T. Christensen and J. Kjeldahl are present, and so is the fermentation physiologist Emil Christian Hansen.

Krøyer's painting of "Men of Industry" dating from 1904 was Hagemann's invention and personal project.[70] Wanting to create a monument for Danish technology and industry he arranged with

---

69. See Lomholt (1954) for the history of the painting.

70. A detailed analysis of the painting in its historical and social context is given in Nielsen and Wistoft (1996) and (1998). The impressive painting was paid by Hagemann and placed in his private home. It was later bought by the Frederiksborg National Museum where it still hangs.

Figure 8.6. Painting in watercolour by Thomsen, dating from his journey to the Alps in 1887. Royal Library, Copenhagen.

Krøyer how to place the select 53 men of industry representing the country's technological progress (Figure 8.8). The scene of the painting is the machine hall of a new electricity plant in Copenhagen dominated by powerful steam engines and electric dynamos. It is all about power, technological and human, with the relaxed industrialists ruling confidently over the machines in the new temple of electricity. In the painting's central part are placed nine older gentlemen listening to chief engineer Ib Windfeld-Hansen, a pioneer of Copenhagen's electricity supply. One of the gentlemen is Thomsen, leaning against a railing, and behind him and half covered by him is S. M. Jørgensen. Given Thomsen's role in the cryolite industry and the Polytechnic College, not to mention Hagemann's admiration for him, it is understandable that he figures prominently in the painting. It is a little more surprising to find Jørgensen among the men of industry as he was not, in fact, engaged in technological or industrial projects. But his position as director of the Carlsberg Foundation may have been the reason to include him. Notice that Hagemann has modestly placed himself in the upper left corner of the painting.

Few Danes (and even fewer foreigners) will today know who Julius Thomsen was. If they do they may associate the name with a street and a square in the Frederiksberg area of Copenhagen, the area where the chemist spent the last part of his life. Julius Thomsen Street and Julius Thomsen Square were so named in 1925. They may also have passed his statue placed in front of the old Polytechnic College which is currently part of the National History Museum of Copenhagen. The statue, revealed on the occasion of Thomsen's 80-year's birthday in 1906 but dating from 1897, was made by he celebrated Danish sculptor August Wilhem Saabye. Several of Saabye's statues and busts can be seen in Copenhagen, the most popular being the statue of Hans Christian Andersen in the Rosenborg Castle Gardens. Some of the country's best known painters used Thomsen as a model. Before he entered Krøyer's paintings he was portrayed by Carl Bloch in 1881 and by August Jerndorff in 1896; the latter painting is presently at the Frederiksborg Castle Museum. Jerndorff also portrayed two of Julius Thomsen's daughters, Anna Sophie Frederikke in 1890 and Johanne in 1894, in both cases ordered by their father.

Figure 8.7. Part of P. S. Krøyer's painting of a meeting in the Royal Danish Academy of Sciences and Letters.

Danish chemists and engineers may have heard of the Julius Thomsen Medal awarded for meritorious achievements in chemical engineering since 1936, but in most cases they will be unaware of the person after whom the medal is named. They will not come across the name in their textbooks and courses not even if they choose to follow a course in history of science. As seen in retrospect Julius Thomsen did not make any discovery of lasting significance and his work did not change the course of chemistry. In current standard books on the history of chemistry his name may turn up but if so only briefly and typical in conjunction with Berthelot as a co-founder of classical thermochemistry.[71] This branch of chemistry came to be regarded as a somewhat dull and peripheral area, but in the second half of the nineteenth century it was at the very heart of chemical front research. To describe Thomsen's work as "foreign to the central concerns of contemporary chemists," as one historian has done, is quite mistaken.[72] Apart from his work in thermochem-

71. Berry (1968), for a time a widely read account of the history of chemistry, dealt in some detail with the Thomsen-Berthelot controversy (on pp. 33-35).

72. Servos (1990), p. 19.

Figure 8.8. Part of P. S. Krøyer's painting of the leaders of Danish industry and technology.

istry, Thomsen's contributions to the periodic system continue to attract some historical attention. His bold interpretation of the system in terms of atomic constitution turned out to be prophetic (Section 7.4), but this was realised only later on. Had Thomsen been more open to the physicists' ideas about electrons as the primary constituents of matter his contribution might have had a greater impact.

Although Thomsen was indeed an experimental chemist who enjoyed and immersed himself in meticulous and often tedious measurements of thermal effects, this was only part of his work and attitude to science. He was not in fact a narrow experimenticist who identified science with data obtained from precise laboratory measurements. The higher theoretical aims of chemistry were always in his mind. As we have seen, on many occasions Thomsen speculated quite freely on more fundamental questions such as the composite nature of atoms and the existence of new elements. Conservative he

Figure 8.9. To the left, Thomsen in his laboratory, probably from late 1890s; to the right, the statue at the Polytechnic College. Image credit: DTU History of Technology.

was in many ways and yet he was not afraid of suggesting ideas that deviated from the consensus view shared by the majority of chemists. For example, his hypothesis of the octahedral structure of benzene was unorthodox and rightly described as "a bold push beyond the beaten tracks of chemistry" (Section 5.2). No less unorthodox was the hypothesis that he entertained at the end of his life, that argon might not be elementary but rather of a compound nature. In both cases his unorthodoxies beyond the the beaten tracks of chemistry proved unfounded.

In the half-century from about 1860 to 1910 Thomsen enjoyed a very high reputation in Danish science and cultural life generally. In 1899 he figured on the front page of *Illustreret Familie-Journal* (Illustrated Family Journal), a popular and widely read borgeois journal founded in 1877 (Figure 8.10). According to the journal Thomsen

Figure 8.10. Thomsen as a public celebrity.

was one of Denmark's greatest sons ever, a man who had "cast his lustre over the country and who the entire nation must look up to in reverence and gratitude."[73] Although Thomsen had no literary ambitions he and his work were considered part of the Danish literary tradition in a broader sense. For example, he was among the scientists included in the massive and widely read *Illustreret Dansk Litteraturhistorie* (Illustrated Danish History of Literature) written by the author H. C. Peter Hansen. The section on Danish chemical literature was due to S. M. Jørgensen, who called attention to Thomsen's recently completed *Untersuchungen*, a publication which he characterised as "unequalled in richness and systematic coverage."[74]

As seen from an international perspective, Thomsen was undoubtedly the best known Danish chemist of his time, and possibly the best known scientist generally. His work was widely known and cited, not only by chemists but also by eminent physicists such as Max Planck and J. J. Thomson. As a somewhat curious illustration, in his *Dialektik der Natur*, a collection of notes written in the period 1872-1882 but only published in 1927, Karl Marx's close collaborator Friedrich Engels referred several times to Thomsen.[75] Much interested in science and its historical development, Engels based in part his system of dialectical natural philosophy on the two conservation laws of matter and energy. When discussing the chemical processes in an electric battery he cited Thomsen's thermochemical data. Engels's sources were not Thomsen's own publications but a textbook in general chemistry written by Alexander Naumann, the German chemist with whom Thomsen had been involved in a minor controversy. Few Danish scientists have been included in the authoritative *Encyclopædia Britannica*, but Thomsen belonged to the exclusive class. He was described in an article in the famous 1911 edition as a scientist whose "name is famous for his researches in thermochemistry" and also as the inventor of the cryolite soda man-

---

73. *Illustreret Familie-Journal*, 5 February 1899.
74. Hansen (1886), vol. 2, p. 759.
75. Engels (1951), pp. 124-141.

ufacturing process.[76] The article even mentioned his two brothers, August and Thomas.

Except for the Nobel Prize, Thomsen received almost all honours a scientist of his time could dream of, the most prestigious of which was perhaps the 1883 Davy Medal from the Royal Society (Section 5.1). Six years earlier he had been elected a foreign honorary member of the Chemical Society of London and in 1891 a similar honour was bestowed upon him by the Royal Institution.[77] Late in life honorary memberships of the American Chemical Society (1905) and the German Chemical Society (1907) followed.[78] Thomsen became a corresponding member of the prestigious Prussian Academy of Science in 1900, at a time when the only other Danish member of the scientific class was the botanist Eugen Warming, who was elected the previous year. In 1902 Thomsen was elected a foreign member of the Royal Society, a rare honour for a Danish scientist and the only time it has fallen on a chemist. Among the previous foreign members were H. C. Ørsted (1821) and J. J. Steenstrup (1863). Only in 1926 was another Danish scientist elected member of the Royal Society. His name was Niels Bohr.

As to foreign academies Thomsen was also a member of the Royal Swedish Academy of Science (1880), the Reale Accademia dei Lincei in Rome (1883), the American Academy of Arts and Sciences in Boston (1884), the Académie Royale des Sciences de Belgique (1887), and the Christiania Scientific Academy in Norway (1891). In addition he received honorary doctorates from the universities of Copenhagen and Uppsala. With regard to the Nobel Prize, by the early twentieth century Thomsen's work in thermochemistry was neither novel nor of the importance it had been twenty years earlier. Neither he nor Berthelot was ever nominated (but his contemporary Mendeleev was, unsuccessfully, nominated nine times between

---

76. *Encyclopædia Britannica*, 1911 edition, p. 817, in which it was wrongly stated that Thomsen retired from active work in 1891.

77. *Nature* **44** (1891): 184. Other distinguished physicists and chemists elected honorary members included Gibbs, Helmholtz, Mendeleev, Berthelot, Bunsen, and Cannizarro.

78. See *Science* **22** (1905): 84 for the American distinction which was also conferred to Nernst, Arrhenius, and the Dutch physical chemist Hendrik Roozeboom.

1905 and 1907). As a professor at the University of Copenhagen, Thomsen had the right to nominate scientists for the Nobel Prize but he did not make use of it.

On the occasion of the twenty-fifth anniversary of the American Chemical Society in 1902 several distinguished chemists were invited to New York to celebrate the anniversary, among them Mendeleev, Berthelot, van't Hoff, Ramsay, James Dewar, and Emil Fischer. Thomsen also received an invitation but he, like most of the European chemists were unable to come and instead sent telegrams of congratulation. "Dear Sir," Thomsen wrote, "I beg to thank you for your invitation to assist at the celebration of the twenty-fifth anniversary of the foundation of the American Chemical Society, which I received to-day, but very much regret to be prevented in accepting the same. I, however, beg to present my best compliments and wishes for the prosperity of your society and remain, dear Sir, yours very truly, Julius Thomsen."[79] While Thomsen was thus highly regarded in Scandinavia, Germany, England, and the United States, he was never honoured with any kind of distinction from France, possibly due to the influence of Berthelot. But then Berthelot was never considered for election as a foreign member of the Royal Danish Academy of Sciences.

As Thomsen and his younger colleague Jørgensen had initiated a new phase in Danish chemistry in the 1860s, so another change of guard, a much needed generational shift, occurred in the first decade of the twentieth century. Thomsen had retired in 1902 and he was followed by Jørgensen in 1908; Thomsen's successor Emil Petersen died in 1907. As a result, a new university professorship in chemistry was announced in 1908. The two 29-year-old candidates, Niels Bjerrum and Johannes Brønsted, were highly promising chemists representing new-style chemistry.[80] After a competition including oral performances, on 17 December 1908 Brønsted was chosen as the new professor (Bjerrum was appointed professor at

---

79. Telegram addressed to Albert C. Hall, secretary of the American Chemical Society. Reproduced in *Twenty-Fifth Anniversary of the American Chemical Society* (Easton, PA: Chemical Publ. Comp., 1902), p. 15.
80. Jerslev (1963), pp. 71-72; Nielsen (2000), pp. 285-290.

the Agricultural College six years later). Thomsen presumably followed the development with interest, but he was not involved in it and we do not know his opinion of it. Still, one may imagine that he was pleased with the appointment of Brønsted, whose research area was in a sense a continuation of Thomsen's.[81] Brønsted had just defended his doctoral thesis on "Studies in Affinity," a subject with more than a little similarity to the one Thomsen had started his scientific career more than fifty years earlier.

---

81. Jerslev (1963), p. 86 notes the continuity between Thomsen's thermochemical investigations of affinity and Brønsted's early work.

Appendices

Appendix A: Thomsen's publications

Julius Thomsen was a prolific author and undoubtedly the most productive Danish nineteenth-century scientist. Even today few chemists can boast of a publication list longer than Thomsen's. The precise number of his publications is a little uncertain but close to 254 of which the large majority were research papers in either Danish or German scientific journals. There exists two major bibliographies, the one due to Thomsen's own hand and the other to the Danish chemist Stig Veibel, a professor in organic chemistry at the Polytechnic College.[1] These are valuable but not quite complete.

As shown by the table below, Thomsen's rather amazing productivity peaked in the 1870s with no less than 132 papers. The publications in the category "other languages" are almost all in German, but in one case they refer to a paper in English and in two cases to papers written in French but published by the Royal Danish Academy of Sciences. Some of the papers published in the journals of the Academy included summaries in French. The table does not include reviews, minor notes and the like, but it does include several short communications some of which relate to the controversies Thomsen was involved in. The distinction between research and non-research papers is somewhat fluid; I have counted all publications in recognised scientific journals as belonging to the first category. All of Thomsen's non-research publications, including books and separately published pamphlets were written in Danish. I have not distinguished between books and papers, and have included his opus major, *Thermochemische Untersuchungen* from 1882-1886, as four separate publications.

It should be pointed out that in many cases Thomsen communicated his results both in Danish and German, meaning that some of his publications are essentially doublets. He quite often published papers in Danish in either *Tidsskrift* or one of the journals of the Academy, and simultaneously or shortly later published a slightly

---

1. Thomsen (1905a); Veibel (1943).

changed version in one of the major German journals. In a few cases he even published three versions of what was basically the same paper, for example in *Tidsskrift*, *Oversigter* and *Chemische Berichte*.

	Danish	Other languages (German)	Total	Non-research
1850-1854	6	4	10	5
1855-1860	8	1	9	6
1861-1865	14	5	19	2
1866-1870	13	22	35	1
1871-1875	46	22	68	0
1876-1880	18	46	64	2
1881-1885	5	11	16	2
1886-1890	3	8	11	2
1891-1900	5	4	9	3
1901-1905	4	8	12	1
1906-1909	0	1	1	0
Total	122	132	254	24

Table 7. Thomsen's publications.

Thomsen published his many research publications primarily in German chemical journals of which *Chemische Berichte*, the journal of the German Chemical Society founded in 1868, was the most popular. The distribution on German journals is as follows: *Chemische Berichte* (97), *Journal für praktische Chemie* (29), *Annalen der Physik und Chemie* (23), *Zeitschrift für anorganische Chemie* (10), *Zeitschrift für physikalische Chemie* (7), *Chemische Centralblatt* (1), and *Annalen der Chemie und Pharmacie* (1). His papers published in Danish journals were distributed as *Tidsskrift for Physik og Chemie* (40), *Skrifter*, Royal Danish Academy (9), and *Oversigt*, Royal Danish Academy (23). Remarkably by modern standards, with a single exception Thomsen's publications were all published by him as the only author. The exception was the 1853 booklet on the causes of the Copenhagen cholera epidemic that he wrote with L. A. Colding.

Like Thomsen, his eleven years younger colleague S. M. Jørgensen was internationally oriented. Although he did not match

Thomsen in productivity, he had a solid publication record in Danish and German periodicals. His publication pattern, based on Veibel's bibliography, was roughly the same as Thomsen's. Of Jørgensen's 92 publications far most were on chemical research and only a single of these was written with a co-author (a paper from 1906 written with S. P. L. Sørensen). As an illustration of Jørgensen's international reputation, in 1907 he was nominated for the Nobel Prize by two French scientists, the mathematician Gaston Darboux and the chemist Henri Moissan, the latter a laureate of 1906. However, he did not receive the coveted prize. It took more than ninety years until a Dane would be awarded with the chemistry prize, and then it would go to a medical doctor and not a chemist (Jens Christian Skou, 1997).

	Danish	Other languages (German)	Total	Non-research
1860-1864	4	0	4	2
1865-1869	10	2	12	2
1870-1874	5	6	11	1
1875-1879	3	11	14	1
1880-1884	2	7	9	2
1885-1889	2	6	8	2
1890-1894	0	10	10	0
1895-1899	2	8	10	1
1900-1904	3	5	8	1
1905-1909	2	4	6	3
1910-1914	0	0	0	0
Total	33	59	92	15

Table 8. S. M. Jørgensen's publications.

Appendix B:
Time-line of Thomsen's life and career

1826	Born 16 February in Copenhagen.
1840-41	Gymnasium student, the von Westen Institute, Copenhagen. No exam.
1841-43	Assistant to prof. E. Scharling.
1843	Passes entrance exam for the Polytechnic College.
1846	Graduation as candidate in "applied science," Polytechnic College.
1847	Attends meeting of Scandinavian scientists, Copenhagen.
1846-53	Assistant to prof. J. G. Forchhammer, Chemical Laboratory, Polytechnic College.
1850	Thomsen's textbook in inorganic chemistry-
1850-54	Teacher in agricultural chemistry, the Polytechnic College.
1851	Meeting of Scandinavian scientists, Stockholm.
---	Ørsted recommends Thomsen for chair in Christiania.
1852	First essay of thermochemistry to the Royal Academy; receives the Academy's silver medal for it.
1853	Patent on method for producing soda from cryolite.
---	Work with L. A. Colding on the causes of the Copenhagen cholera.
1853-54	Study travel abroad, mainly to Germany and France.
1856	Meeting of Scandinavian Scientists, Christiania; paper on the heat equivalent of electromotive force (published 1858).
---	Invention of polarisation battery.
---	Popular book (*Excursions into the Landscape of Science*).
1856-59	Adjuster of weights and measures, Copenhagen.
1857	Marriage to Elmine Hansen.
1859	Inauguration of the cryolite factory Øresund.
1859-66	Physics teacher at the Royal Military High School.

1860	Member of the Royal Danish Academy of Sciences and Letters.
1861	Permanent member of the Royal Danish Agricultural Society.
1861-94	City Councillor, Copenhagen municipality.
1862	Titular professor.
---	Visit to Paris to study gas light.
1862-78	Co-founder and co-editor (with A. Thomsen) of *Tidsskrift for Fysik og Chemi*.
1863	Meeting of Scandinavian Scientists, Stockholm; first paper on the mechanical equivalent of light.
1864-65	Chemistry teacher, University of Copenhagen.
1865	Formation of Cryolite Mining and Trading Company Ltd (Kryolit, Mine- og Handelsselskabet, KMHS).
1866-1901	Professor of chemistry, University of Copenhagen.
1869	Offer to become professor in Leipzig.
---	Confirmation of Guldberg-Waage law of mass action.
1870	Dispute with A. Naumann on the status of Avogadro's law.
1872	Thomsen attacks M. Berthelot's thermochemistry; beginning of Thomsen-Berthelot controversy.
1876	Foreign Honorary Member of the Chemical Society, London.
1877	Honorary Doctor (dr.phil.) at Uppsala University, Sweden.
1879	Honorary Doctor (dr.med.) at the University of Copenhagen.
1880	Foreign member of the Royal Swedish Academy of Science.
---	Thomsen's model of the benzene molecule.
1881	Member of the Physiographic Society, Lund.
1882-86	*Thermochemische Untersuchungen*, 4 vols., Leipzig.
1882-92	Member of Copenhagen Harbour Council.
1882-1903	Director of the Polytechnic College.
1883	Recipient of the Davy medal, Royal Society.
---	Member of Reale Accademia dei Lincei, Rome.

1884	Member of the American Academy of Arts and Sciences, Boston.
---	Member of Reale Accademia delle Scienze, Turin.
1886-87	Rector of University of Copenhagen (also 1891-92).
1887	University programme on chemical elements and the unity of matter.
---	Member of Académie Royale des Sciences de Belgique.
1888-1909	President of the Royal Danish Academy.
1891	Honorary member of the Royal Institution.
1895	New version of the periodic table; precise atomic weight determinations.
---	Prediction of group of inert gases.
1896	Grand Cross of the Dannebrog Order.
1898	Thomsen finds helium in fluorite mineral from Greenland.
1900	Corresponding member of the Prussian Academy of Science.
1902	Retires from professorship at Copenhagen University, 31 January.
1902	Privy Councillor (Gehejmekonferensraad); statue of Thomsen.
---	Foreign member of the Royal Society, London.
1905	*Thermokemiske Resultater*, Copenhagen.
1907	Honorary member of the German Chemical Society.
1908	Work on argon's chemistry (unpublished).
1909	Death on 13 February.

Appendix C: Family relations

Hans Peter Jürgen Julius Thomsen (1826-1909)

Grandfather: Jürgen Thomsen

Parents: Thomas Thomsen (1787-1862) & Jensine Frederikke Thomsen (née Lund; 1798-1862)

Siblings: Franziska Helleonora Marie Thomsen (b. 1824)
Meta Catharine Thomsen (b. 1828)
Caroline Auguste Thomsen (b. 1830)
Sigismund Gotthelf Thomsen (1831-1903)
Carl August Thomsen (1834-1894)
Carl Aage Thomsen (b. 1834) ?
Thomas Gottfried Thomsen (1841-1901)

Wife: Elmine Thomsen (née Hansen; 1832-1890)

Children of Julius and Elmine Thomsen:
Anna Sophie Frederikke Thomsen (1859-1950)
Marie Franziska Thomsen (1862-1939)
Ellen Thomsen (1865-1958)
Johanne Thomsen (1867-1963)
Julius Thomsen (d. 1883)

Nephew: Thomas Marius Thomsen (1870-1941)

Bibliography

Aaserud, Finn and John Heilbron (2013). *Love, Literature, and the Quantum Atom.* Oxford: Oxford University Press.

Abildgaard, Peter C. (1800). "Om norske titanertser og om en nye Steenart fra Grönland." *Kgl. Danske Videnskabernes Selskabs Skrifter*: 305-316.

Allan, Thomas (1813). "On a collection of minerals from Greenland." *Annals of Philosophy* **1**: 99-110.

Andersen, Sigurd, ed. (1985). *P. C. Abildgaard (1740-1801): Biography & Bibliography.* Copenhagen: Kandrup.

Andrada, Jozé B. de (1800). "Kurze Angabe der Eigenschaften und Kennzeichen einiger neuen Fossilien aus Schweden und Norwegen." *Scherer's Allgemeine Journal der Chemie* **4**: 28-39.

Armstrong, Henry E. (1887). "The determination of the constitution of carbon compounds from thermochemical data." *Philosophical Magazine* **23**: 73-109.

Armstrong, Henry E. (1927). "Marcelin Berthelot." *Nature* **120**: 659-663.

Arrhenius, Svante (1912). *Theories of Solution.* New Haven: Yale University Press.

Bak, Tor A. (1974). "The history of physical chemistry in Denmark." *Annual Review of Physical Chemistry* **25**: 1-10.

Barkan, Diana K. (1992). "A usable past: Creating disciplinary space for physical chemistry." In: *The Invention of Physical Science*, eds. Mary Jo Nye, Joan Richards, and Roger Stuewer, pp. 175-202. Dordrecht: Kluwer Academic.

Barkan, Diana K. (1999). *Walther Nernst and the Transition to Modern Physical Science.* Cambridge: Cambridge University Press.

Bensaude-Vincent, Bernadette (1999). "Atomism and positivism: A legend about French chemistry." *Annals of Science* **56**: 81-94.

Bergsøe, Paul (1945). *De Tre Vinduer: Erindringer og Tidsbilleder.* Copenhagen: Thanning & Appel.

Berry, A. J. (1968). *From Classical to Modern Chemistry: Some Historical Sketches.* New York: Dover Publications.

Berthelot, Marcellin (1873a). "Sur la chaleur de combinaison rapporté à l'état solide." *Comptes Rendus* **77**: 24-32.

Berthelot, Marcellin (1873b). "Sur la réclamation de priorité élevée par M. J. Thomsen relativement aux principes de la thermochimie." *Bulletin de Société Chimiques de Paris* **19**: 485-489.

Berthelot, Marcellin (1875a). "Recherches thermique sur le chlore et sur les agents d'oxydation et de reduction." *Annales de Chimie et de Physique* **5**: 318-356.

Berthelot, Marcellin (1875b). "Sur les principes géneraux de la thermochimie." *Annales de Chimie et de Physique* **4**: 5-131, 141-213.

Berthelot, Marcellin (1878). "Sur la chaleur de dissolution du sulfate de soude." *Annales de Chimie et de Physique* **14**: 445-452.

Berthelot, Marcellin (1879). *Essai de Mécanique Chimique Fondée sur la Thermochimie*, 2 vols. Paris: Dunod.

Berthelot, Marcellin (1894). "Le principe du travail maximum et l'entropie." *Comptes Rendus* **118**: 1378-1392.

Berzelius, Jöns J. (1824). *Undersökning af Flusspatsyran och dess Märkvärdigaste Föreningar*. Stockholm: Kongliga Svenska Vetenskaps-Akademien.

Bjerrum, Niels (1907). "Nyere og ældre anskuelser om grundstoffernes natur." *Fysisk Tidsskrift* **6**: 71-85.

Bjerrum, Niels (1909). "Julius Thomsen." *Berichte der Deutschen Chemischen Gesellschaft* **42**: 4971-4988.

Blake, William P., ed. (1870). *Reports of the United States Commissioners to the Paris Universal Exposition 1867*, vol. 3. Washington D.C.: Government Printing Office.

Boisbaudran, P. E. Lecoq de (1895). "Remarques sur les poids atomique." *Comptes Rendus* **120**: 361-362.

Bostrup, Ole (1996). *Dansk Kemi 1770-1807: Den Kemiske Revolution*. Copenhagen: Teknisk Forlag.

Bottone, S. R. (1902). *Galvanic Batteries: Their Theory, Construction and Use*. New York: Whittaker & Co.

Brock, William H. (1985). *From Protyle to Proton: William Prout and the Nature of Matter, 1785-1985*. Bristol: Adam Hilger.

Brønsted, Johannes N. (1909). "Mindefest for Julius Thomsen." *Fysisk Tidsskrift* **7**: 132-141.

Brønsted, Johannes N. (1932). "Julius Thomsen." In: *Prominent Danish Scientists Through the Ages*, ed. V. Meisen, pp. 143-147. Copenhagen: Levin & Munksgaard.

Brown, Robert (1894). "Dr. Hendrik Rink." *Geographical Journal* **3**: 65-67.

Brühl, Julius W. (1887). "Kritik der Grundlagen und Resultate der sogenannten Theorie der Bildungswärme organischer Körper." *Journal für Praktische Chemie* **35**: 181-204, 209-236.

Brühl, Julius W. (1894). "Neue Beiträge zur Frage nach der Constitution des Benzols." *Berichte der Deutschen Chemischen Gesellschaft* **27**: 1065-1083.

Brush, Stephen G. (1986). *The Kind of Motion We Call Heat: A History of the Kinetic Theory of Gases in the 19th Century*. Amsterdam: North-Holland.

Brush, Stephen G. (1987). "The nebular hypothesis and the evolutionary world view." *History of Science* **25**: 245-278.

Brush, Stephen G. (1999). "Dynamics of theory change in chemistry, Part 1: The Benzene problem." *Studies in History and Philosophy of Science* **30**: 21-79.

Callisen, Heinrich (1807). *Physisk Medizinske Betragtninger over Kjöbenhavn.* Copenhagen: J. F. Schulz.

Caneva, Kenneth (1997). "Colding, Ørsted, and the meaning of force." *Historical Studies in the Physical and Biological Sciences* **28**: 1-138.

Cardwell, Donald (1989). *From Watt to Clausius: The Rise of Thermodynamics in the Early Industrial Age.* Ames: Iowa State University Press. Johnson Reprint Corporation.

Christensen, Dan C. (2013). *Hans Christian Ørsted: Reading Nature's Mind.* Oxford: Oxford University Press.

Christensen, Odin T. (1880). "Nogle i de senere aar opdagede grundstoffer." *Tidsskrift for Populære Fremstillinger af Naturvidenskaben* **27**: 417-429.

Christensen, Odin T. (1910). "Julius Thomsen." *Nordisk Tidskrift för Vetenskap, Konst och Industri*: 155-170.

Christiansen, Jens A. (1929). "Julius Thomsen." *Polyteknisk Tidsskrift* **10**: 220-224.

Clarke, Frank W. (1896). "Third annual report of committee on atomic weights." *Journal of the American Chemical Society* **18**: 197-310.

Clarke, Frank W. (1903). "A new law in thermochemistry." *Proceedings of the Washington Academy of Science* **5**: 1-37.

Colding, L. August (1856). "Naturvidenskabelige betragtninger over slægtskabet mellem det aandelige livs virksomheder og de almindelige naturkræfter." *Kgl. Da. Vid. Selsk. Forhandl., Oversigt*: 136-168.

Colding, L. August and Julius Thomsen (1853). *Om de Sandsynlige Aarsager til Choleraens Ulige Styrke i de Forskjellige Dele af Kjöbenhavn.* Copenhagen: C. A. Reitzel.

Cooke, Josiah P. (1881). "Notice of Julius Thomsen's thermochemical investigation of the molecular structure of the hydrocarbon compounds." *American Journal of Science* **21**: 87-98.

Crawford, Elisabeth (1996). *Arrhenius: From Ionic Theory to the Greenhouse Effect.* Canton, MA: Science History Publications.

Crookes, William (1886). "On the nature and origin of the so-called elements." *Report of the British Association for the Advancement of Science*: 558-576.

Crosland, Maurice P. (1970-1980). "Berthelot, Pierre Eugène Marcellin." In: *Dictionary of Scientific Biography*, ed. C. G. Gillispie, vol. 2, pp. 63-72. New York: Charles Scribner's Sons.

Crosland, Maurice P. (1978). *Gay-Lussac: Scientist and Bourgeois.* Cambridge: Cambridge University Press.

Crowe, Michael J. (1999). *The Extraterrestrial Life Debate 1750-1900*. Mineola, NY: Dover Publications.

Dahl, Per F. (1972). *Ludvig Colding and the Conservation of Energy Principle*. New York: Johnson Reprint Corporation.

Davies, Alwyn G. (2012). "Sir William Ramsay and the noble gases." *Science Progress* **95**: 23-49.

De Milt, Clara (1948). "Carl Weltzein and the congress at Karlsruhe." *Chymia* **1**: 153-169.

Deville, Henri E. Sainte-Claire (1856). "Mémoire sur la fabrication du sodium et de l'aluminium." *Annales de Chimie et de Physique* **46**: 415-458.

Deville, Henri E. Sainte-Claire (1859). *De l'Aluminium: Ses Propriétés, sa Fabrication et se Applications*. Paris: Mallet-Bachelier.

Dick, Allan (1855). "On the preparation of aluminium from Kryolite." *Philosophical Magazine* **10**: 364-365.

Dolby, R. G. A. (1976a). "Debates over the theory of solution: A study of dissent in physical chemistry in the English-speaking world in the late nineteenth and early twentieth centuries." *Historical Studies in the Physical Sciences* **7**: 297-404.

Dolby, R. G. A. (1976b). "The case of physical chemistry." In: *Perspectives on the Emergence of Scientific Disciplines*, eds. Gerard Lemaine et al., pp. 63-73. The Hague: W. De Gruyter.

Dolby, R. G. A. (1984). "Thermochemistry versus thermodynamics: The nineteenth century controversy." *History of Science* **22**: 375-400.

Drechsel, Johanne (1926). *En Fader og en Moder*. Copenhagen: Privately printed.

Duhem, Pierre (1893). *Introduction à la Mécanique Chimique*. Paris: C. Naud.

Duhem, Pierre (1897). "Thermochimie." *Revue des Questions Scientifique* **12**: 361-392.

Duhem, Pierre (2002). *Mixture and Chemical Combination and Related Essays*, ed. Paul Needham. Dordrecht: Kluwer Academic.

Duncan, Alistair M. (1996). *Laws and Order in Eighteenth-Century Chemistry*. Oxford: Clarendon Press.

Ehrenfeld, Richard (1906). *Grundriss einer Entwicklungsgeschichte der Chemischen Atomistik*. Heidelberg: Winter's Universitätsbuchhandlung.

Elkana, Yehuda (1974). *The Discovery of the Conservation of Energy*. Cambridge, MA: Harvard University Press.

Emsmann, August (1870). "Die thermochemischen Untersuchungen von Julius Thomsen in Kopenhagen." *Gaea* **6**: 45-50.

Engelhardt, H. Tristram and Arthur L. Caplan, eds. (1984). *Scientific Controversies: Case Studies in the Resolution and Closure of Disputes in Science and Technology*. Cambridge: Cambridge University Press.

Engels, Friedrich (1951). *Dialektik der Natur*. Berlin: Dietz Verlag.

Eriksson, Nils (1991). *"I Andans Kraft, på Sannings Stråt…": De Skandinaviska Naturforskarmötene*. Gothenburg: Acta Universitatis Gothenburgensis.

Erslew, Thomas H. (1868). *Supplement til Almindeligt Forfatter-Lexicon*, bd. 3. Copenhagen: Forlagsforeningens Forlag.

Espmark, Kristina and Christer Nordlund (2012). "Married for science, divorced for love: Success and failure in the collaboration between Astrid Cleve and Hans von Euler-Chelpin." In: *For Better or For Worse? Collaborative Couples in the Sciences*, eds. Annette Lykknes, Donald Opitz and Brigitte Van Tiggelen, pp. 81-102. Basel: Birkhäuser.

Farmer, Moses G. (1866). "Note on the mechanical equivalent of light." *Philosophical Magazine* **31**: 403-404.

Farrar, W. V. (1965). "Nineteenth-century speculations on the complexity of the chemical elements." *British Journal for the History of Science* **2**: 297-323.

Favre, Pierre A. (1874). "Réclamation relative a une note de M. J. Thomsen." *Bulletin de Societé Chimiques de Paris* **21**: 487.

Favre, Pierre A. and Johann T. Silbermann (1853). "Recherches sur les quantités de chaleur dégagées dans les actions chimiques et moléculaires." *Annales de Chimie et de Physique* **37**: 406-508.

Fisher, Nicholas (1982). "Avogadro, the chemists, and historians of chemistry." *History of Science* **20**: 77-94, 213-231.

Flavitsky, Flavian M. (1896). "Ueber eine Funktion, welche der Periodizität der Eigenschaften der chemischen Elemente entspricht." *Zeitschrift für Anorganische Chemie* **11**: 264-267.

Fleury, August (1875). *Kortfattet Vejledning til Kvantitativ Bestemmelse af de i Hygiejnisk Henseende Vigtige Stoffer i Vandet*. Copenhagen: Høst og Søn.

Forchhammer, Johann Georg (1842). *Lærebog i Stoffernes Almindelige Chemie*. Copenhagen: C. A. Reitzel.

Foster, Peter Le Neve (1859). "On aluminium." *Journal of the Society of Arts* **7**: 162-169.

Freund, Ida (1904). *The Study of Chemical Composition: An Account of Its Method and Historical Development*. Cambridge: Cambridge University Press.

Gad, Finn (1976). *Grønlands Historie, 1782-1808*. Copenhagen: Nyt Nordisk Forlag.

Garboe, Axel (1959-1961). *Geologiens Historie i Danmark*, 2 vols. Copenhagen: C. A. Reitzel.

Garboe, Axel (1966). "Farmaceuten, kemikeren, naturforskeren B. Levy (Lewy)." *Theriaca* **11**: 5-73.

Giesecke, Karl L. (1822). "On cryolite." *The Edinburgh Philosophical Journal* **6**: 141-144.

Giesecke, Karl L. (1910). "Bericht über einer mineralogischen Reise in Grönland." *Meddelelser om Grønland* **35**: 1-478.

Giunta, Carmen (2001). "Argon and the periodic system: The piece that would not fit." *Foundations of Chemistry* **3**: 105-128.

Glamann, Kristof (2003). *The Carlsberg Foundation: The Early Years*. Copenhagen: The Carlsberg Foundation.

Glamann, Kristof and Kirsten Glamman (2004). *Nordens Pasteur: Fortællingen om Emil Chr. Hansen*. Copenhagen: Gyldendal.

Goldschmidt, Meir A. (1857). "Et industrielt billede fra Danmark." *Nord og Syd* **2**: 249-261.

Gregory, Frederic (1984). "Romantic Kantianism and the end of the Newtonian dream in chemistry." *Archive Internationales d'Histoire des Sciences* **34**: 108-123.

Grossman, Louis and Marianne M. Jennings (2002). *Through Good Times and Bad*. Westport, CT: Quorum Books.

Guldberg, Cato M. and Peter Waage (1867). *Études sur les Affinités Chimique*. Oslo: Brögger and Christie.

Haber, L. F. (1958). *The Chemical Industry during the Nineteenth Century*. Oxford: Clarendon Press.

Hagemann, Gustav A. (1866a). "On some minerals associated with the cryolite in Greenland." *American Journal of Science and Arts* **42**: 93-94.

Hagemann, Gustav A. (1866b). "On crystallized cryolite." *American Journal of Science and Arts* **42**: 268-273.

Hagemann, Gustav A. (1886). *Studien über das Molekularvolumen einiger Körper*. Berlin: R. Friedländer & Sohn.

Hagemann, Gustav A. (1887). *Einige Kritische Bemerkungen zur Aviditätsformel*. Berlin: R. Friedländer & Sohn.

Hagemann, Gustav A. (1888). *Die Chemischen Kräfte*. Berlin: R. Friedländer & Sohn.

Hagemann, Gustav A. (1909). "Mindefest for Julius Thomsen." *Fysisk Tidsskrift* **7**: 129-132.

Hall, M. Boas (1976). "The strange case of aluminium." *History of Technology* **1**: 143-158.

Halland, Alfred S. (1911). "Cryolite and its industrial applications." *Journal of Industrial and Engineering Chemistry* **3**: 63-66.

Hamerla, Richard (2003). "Edward Williams Morley and the atomic weight of oxygen: The death of Prout's hypothesis revived." *Annals of Science* **60**: 341-372.

Hamlin, Christopher (1990). *A Science of Impurity: Water Analysis in Nineteenth Century Britain*. Bristol: IOP Publishing.

Hansen, H. C. Peter (1886). *Illustreret Dansk Litteraturhistorie*. Copenhagen: P. G. Philipsens Forlag.

Harding, Marius C., ed. (1920). *Correspondance de H. C. Örsted avec Divers Savants*. Copenhagen: H. Aschehoug & Co.

Harding, Marius C. (1924). *Selskabet for Naturlærens Udbredelse*. Copenhagen: Gjellerups Forlag.

Hargrove, James L. (2006). "History of the calorie in nutrition." *Journal of Nutrition* **136**: 2958-2961.

Harnow, Henrik (1993). "The Danish engineer in transition: The reformation of Danish engineering education ca. 1890-1933." In: *European Historiography of Technology*, ed. Dan Ch. Christensen, pp. 164-174. Odense: Odense University Press.

Harnow, Henrik (1998). *Den Danske Ingeniørs Historie 1850-1920*. Herning: Systime.

Hartley, Walther N. (1881). "Researches on the relation between molecular structure of carbon compounds, and their absorption spectra." *Journal of the Chemical Society, Transactions* **39**: 153-168.

Hartley, Walther N. (1898). "The thermo-chemistry of the Bessemer process." *Journal of the Society of Arts* **46**: 721-732.

Helmholtz, Hermann (1995). *Science and Culture: Popular and Scientific Essays*, ed. David Cahan. Chicago: University of Chicago Press.

Hevesy, George von (1927). *Die Seltenen Erden vom Standpunkt des Atombaues*. Berlin: Julius Springer.

Hevesy, George von and V. Thal Jantzen (1927). "Über den Hafniumgehalt einiger historischer Zirkonpräparate." *Naturwissenschaften* **12**: 729-732.

Hinrichs, Gustavus D. (1866). "On the spectra and composition of the elements." *American Journal of Science and Arts* **42**: 350-368.

Hinrichs, Gustavus D. (1894). *The True Atomic Weights of the Chemical Elements and the Unity of Matter*. St. Louis: G. D. Hinrichs.

Hjermitslev, Hans H. (2004). "Da videnskaben blev populær." *Aktuel Naturvidenskab* nr. 4: 32-34.

Holmes, Frederick L. (1962). "From elective affinities to chemical equilibria: Berthollet's law of mass action." *Chymia* **8**: 105-145.

Holst, Helge (1908). "Julius Thomsen." *Frems Aarbog* **1**: 215-219.

Hornemann, Emil (1856). "Choleraudbruddet paa Kastelsveien i Efteraaret 1854." *Hygiejniske Meddelelser og Betragtninger* **1**: 1-19.

Hornemann, Emil (1857). "Undersøgelsen af brøndvandet i omegnen af Frederiksberg Kirkegaard." *Hygiejniske Meddelelser og Betragtninger* **2**: 1-24.

Hyldtoft, Ole (1994). *Den Lysende Gas: Etableringen af det Danske Gassystem 1800-1890*. Herning: Systime.

Ibsen, Eugen (1865). "Thomsens polarisationsbatteri." *Ugeskrift for Læger* **43**: 33-42.

Ihde, Aaron J. (1961). "The Karlsruhe congress: A centennial retrospective." *Journal of Chemical Education* **38**: 83-86.

Jacobsen, Lis, ed. (1925). *Kvindelige Akademikere 1875-1925*. Copenhagen: Gyldendal.

Jahn, Hans (1882). *Die Grundsätze der Thermochemie und Ihre Bedeutung für die Theoretiche Chemie*. Vienna: Alfred Hölder.

Jaki, Stanley (1984). *Uneasy Genius: The Life and Work of Pierre Duhem*. The Hague: Martinus Nijhoff.

Jarl, Carl F. (1909a). *Fabrikken Øresund 1859-1909: Kryolitindustriens Historie og Udvikling*. Copenhagen: J. Jørgensen.

Jarl, Carl F. (1909b). "Julius Thomsen." *Ingeniøren* no. 8 (20 February): 53-54.

Jensen, Kai A. (1983). "Kemi." In: *Københavns Universitet*, vol. 12, ed. Mogens Pihl, pp. 427-580. Copenhagen: Gads Forlag.

Jensen, William B. (2002). *Mendeleev on the Periodic Law: Selected Writings, 1869-1905*. Mineola, NY: Dover Publications.

Jensen, William B. (2009). "August Horstmann and the origins of chemical thermodynamics." *Bulletin for the History of Chemistry* 34: 83-91.

Jerslev, Bodil, ed. (1963). *Kemien i Danmark, III: Danske Kemikere*. Copenhagen: Nyt Nordisk Forlag.

Jespersen, Knud J. V. (2004). *A History of Denmark*. Houndmills, Hampshire: Palgrave.

Jessen-Hansen, Hans (1910). "Julius Thomsen." *Højskolebladet*: 346-352.

Jevons, William Stanley (1973). *Papers and Correspondence of William Stanley Jevons*, ed. R. D. Collison Black, vol. 2. London: Macmillan.

Jewess, Michael (2010). "What happened in Thomsen's kiln? A detective story." *RSC Historical Group Newsletter* (August): 15-17.

Jørgensen, Sophus M. (1860). "Om atomer og atomtheorien." *Tidsskrift for Populære Fremstillinger af Naturvidenskaben* 2: 191-219.

Jørgensen, Sophus M. (1909). "Mindeord om Julius Thomsen." *Kgl. Da. Vid. Selsk. Forhandl., Oversigt*: 27-31.

Jones, Harry C. (1903). *A New Era in Chemistry*. New York: Van Nostrand.

Kant, Immanuel (2004). *Metaphysical Foundations of Natural Science*, ed. Michael Friedman. Cambridge: Cambridge University Press.

Kauffman, George B., ed. (1976). *Classics in Coordination Chemistry, Part II*. New York: Dover Publications.

Kauffman, George B. (1977). "Christian Wilhelm Blomstrand (1826-1897) and Sophus Mads Jørgensen (1837-1914): Their correspondence from 1870 to 1897." *Centaurus* 21: 44-63.

Kipnis, Alexander (1997). *August Friedrich Horstmann und die Physikalische Chemie*. Berlin: ERS Verlag.

Kjølsen, Hans (1965). *Fra Skidenstræde til H. C. Ørsted Instituttet*. Copenhagen: Gjellerup.

Klaproth, Martin H. (1801). "Chemische Untersuchung des Kryoliths." *Neue Schriften der Gesellschaft Naturforschender Freunde zu Berlin* 3: 322-328.

Koefoed, Emil (1974). *Emil Koefoeds Erindringer.* Copenhagen: Dansk Farmaci-historisk Selskab.

Koefoed, Rudolph G. (1885). "Den periodiske lov." *Tidsskrift for Physik og Chemi* 24: 161-174.

Körber, H.-G. (1961). *Aus dem wissenschaftlichen Briefwechsel Wilhelm Ostwald.* Berlin: Akademie Verlag.

Kragh, Helge (1982). "Julius Thomsen and 19th-century speculations on the complexity of atoms." *Annals of Science* 39: 37-60.

Kragh, Helge (1984). "Julius Thomsen and classical thermochemistry." *British Journal for the History of Science* 17: 255-272.

Kragh, Helge (1989). "The aether in late nineteenth century chemistry." *Ambix* 36: 49-65.

Kragh, Helge (1991). "Ludvig Lorenz and nineteenth century optical theory: The work of a great Danish scientist." *Applied Optics* 30: 4688-4695.

Kragh, Helge (1993). "Between physics and chemistry: Helmholtz's route to a theory of chemical thermodynamics." In: *Hermann von Helmholtz and the Foundations of Nineteenth-Century Science*, ed. David Cahan, pp. 401-429. Berkeley: University of California Press.

Kragh, Helge (1995). "From curiosity to industry: The early history of cryolite soda manufacture." *Annals of Science* 52: 285-301.

Kragh, Helge (1996). "Elements no. 70, 71 and 72: Discoveries and controversies." In: *Episodes from the History of Rare Earth Elements*, ed. C. H. Evans, pp. 67-90. Dordrecht: Kluwer Academic.

Kragh, Helge (1998). "Out of the shadow of medicine: Themes in the development of chemistry in Denmark and Norway." In: *The Making of the Chemist: The Social History of Chemistry in Europe 1789-1914*, eds. David Knight and Helge Kragh, pp. 345-364. Cambridge: Cambridge University Press.

Kragh, Helge (2000). "Confusion and controversy: Nineteenth-century theories of the Voltaic pile." In: *Nuova Voltiana: Studies on Volta and his Times*, vol. 1, eds. Fabio Bevilacqua and Lucio Fregonese, pp. 133-157. Milan: Hoepli.

Kragh, Helge (2002). "The first subatomic explanations of the periodic system." *Foundations of Chemistry* 3: 129-143.

Kragh, Helge (2009a). "Conservation and controversy: Ludvig Colding and the imperishability of 'forces'." *RePoss* no. 4 (Research Publications on Science Studies). http://www.ivs.au.dk/reposs.

Kragh, Helge (2009b). "Uorganisk darwinisme: Udviklingstanken i de fysiske videnskaber." *Slagmark* no. 54: 63-76.

Kragh, Helge (2010a). "Aspekter af tidlig dansk miljøkemi." In: *Aspekter af Dansk Miljøkemis Historie*, ed. Anita Kildebæk Nielsen, pp. 7-22. Copenhagen: Dansk Selskab for Historisk Kemi.

Kragh, Helge (2010b). "Auroral chemistry: The riddle of the green line." *Bulletin for the History of Chemistry* **35**: 97-104.

Kragh, Helge (2012). *Niels Bohr and the Quantum Atom: The Bohr Model of Atomic Structure 1913-1925*. Oxford: Oxford University Press.

Kragh, Helge (2013). "Superheavy elements and the upper limit of the periodic table: Early speculations." *European Physical Journal H* **38**: 411-438.

Kragh, Helge (2015a). "Reception and early use of the periodic system: The case of Denmark." In: *Early Responses to the Periodic System*, eds. Masanori Kaji, Helge Kragh and Gábor Palló, pp. 171-190. New York: Oxford University Press.

Kragh, Helge (2015b). "From Ørsted to Bohr: The sciences and the Danish university system, 1800-1920." In: *The Sciences in the Universities of Europe, Nineteenth and Twentieth Centuries*, eds. Ana Simões, Maria Diogo and Kostas Gavroglu, pp. 31-48. Dordrecht: Springer.

Kragh, Helge and Jørgen S. Petersen (1995). *En Nyttig Videnskab: Episoder fra den Tekniske Kemis Historie i Danmark*. Copenhagen: Gyldendal.

Kragh, Helge and Stephen J. Weininger (1996). "Sooner silence than confusion: The tortuous entry of entropy into chemistry." *Historical Studies in the Physical and Biological Sciences* **27**: 91-130.

Kragh, Helge et al. (2008). *Science in Denmark: A Thousand-Year History*. Aarhus: Aarhus University Press.

Kruse, Poul R. and Helmer Kofod, eds. (1992). *Danmarks Farmaceutiske Højskole 1892-1992*. Copenhagen: Lægeforeningens Forlag.

Kubbinga, Henk (2001). *L'Histoire du Concept de "Molecule."* Paris: Springer.

Lagerlöf, Daniel (1905). "Antwort an Herrn Julius Thomsen hinsichtlich seiner Beurteilung." *Journal für Praktische Chemie* **72**: 80-104.

Laidler, Keith J. (1993). *The World of Physical Chemistry*. Oxford: Oxford University Press.

Landolt, Hans H. (1893). "Untersuchungen über etwaige Aenderungen des Gesamtgewichtes chemisch sich umsetzender Körper." *Zeitschrift für Physikalische Chemie* **12**: 1-32.

Langberg, L. Christian (1845). "Den ved de forskjellige svovlsyrehydraters forbindelse med vand frembragte volumforminskelse, og dennes forhold til den frigjorte varme." *Nyt Magazin for Naturvidenskaberne* **5**: 319-335.

Lange, Ole (2006). *Stormogulen. C. F. Tietgen – en Finansmand, Hans Imperium og Hans Tid 1829-1901*. Copenhagen: Gyldendal.

Larsen, Absalon (1950). *Telegrafonen og den Traadløse, og Opfinderparret Valdemar Poulsen og P. O. Pedersen*. Copenhagen: Teknisk Forlag.

Leenson, I. A. (2004). "Sulfuric acid and water: Paradoxes of dilution."
  Journal of Chemical Education **81**: 991-994.

Lehmann, Alfred (1920). "Det psykofysiske laboratorium." *Naturens Verden* **4**:
  118-129.

Leone, Matteo and Nadia Robotti (2003). "Are the elements elementary?
  Nineteenth-century chemical and spectroscopical answers." *Physics in
  Perspective* **5**: 360-383.

Leicester, Henry M. (1951). "Germain Henri Hess and the foundations of
  thermochemistry." *Journal of Chemical Education* **28**: 581-583.

Leicester, Henry M. and Herbert S. Klickstein, eds. (1968). *A Source Book in
  Chemistry, 1400-1900*. Cambridge, MA: Harvard University Press.

Levere, Trevor L. (1971). *Affinity and Matter: Elements of Chemical Philosophy
  1800-1865*. Oxford: Oxford University Press.

Lewicki, Wilhelm, ed. (1982). *Wöhler und Liebig: Briefe von 1829-1873*. Göttingen:
  Jürgen Cromm Verlag.

Liebig, Justus (1846). *Chemiske Breve*. Copenhagen: P. G. Philipsen.

Lindauer, Maurice W. (1962). "The evolution of the concept of chemical
  equilibrium from 1775 to 1923." *Journal of Chemical Education* **39**: 384-390.

Lindegaard, Hanne (2001). *Ud af Røret? Planer, Processer og Paradokser Omkring det
  Københavnske Kloaksystem 1840-2001*. Ph.D. thesis, Denmark's Technical
  University.

Lomholt, Asger (1942-1973). *Det Kongelige Danske Videnskabernes Selskab 1742-1942*,
  vol. 1-5. Copenhagen: Munksgaard.

Lomholt, Asger (1954). *Et Møde i Videnskabernes Selskab: P. S. Krøyers Maleri og dets
  Tilblivelse*. Copenhagen: Munksgaard.

Lorenz, Ludvig V. (1877). *Læren om Varmen*. Copenhagen: C. A. Reitzel.

Lund, E. W. (1965). "Guldberg and Waage and the law of mass action."
  Journal of Chemical Education **42**: 548-550.

Lundbye, Johan T. (1929). *Den Polytekniske Læreanstalt 1829-1929*. Copenhagen:
  G. E. C. Gad.

Mach, Ernst (1926). *Die Prinzipien der Physikalischen Optik Historisch und Erkenntnis-
  psychologischen Entwickelt*. Leipzig: J. A. Barth.

Magie, William F. (1907). "The association theory of solutions." *Proceedings of
  the American Philosophical Society* **46**: 138-145.

Mahoney, Dennis W. et al. (1981). "A continuous variation study of heats of
  neutralization." *Journal of Chemical Education* **58**: 730-732.

Marsh, James E. (1888). "Van't Hoff's hypothesis and the constitution of
  benzene." *Philosophical Magazine* **26**: 426-434.

Marstrand, Jacob (1928). *Tilbageblik Gennem et Langt Liv*. Copenhagen: Gyldendal.

Marstrand, Vilhelm V. (1929). "Ingeniøren og fysikeren Ludvig August
  Colding." *Ingeniørvidenskabelige Skrifter A* **20**: 1-61.

Médard, Louis and Henri Tachoire (1994). *Histoire de la Thermochimie: Prélude à la Thermodynamique Chimique*. Aix-en-Provence: Publications de l'Université de Provence.

Mendeleev, Dmitri I. (1882). "Ueber die Verbrennungswärme der Kohlenwasserstoffe." *Berichte der Deutschen Chemischen Gesellschaft* **15**: 1555-1559.

Mendeleev, Dmitri I. (1889). "The periodic law of the chemical elements." *Journal of the Chemical Society* **55**: 634-656.

Mendeleev, Dmitri I. (1892). *Grundlagen der Chemie*. Leipzig: Carl Ricker.

Mendeleev, Dmitri I. (1895). "Professor Mendeléef on argon." *Nature* **151**: 543.

Merz, John T. (1904-1912). *A History of European Thought in the Nineteenth Century*, 4 vols. Edinburgh: Wm. Blackwood and Sons.

Meyer, Lothar (1871). "Ueber die Hypothese Avagadro's." *Berichte der Deutschen Chemischen Gesellschaft* **4**: 25-32.

Meyer, Lothar (1872). *Die Modernen Theorien der Chemie und Ihre Bedeutung für die Chemische Statik*. Breslau: Maruscke & Berendt.

Meyer, Lothar (1873). "Zur Systematik der anorganischen Chemie." *Berichte der Deutschen Chemischen Gesellschaft* **6**: 101-106.

Meyer, Lothar (1887). "Die bisherigen Entwickelung der Affinitätslehre." *Zeitschrift für Physikalische Chemie* **1**: 134-144.

Meyer, Lothar and Karl Seubert (1885). "On the unit adopted for the atomic weights." *Journal of the Chemical Society, Transactions* **47**: 426-433.

Meyer, Lothar and Karl Seubert (1894). "Über das Verhältnis der Atomgewichte des Wasserstoffes und des Sauerstoffes." *Berichte der Deutschen Chemischen Gesellschaft* **27**: 2770-2773.

Meyer, Victor (1895). "Probleme der Atomistik." *Verhandlungen der Gesellschaft Deutscher Naturforscher und Ärtzte* **67**: 95-110.

Miller, Alexander K. (1887). "Notes on the recent papers by A. von Baeyer and Julius Thomsen on the constitution of benzene." *Journal of the Chemical Society, Transactions* **51**: 208-215.

Mohr, Friedrich (1868). *Mechanische Theorie der Chemischen Affinität und die Neuere Chemie*. Braunschweig: Vieweg und Sohn.

Mond, Ludwig, William Ramsay, and John Shields (1898). "On the occlusion of hydrogen and oxygen by palladium." *Philosophical Transactions of the Royal Society A* **191**: 105-121.

Morel, Patrice, ed. (1992). *Histoire Technique de la Production d'Aluminium*. Grenoble: Presses Universitaires de Grenoble.

Morrisson, Mark S. (2007). *Modern Alchemy: Occultism and the Emergence of Atomic Theory*. Oxford: Oxford University Press.

Muir, Matthew M. Pattison (1884). *A Treatise on the Principles of Chemistry*. Cambridge: Cambridge University Press.

Muir, Matthew M. Pattison (1885). *The Elements of Thermal Chemistry*. London: Macmillan and Co.

Muir, Matthew M. Pattison (1907). *A History of Chemical Theories and Laws*. New York: John Wiley & Sons.

Muir, Matthew M. Pattison (1909). "Prof. Julius Thomsen." *Nature* **80**: 46-47.

Nath, Biman B. (2013). *The Story of Helium and the Birth of Astrophysics*. New York: Springer.

Naumann, Alexander (1869a). "Das Avogadro'sche Gesetz abgeleitet aus der Grundvorstellung der mechanischen Gastheorie." *Berichte der Deutschen Chemischen Gesellschaft* **2**: 690-693.

Naumann, Alexander (1869b). *Grundriss der Thermochemie*. Braunschweig: Vieweg und Sohn.

Nernst, Walther (1888). "Über die Bildungswärme der Quecksilberverbindungen." *Zeitschrift für Physikalische Chemie* **2**: 23-28.

Nernst, Walther (1904). *Theoretical Chemistry from the Standpoint of Avogadro's Rule and Thermodynamics*. London: Macmillan and Co.

Nichols, Edward L. and William W. Coblentz (1903). "On methods of measuring radiant efficiencies." *Physical Review* **17**: 267-276.

Nielsen, Anita Kildebæk (2000). *The Chemists: Danish Chemical Communities and Networks, 1900-1940*. Ph.D. dissertation. Aarhus: History of Science Department.

Nielsen, Anita Kildebæk (2001). "En disciplins demarkation og selvforståelse: Dansk kemisk tidsskriftlitteratur i 1800-tallet." *Rotunden* **16**: 25-53.

Nielsen, Anita Kildebæk, ed. (2002). *Dansk Landbrugskemi i Historisk Perspektiv 1750-1930*. Copenhagen: Dansk Selskab for Historisk Kemi.

Nielsen, Anita Kildebæk, ed. (2004). *Niels Bjerrum (1879-1958): Liv og Værk*. Copenhagen: Dansk Selskab for Historisk Kemi.

Nielsen, Anita Kildebæk (2007). "Fashioning and demarcation of the Danish chemical community in the 19th century." *Centaurus* **49**: 199-226.

Nielsen, Anita Kildebæk (2008). "Denmark: Creating a Danish identity in chemistry between pharmacy and engineering, 1879-1914." In: *Creating Networks in Chemistry: The Founding and Early History of Chemical Societies in Europe*, eds. Anita K. Nielsen and Soňa Štrbáňová, pp. 75-90. London: Royal Society of Chemistry.

Nielsen, Anita Kildebæk and Helge Kragh (1997). "An institute for dollars: Physical chemistry in Copenhagen between the world wars." *Centaurus* **39**: 311-331.

Nielsen, Hans Toftlund (1994). "Kemikeren J. G. Forchhammer." *Dansk Kemi* no. 12: 20-24.

Nielsen, Henry and Birgitte Wistoft (1996). *Industriens Mænd: Et Krøyer-Maleris Tilblivelse og Industrihistoriske Betydning*. Aarhus: Klim.

Nielsen, Henry and Birgitte Wistoft (1998). "Painting technological progress: P. S. Krøyer's *The Industrialists.*" *Technology and Culture* **39**: 408-433.

Nielsen, Rasmus (1873). *Natur og Aand: Bidrag til En med Physiken Stemmende Naturphilosophie.* Copenhagen: J. H. Schubote.

Noyes, William A. (1896). "A new determination of the relative densities of oxygen and hydrogen and of the ratio of their atomic weights." *Science* **4**: 26-27.

Nye, Mary Jo (1981). "Berthelot's anti-atomism: A 'matter of taste'?" *Annals of Science* **38**: 585-590.

Nye, Mary Jo (1993). *From Chemical Philosophy to Theoretical Chemistry: Dynamics of Matter and Dynamics of Disciplines 1800-1950.* Berkeley: University of California Press.

Ostwald, Wilhelm (1902). *Lehrbuch der Allgemeinen Chemie.* Leipzig: W. Engelmann.

Ostwald, Wilhelm (1980). *Electrochemistry: History and Theory.* New Delhi: Amerind Publishing Co.

Ostwald, Wilhelm (2013). *Lebenslinien. Eine Selbstbiographie.* Berlin: Hofenberg.

Pais, Abraham (1991). *Niels Bohr's Times, in Physics, Philosophy, and Polity.* Oxford: Clarendon Press.

Paneth, Fritz (1923). "Über das Element 72 (Hafnium)." *Ergebnisse der Exakten Naturwissenschaften* **2**: 163-176.

Parks, George S. (1949). "Some notes on the history of thermochemistry." *Journal of Chemical Education* **26**: 262-266.

Partington, James R. (1964). *A History of Chemistry,* vol. 4. London: Macmillan & Co.

Pauly, Hans (1986). "Cryolithionite and Li in the cryolite deposit Ivigtut, South Greenland." *Kgl. Da. Vid. Selskab, Matematisk-Fysiske Meddelelser* **42** (1).

Pedersen, Bjørn (2007). *Syv Bidrag til Norsk Kjemihistorie.* Oslo: Universitetet i Oslo.

Pedersen, Olaf (1992). *Lovers of Learning: A History of the Royal Danish Academy of Sciences and Letters 1742-1992.* Copenhagen: Munksgaard.

Pedersen, Peder O. (1941). "Julius Thomsen." In: P. O. Pedersen, *Livsskildringer: Radioforedrag 1930-1941,* pp. 159-175. Copenhagen: G. E. C. Gads Forlag.

Pell, Morris B. (1872). "On the constitution of matter." *Philosophical Magazine* **43**: 161-185.

Petersen, Emil (1889). "Neutralisationswärme der Fluoride." *Zeitschrift für Physikalische Chemie* **4**: 384-412.

Petersen, Emil (1890). "Grundstoffernes natur." *Naturen og Mennesket* **4**: 13-32.

Petersen, Emil (1893). "Über die Dissociationswärme einiger Säuren." *Zeitschrift für Physikalische Chemie* **11**: 174-184.

Petersen, Emil (1895). "Om argon." *Nordisk Farmaceutisk Tidsskrift* **2**: 233-240.

Petersen, Niels (1993). "Københavns Universitet 1848-1902." In: *Københavns Universitet 1479-1979*, vol. 2, eds. Leif Grane and Kai Hørby, pp. 269-454. Copenhagen: Gads Forlag.

Pickering, Spencer U. (1888). "On thermochemical constants." *Philosophical Magazine* **26**: 53-62.

Pickering, Spencer U. (1889). "The principles of thermochemistry." *Journal of the Chemical Society, Transactions* **55**: 14-33.

Pickering, Spencer U. (1890). "The nature of solutions, as elucidated by a study of (…) sulphuric acid solutions." *Journal of the Chemical Society, Transactions* **57**: 64-184.

Planck, Max (1897). *Vorlesungen über Thermodynamik*. Leipzig: Veit & Comp.

Poulsen, Valdemar (1930). "Nogle erindringsstrejf og smaabetragtninger gennem fire decennier." *Politiken*, 16 May.

Raestad, Christen (2000). "Drinking water link to cholera in Denmark in 1853." *Health and Hygiene* **21**: 148.

Ramsay, William (1907). "Radium emanation." *Nature* **76**: 269.

Rathke, Bernhard (1882). "Ueber die Principien der Thermochemie." *Abhandlungen der Naturforschenden Gesellschaft zu Halle* **15**: 197-224.

Rayleigh, Lord (1875). "On the dissipation of energy." *Nature* **11**: 454-455.

Rayleigh, Lord (1892). "On the relative densities of hydrogen and oxygen." *Nature* **46**: 101-104.

Rayner-Canham, Marelene and Geoff Rayner-Canham (2008). *Chemistry Was Their Life: Pioneer British Women Scientists 1880-1949*. London: Imperial College Press.

Reed, C. J. (1895). "A prediction of the discovery of argon." *Chemical News* **71**: 213-215.

Richards, Joseph W. (1887). *Aluminium: Its History, Occurrence, Properties, Metallurgy and Applications, Including Its Alloys*. Philadelphia: Henry C. Baird & Co.

Riis Larsen, Børge (1991). *Naturvidenskab og Dannelse*. Copenhagen: Dansk Selskab for Historisk Kemi.

Riis Larsen, Børge (1998). *Otte Kapitler af Kemiundervisningens Historie*. Copenhagen: Dansk Selskab for Historisk Kemi.

Rocke, Alan J. (1984). *Chemical Atomism in the Nineteenth Century: From Dalton to Cannizzaro*. Columbus: Ohio State University Press.

Rocke, Alan J. (1985). "Hypothesis and experiment in the early development of Kekulé's benzene theory." *Annals of Science* **42**: 355-381.

Rode, Jørgen, ed. (1942). *Københavns Elektricitetsværker*. Copenhagen: Københavns Belysningsvæsen.

Rose, Heinrich (1855). "Ueber eine neue und vortheilhafte Darstellung des Aluminiums." *Annalen der Physik und Chemie* **96**: 152-163.

Rubin, Mordecai B. (2009). "The mythical spawn of ozone: Antozone, oxozone, and ozohydrogen." *Bulletin for the History of Chemistry* **34**: 39-49.

Rud Nielsen, Jens, ed. (1972). *Niels Bohr. Collected Works*, vol. 1. Amsterdam: North-Holland.

Russell, Colin A. (1971). *The History of Valency*. New York: Humanities Press.

Scerri, Eric R. (2007). *The Periodic Table: Its Story and Its Significance*. Oxford: Oxford University Press.

Scharling, Edvard A. (1843). "Versuche über die Quantität der, von einem Menschen in 24 Stunden ausgeathmeten, Kohlensäure." *Annalen der Chemie und Pharmacie* **44**: 214-242.

Schelar, Virginia M. (1966). "Thermochemistry and the third law of thermodynamics." *Chymia* **11**: 99-124.

Schumacher, Heinrich C. F. (1795). "Fortegnelse og beskrivelse over nogle grøndlandske mineralier." *Skrivter af Naturhistorie-Selskabet* **4** (2): 206-233.

Scott, Alexander (1896). "Notes on atomic weights." *Science Progress* **5**: 202-214.

Sebelien, John (1884a). "Nogle bemærkninger om de chemiske hypotheser." *Tidsskrift for Physik og Chemi* **5**: 65-76.

Sebelien, John (1884b). *Beiträge zur Geschichte der Atomgewichte*. Braunschweig: Vieweg und Sohn.

Servos, John W. (1990). *Physical Chemistry from Ostwald to Pauling: The Making of a Science in America*. Princeton: Princeton University Press.

Siegel, Daniel M. (1978). "Classical electromagnetic and relativistic approaches to the problem of nonintegral atomic masses." *Historical Studies in the Physical Sciences* **9**: 323-360.

Smith, Fritze (1950). *Doktordisputatsens Historie ved Københavns Universitet*. Copenhagen: Munksgaard.

Snelders, Harry A. M. (1977). "Dissociation, Darwinism and Entropy: A case-study from the history of physical chemistry." *Janus* **64**: 51-75.

Sørensen, Søren P. L. (1896). "Om argon og helium." *Nyt Tidsskrift for Fysik og Kemi* **1**: 1-18.

Stachel, John, ed. (1989). *The Collected Papers of Albert Einstein*, vol. 2. Princeton: Princeton University Press.

Stallo, John (1960). *The Concepts and Theories of Modern Physics*. Cambridge, MA: Belknap Press.

Stohmann, Friedrich (1886). "Entgegnung zu vorstehender Abhandlung des Herrn Thomsen." *Journal für Praktische Chemie* **33**: 568-576.

Stohmann, Friedrich (1887). "Zur weiteren Beleuchtung der Untersuchungen des Herrn Thomsen." *Journal für Praktische Chemie* **35**: 136-141.

Stohmann, Friedrich, P. Rodatz, and H. Herzberg (1886). "Ueber den Wärmewerth des Benzols." *Journal für Praktische Chemie* **33**: 241-260.

Strand, K. Aa. (1968). "Ejnar Hertzsprung." *Proceedings of the Astronomical Society of the Pacific* **80**: 51-56.

Strecker, Adolph (1859). *Theorien und Experimente zur Bestimmung der Atomgewichte der Elemente.* Braunschweig: Vieweg und Sohn.

Strutt, Robert J. (1907). "Note on the association of helium and thorium in minerals." *Proceedings of the Royal Society A* **80**: 56-57.

Sutton, Mike (2009). "Wealth from Greenland, honour from London." *Chemistry World* (September): 56-58.

Sveistrup, P. P. (1956). "Kryoliten og staten." *Tidsskriftet Grønland* no. 2, 41-49.

Thomsen, August (1860). "Solvarmen som bevægende kraft." *Tidsskrift for Populære Fremstillinger af Naturvidenskaben* **2**: 271-278.

Thomsen, August (1877). "Belysningsgassen fra Kjøbenhavns gasværk med hensyn til forbrændingsproducter, indhold af svovl og brændværdi." *Tidsskrift for Physik og Chemi* **16**: 289-295.

Thomsen, August (1879). *Naturkræfterne i Menneskets Tjeneste: Belærende Underholdninger paa Videnskabens og Industriens Gebet.* Copenhagen: Philipsen.

Thomsen, August (1883). *Forelæsninger over Technisk Chemi ved den Polytekniske Læreanstalt.* Copenhagen: Reitzel.

Thomsen, Julius (1850a). *Kortfattet Lærebog i den Uorganiske Chemie.* Copenhagen: C. A. Reitzel.

Thomsen, Julius (1850b). *Om Vexelvirkningen Imellem Planten og dens Omgivelser.* Copenhagen: S. Trier.

Thomsen, Julius (1852). "Bidrag til et thermochemisk system." *Kgl. Danske Videnskabernes Selskab, Skrifter* **3**: 115-165.

Thomsen, Julius (1853a). *Vejledning i den Præparative Chemie.* Copenhagen: C. A. Reitzel.

Thomsen, Julius (1853b). *Et Forsøg paa en Almeenfattelig Fremstilling af Chemiens Vigtigste Resultater.* Copenhagen: Thaarup.

Thomsen, Julius (1853c). "Die Grundzüge eines thermochemischen Systems; I, II." *Annalen der Physik und Chemie* **88**: 349-362.

Thomsen, Julius (1854a). "Die Grundzüge eines thermochemischen Systems; III." *Annalen der Physik und Chemie* **90**: 261-288.

Thomsen, Julius (1854b). "Die Grundzüge eines thermochemischen Systems; IV." *Annalen der Physik und Chemie* **91**: 83-104.

Thomsen, Julius (1854c). "Die Grundzüge eines thermochemischen Systems; V." *Annalen der Physik und Chemie* **92**: 34-57.

Thomsen, Julius (1855a). "Om naturkræfternes gjensidige forhold." *Tidsskrift for Populære Fremstillinger af Naturvidenskaben* **1**: 227-240, 360-374.

Thomsen, Julius (1855b). *Om den Naturvidenskabelig-Mathematiske Underviisning: Den Polytekniske Læreanstalt.* Copenhagen: C. A. Reitzel.

Thomsen, Julius (1856). *Vandringer paa Naturvidenskabens Gebeet*. Copenhagen: C. A. Reitzel.

Thomsen, Julius (1858). "Den elektromotoriske kraft udtrykt i varmeenheder." *Kgl. Danske Videnskabernes Selskab, Skrifter* 5: 155-175.

Thomsen, Julius (1860a). "Laskaris." *Illustreret Tidende* 2, no. 66: 106.

Thomsen, Julius (1860b). "Die konstante Kupfer-Kohlenkette." *Annalen der Physik und Chemie* 187: 192.

Thomsen, Julius (1861). "Om de chemiske processers almindelige charakteer og en paa denne bygget Affinitetslære." *Kgl. Da. Vid. Selsk. Forhandl., Oversigt*: 100-134.

Thomsen, Julius (1862a). "Nogle meddelelser angaaende kryolithindustrien." *Tidsskrift for Physik og Chemi* 1: 321-332.

Thomsen, Julius (1862b). "Zur Geschichte der Kryolithindustrie." [Dingler's] *Polytechnisches Journal* 166: 441-443.

Thomsen, Julius (1862c). "Om det Steinheilske spectralapparat." *Kgl. Da. Vid. Selsk. Forhandl., Oversigt*: 94-95.

Thomsen, Julius (1863a). "Der Kryolith-Sodaofen." [Dingler's] *Polytechnisches Journal* 167: 362-370.

Thomsen, Julius (1863b). "Quantitativ bestemmelse of leerjord ved titrering." *Tidsskrift for Physik og Chemi* 2: 225-233.

Thomsen, Julius (1863c). "Om gasbelysningsapparaternes tilstand i Kjøbenhavn." *Tidsskrift for Physik og Chemi* 2: 65-82.

Thomsen, Julius (1863d). "Lysets mechaniske æqvivalent." *Tidsskrift for Physik og Chemi* 2: 193-206.

Thomsen, Julius (1863e). "Om de galvaniske apparaters natur og deres caloriske virkninger." *Tidsskrift for Physik og Chemi* 2: 321-337.

Thomsen, Julius (1864). "Polarisationsbatteriet." *Tidsskrift for Physik og Chemi* 3: 193-217.

Thomsen, Julius (1865a). "Om de saakaldte grundstoffers natur." *Tidsskrift for Physik og Chemi* 4: 65-80, 97-115.

Thomsen, Julius (1865b). "On the mechanical equivalent of light." *Philosophical Magazine* 30: 246-249.

Thomsen, Julius (1865c). "Die Polarisationsbatterie." *Annalen der Physik und Chemie* 200: 498-500.

Thomsen, Julius (1865d). "Einige Bemerkungen bezüglich der Polarisationsbatterie." *Annalen der Physik und Chemie* 201: 163-165.

Thomsen, Julius (1867). "En række dobbeltchlorider, henhørende til platinbasernes gruppe." *Kgl. Da. Vid. Selsk. Forhandl., Oversigt*: 225-233.

Thomsen, Julius (1868). "Om thermometres følsomhed." *Kgl. Da. Vid. Selsk. Forhandl., Oversigt*: 25-31.

Thomsen, Julius (1869a). "Thermochemische Untersuchungen, I." *Annalen der Physik und Chemie* **138**: 65-102, 201-213.

Thomsen, Julius (1869b). "Thermochemiske undersøgelser over affinitetsforholdende imellem syrer og baser i vandig opløsning." *Tidsskrift for Physik og Chemi* **8**: 1-4, 129-152, 223-242.

Thomsen, Julius (1869c). "Über die Ungenauigkeit der von Favre und Silbermann mittels des Quecksilbercalorimeters gemachten thermochemischen Bestimmungen." *Berichte der Deutschen Chemischen Gesellschaft* **2**: 701-703.

Thomsen, Julius (1869d). "Thermochemische Untersuchungen, III." *Annalen der Physik und Chemie* **138**: 497-514.

Thomsen, Julius (1869e). "Chloralhydratets fremstilling og egenskaber." *Tidsskrift for Physik og Chemi* **8**: 243.

Thomsen, Julius (1869f). "Über Berechnung der Verbrennungswärme organischer Verbindungen." *Berichte der Deutschen Chemischen Gesellschaft* **2**: 482-489.

Thomsen, Julius (1870a). "Über einige Konstante des Wasserstoffs und der Sauerstoffs." *Berichte der Deutschen Chemischen Gesellschaft* **3**: 927-930.

Thomsen, Julius (1870b). "Über die angebliche Ableitung des Avogadro'schen Gesetzes aus der mechanischen Wärmetheorie." *Berichte der Deutschen Chemischen Gesellschaft* **3**: 828-829.

Thomsen, Julius (1870c). "Ueber das Avogadrosche Gesetz." *Berichte der Deutschen Chemischen Gesellschaft* **3**: 949-955.

Thomsen, Julius (1870d). "Einige Vorlesungsversuche." *Berichte der Deutschen Chemischen Gesellschaft* **3**: 927-930.

Thomsen, Julius (1871a). "Chemiske theorier." *Tidsskrift for Physik og Chemi* **10**: 1-9, 65-74, 289-303.

Thomsen, Julius (1871b). "Über die Ungenauigkeit der von Favre und Silbermann mittels des Quecksilbercalorimeters gemachten thermochemischen Bestimmungen." *Berichte der Deutschen Chemischen Gesellschaft* **4**: 591-594.

Thomsen, Julius (1872a). "Die völlige Ungültigkeit der von Berthelot (...) berechneten Zahlenwerte." *Berichte der Deutschen Chemischen Gesellschaft* **5**: 181-185.

Thomsen, Julius (1872b). "Über die Angaben des Quecksilbercalorimeters." *Berichte der Deutschen Chemischen Gesellschaft* **5**: 614-620.

Thomsen, Julius (1872c). "Iagttagelser, som tyde paa, at affinitetens størrelse i forskellige chemiske processer maa opfattes som multipla af fælles constanter." *Tidsskrift for Physik og Chemi* **11**: 161-175.

Thomsen, Julius (1873a). "Eine Prioritätsfrage bezüglich einiger Grundsätze der Thermochemie." *Berichte der Deutschen Chemischen Gesellschaft* **6**: 423-428.

Thomsen, Julius (1873b). "Über die Basicität und Constitution der Überjodsäure." *Berichte der Deutschen Chemischen Gesellschaft* **6**: 368-404.

Thomsen, Julius (1873c). "Über den Einfluss der Temperatur auf die chemische Wärmetönung." *Berichte der Deutschen Chemischen Gesellschaft* 6: 1330-1345.

Thomsen, Julius (1873d). "Thermochemische Untersuchungen, XI." *Annalen der Physik und Chemie* 148: 177-202, 368-404.

Thomsen, Julius (1874). "Ueber die Darstellung von Wasserstoffhyperoxid." *Berichte der Deutschen Chemischen Gesellschaft* 7: 73-74.

Thomsen, Julius (1876a). "Darstellung und Eigenschaften der Chlor- und Bromverbindungen und des Oxyds des Golds." *Journal für Praktische Chemie* 13: 337-347.

Thomsen, Julius (1876b). "Zur Geschichte des Einflusses der Temperatur auf die chemische Wärmetönung." *Berichte der Deutschen Chemischen Gesellschaft* 9: 307-308.

Thomsen, Julius (1877). " Thermochemische Untersuchungen, XIV." *Journal für Praktische Chemie* 15: 435-473.

Thomsen, Julius (1878). "Ueber Genauigkeit thermochemischer Zahlenresultate." *Berichte der Deutschen Chemischen Gesellschaft* 11: 2183-2188.

Thomsen, Julius (1879a). "Das Sinusmanometer, ein Apparat zum Messen kleiner Luftdruckdifferenzen." *Annalen der Physik und Chemie* 242: 451-455.

Thomsen, Julius (1879b). "H. P. J. J. Thomsen." In: *Levnedsbeskrivelser af de ved Kjøbenhavns Universitets Firehundredaarsfest Promoverede Doktorer og Licentiater*, pp. 20-22. Copenhagen: J. H. Schultz.

Thomsen, Julius (1880a). "Die Constitution des Benzols." *Annalen der Chemie und Pharmacie* 205: 133-138.

Thomsen, Julius (1880b). "Die Verbrennungswärme des Benzols." *Berichte der Deutschen Chemischen Gesellschaft* 13: 1806-1807.

Thomsen, Julius (1880c). "Thermochemische Untersuchungen über die Theorie der Kohlenstoffverbindungen." *Berichte der Deutschen Chemischen Gesellschaft* 13: 1321-1334.

Thomsen, Julius (1880d). "Zur Benzolformel." *Berichte der Deutschen Chemischen Gesellschaft* 13: 2166-2168.

Thomsen, Julius (1880e). "Thermochemische Untersuchungen, XXXI: Die Bildungswärme der Salpetersäure (…) und der Nitrate." *Journal für Praktische Chemie* 21: 449-478.

Thomsen, Julius (1880f). "Erindringer fra det tolvte skandinaviske naturforskermøde i Stockholm." *Nordisk Tidskrift för Vetenskap, Konst och Industri* 3: 526-534.

Thomsen, Julius (1880g). "Chemische Energie und elektromotorische Kraft verschiedener galvanischen Combinationen." *Annalen der Physik und Chemie* 11: 246-269.

Thomsen, Julius (1882-1886). *Thermochemische Untersuchungen*, 4 vols. Leipzig: Barth.

Thomsen, Julius (1884). *Om Molekyler og Atomer*. Copenhagen: Copenhagen University.

Thomsen, Julius (1885). "Ueber das Molekulargewicht des flüssigen Wassers." *Berichte der Deutschen Chemischen Gesellschaft* **18**: 1088.

Thomsen, Julius (1886a). "Om benzolmolekylets konstitution." *Kgl. Da. Vid. Selsk. Forhandl., Oversigt*: 179-186.

Thomsen, Julius (1886b). "Ueber die Verbrennungswärme des Benzols." *Journal für Praktische Chemie* **33**: 564-567.

Thomsen, Julius (1887a). *Om Materiens Enhed*. Copenhagen: Copenhagen University.

Thomsen, Julius (1887b). "Über G. A. Hagemanns kritische Bemerkungen zur Aziditätsformel." *Berichte der Deutschen Chemischen Gesellschaft* **20**: 1155-1157.

Thomsen, Julius (1888a). "Über die Reaktion zwischen Gold und Chlor. Darstellung der Verbindung Au_2Cl_4." *Journal für Praktische Chemie* **37**: 105-108.

Thomsen, Julius (1888b). "Über die Bildungswärme der Quecksilberverbindungen." *Zeitschrift für Physikalische Chemie* **2**: 21-23.

Thomsen, Julius (1891). "Über die Beziehung zwischen der Verbrennungswärme organischer Verbindungen und der Konstitution derselben." *Zeitschrift für Physikalische Chemie* **7**: 55-70.

Thomsen, Julius (1892a). [Untitled opening speech]. *Forhandlingerne ved de Skandinaviske Naturforskeres 14. Møde*. Copenhagen: G. C. Ursins Eftf.

Thomsen, Julius (1892b). *Nogle Ord om Land og Folk*. Copenhagen: Berlingske Bogtrykkeri.

Thomsen, Julius (1894a). "Undersøgelser over, hvorvidt hypothesen om materiens enhed kan bringes i samklang med theorien om atomernes relative vægt." *Kgl. Da. Vid. Selsk. Forhandl., Oversigt*: 306-324.

Thomsen, Julius (1894b). "Relations remarquables entre les poids atomique des éléments chimiques." *Kgl. Da. Vid. Selsk. Forhandl., Oversigt*: 325-343.

Thomsen, Julius (1894c). "Experimentelle Untersuchung zur Feststellung des Verhältnisses zwischen den Atomgewichten des Sauerstoffes und Wasserstoffes." *Zeitschrift für Physikalische Chemie* **13**: 398-406.

Thomsen, Julius (1894d). "Über den wahrscheinlichsten Wert der aus den von Stas durchgeführten Untersuchungen sich ableitenden Atomgewichte." *Zeitschrift für Physikalische Chemie* **13**: 726-735.

Thomsen, Julius (1894e). "Nogle træk af de physiske videnskabers historie fra slutningen af det 18de aarhundrede." *Tidsskrift for Physik og Chemi* **33**: 225-234.

Thomsen, Julius (1895a). "Experimentelle Untersuchung über das Atomverhältnis zwischen Sauerstoff und Wasserstoff." *Zeitschrift für Anorganische Chemie* **11**: 14-30.

Thomsen, Julius (1895b). "Classifications des corps simples." *Kgl. Da. Vid. Selsk. Forhandl., Oversigt*: 132-136.

Thomsen, Julius (1895c). "Über die mutmasslische Gruppe inaktiver Elemente." *Zeitschrift für Anorganische Chemie* **8**: 283-288.

Thomsen, Julius (1895d). "Über die Farbe der Ionen als Funktion der Atomgewichte." *Zeitschrift für Anorganische Chemie* **10**: 155.

Thomsen, Julius (1896). "Experimentelle Untersuchung über die Dichte des Wasserstoffes und des Sauerstoffes." *Zeitschrift für Anorganische Chemie* **12**: 1-15.

Thomsen, Julius (1898a). "Über Abtrennung von Helium aus einer natürlichen Verbindung unter starker Licht- und Wärmeentwicklung." *Zeitschrift für Physikalische Chemie* **28**: 112-114.

Thomsen, Julius (1898b). "En transformator for elektriske strømme." *Kgl. Da. Vid. Selsk. Forhandl., Oversigt*: 97-110.

Thomsen, Julius (1899). "Nogl resultater af de seneste aars naturforskning." *Beretning om det 15. Skandinaviske Naturforskermøde*: 66-77.

Thomsen, Julius (1900). "Vinklens tredeling." *Nyt Tidsskrift for Matematik*, A, **11**: 12-14.

Thomsen, Julius (1903). "F. W. Clarkes neues thermochemisches Gesetz." *Zeitschrift für Physikalische Chemie* **43**: 487-493.

Thomsen, Julius (1904). "Om de i nogle grønlandske mineralier indeholdte luftarter." *Kgl. Da. Vid. Selsk. Forhandl., Oversigt*: 53-57.

Thomsen, Julius (1905a). "Fortegnelse over afhandlinger og skrifter af Julius Thomsen." *Kgl. Da. Vid. Selsk. Forhandl., Oversigt*: 489-503.

Thomsen, Julius (1905b). *Systematisk Gennemførte Termokemiske Undersøgelsers Numeriske og Teoretiske Resultater*. Copenhagen: Royal Danish Academy.

Thomsen, Julius (1905c). "Relativer Wert der zur Bestimmung der Verbrennungswärme flüchtiger organischen Verbindungen benutzten kalorimetrischen Methoden." *Zeitschrift für physikalische Chemie* **51**: 657-672.

Thomsen, Julius (1905d). "Allgemeine Theorie der Verbrennungs- und Bildungswärme der Kohlenwasserstoffe in gas- oder dampfförmigen Zustände (etc.)." *Journal für Praktische Chemie* **71**: 164-181.

Thomsen, Julius (1905e). "Herrn Daniel Lageröfs 'Antwort'." *Journal für Praktische Chemie* **72**: 341-342.

Thomsen, Julius (1908). *Thermochemistry*, translated by Katharine A. Burke. London: Longmans, Green, & Co.

Thomson, Joseph J. (1888). *Applications of Dynamics to Physics and Chemistry*. London: Macmillan & Co.

Thorpe, Edward (1910). "Thomsen memorial lecture." *Journal of the Chemical Society* **97**: 161-173.

Thyssen, Pieter and Koen Binnemans (2011). "Accomodation of the rare earths in the periodic table: A historical analysis." In: *Handbook on the Physics and Chemistry of Rare Earths*, vol. 41, ed. Karl A. Gschneidner, pp. 1-94. Burlington: Academic Press.

Tietgen, Carl F. (1904). *Erindringer og Optegnelser*, ed. O. C. Molbech. Copenhagen: Gyldendal.

Tilden, William A. (1899). *A Short History of the Progress of Scientific Chemistry in Our Own Times*. London: Longmans, Green, and Co.

Topp, Niels-Henrik (1990). *Kryolitindustriens Historie 1847-1900*. Copenhagen: Kryolitselskabet Øresund A/S.

Trenn, Thaddeus J. (1974). "The justification of transmutation: Speculations of Ramsay and experiments of Rutherford." *Ambix* **21**: 53-77.

Tumlirz, Ottokar (1889). "Das mechanische Äquivalent des Lichtes." *Annalen der Physik* **274**: 640-662.

Van der Kolk, Schröder (1864). "Ueber die mechanische Energie der chemischen Wirkungen." *Annalen der Physik und Chemie* **122**: 439-454.

Van Laar, Johannes J. (1901). *Lehrbuch der Mathematischen Chemie*. Leipzig: Barth.

Van Spronsen, Jan W. (1969). *The Periodic System of Chemical Elements: A History of the First Hundred Years*. Amsterdam: Elsevier.

Van't Hoff, Jacobus H. (1905). "The relations of physical chemistry to physics and chemistry." *Journal of Physical Chemistry* **9**: 81-89.

Varberg, Rudolph (1868). *Udflugter paa Naturvidenskabens Enemærker*. Copenhagen: Philipsens Forlag.

Vaubel, Wilhelm (1903). *Lehrbuch der Theoretischen Chemie*. Berlin: J. Springer.

Vauqelin, Louis N. (1800). "Sur la presence de la soude dans la chryolithe du Groënland, annoncé par M. Klaproth." *Annales de Chimie* **37**: 89-93.

Venable, Francis P. (1896). *The Development of the Periodic Law*. Easton, PA: Chemical Publishing Co.

Veibel, Stig (1939). *Kemien i Danmark, I: Kemiens Historie i Danmark*. Copenhagen: Arnold Busck.

Veibel, Stig (1943). *Kemien i Danmark, II: Dansk Kemisk Bibliografi 1800-1935*. Copenhagen: Arnold Busck.

Vestergaard, Lene (1999). *Kemi og Kolera: En Analyse af "Om de Sandsynlige Aarsager til Choleraens Ulige Styrke."* Unpublished Master Thesis, Aarhus University.

Vinding, Poul (1941). "Julius Thomsen." In: P. Vinding, *Dansk Teknik Gennem Hundrede Aar*, pp. 29-47. Copenhagen: Gads Forlag.

Vinding, Poul (1942). *G. A. Hagemann: En Dansk Ingeniørs Levnedsløb*. Copenhagen: Gad.

Voigt, Julius J. (1890). *Den Polytekniske Læreanstalts Kandidater*. Copenhagen: J. H. Schultz.

Wagman, Donald D. (1992). "Data bases: Past, present and future." *Pure & Applied Chemistry* **64**: 37-48.

Wagner, J. R. (1862). "Alaun und andere Thonerdesalze." *Jahres-Bericht der Chemischen Technologie* **8**: 291-304.

Wagner, Michael F. (1999). *Det Polytekniske Gennembrud*. Aarhus: Aarhus Universitetsforlag.

Walker, James (1905). "General and physical chemistry." *Annual Reports on the Progress of Chemistry* **2**: 1-29.

Wedell-Wedellborg, Peer S. (1897). *Julius Thomsen's Dualismus der Chemischen Masse Beleuchtet Durch Aufstellung einer Neuen Wärmetheorie*. Copenhagen: Høst og Søn.

Weeks, Mary E. and Henry M. Leicester (1968). *Discovery of the Elements*. Easton, PA: Journal of Chemical Education.

Whittaker, Alfred (1991). "The Magic Flute cast: Geological correlations with Mozart." *Terra Nova* **3**: 9-16.

Whittaker, Alfred (2007). "The travels and travails of Sir Charles Lewis Giesecke." In: *Four Centuries of Geological Travel: The Search for Knowledge on Foot, Bicycle, Sledge and Camel*, ed. Patrick N. Wyse Jackson, pp. 149-160. London: Geological Society.

Williams, Trevor I. (1993). "Aluminium: Latecomer to the metal industry." *Endeavour* **17**: 89-93.

Wöhler, Friedrich (1856). "Über die Reduction des Aluminiums aus Kryolith." *Annalen der Chemie und Pharmacie* **99**: 255-256.

Woods, Thomas (1852). "On the heat of chemical combination." *Philosophical Magazine* **3**: 43-53, 299-303.

Wurtz, Adolphe, ed. (1869-1870). *Dictionnaire de Chimie Pure et Appliquée*. Paris: Libraire Hachette.

Zapffe, C. A. (1969). "Gustavus Hinrichs, precursor of Mendeleev." *Isis* **60**: 461-476.

Index